John Lycett

A Monograph of the British Fossil Trigoniæ

ISBN/EAN: 9783337419608

Printed in Europe, USA, Canada, Australia, Japan

Cover: Foto ©berggeist007 / pixelio.de

More available books at **www.hansebooks.com**

A MONOGRAPH

OF THE

BRITISH FOSSIL TRIGONIÆ.

BY

JO███████TT, L.R.C.P.E., M.R.C.S.Engl.,

MEDICAL O███████L NORTHERN SEA-BATHING INFIRMARY, SCARBOROUGH; HONORARY
███████ THE COTTESWOLD NATURALISTS' FIELD-CLUB.

LONDON:
PRINTED FOR THE PALÆONTOGRAPHICAL SOCIETY.
1872—1879.

PREFATORY.

The introductory portion of this Monograph contains descriptions of all the sectional divisions of the genus with which the British Trigoniæ are connected. The sketch of species in *italics*[1] appended to these sectional divisions has been reconstructed and amended in the Stratigraphical Table[2] at the end of the Monograph, and to this the reader is referred. One division not referred to in the introductory portion is the living section of the genus, the "Pectinidæ" of Agassiz, a section which is special to one of our colonial possessions, being known only in Australia. This particular section will be found alluded to in the last few concluding pages of the Monograph.[3]

The importance of the Trigoniæ both zoologically and stratigraphically appears only lately to have been sufficiently estimated by either the naturalist or the geologist. The gradually increasing occurrence of these forms, and their relationship to the zoological assemblages with which they are connected, may be considered as so many features of constantly increasing interest and importance to science, whatever may be determined upon as to their status whether as species or varieties. It is only within the last few years that Tertiary Trigoniæ of the Pectinidæ group have been discovered in Australia. The Tertiary formations of the other continents are, as far as we yet know, entirely destitute of the genus.

JOHN LYCETT.

Scarborough;
    10*th February*, 1879.

[1] Pages 5—13.   [2] Pages 235—239.   [3] Pages 231—234.

A MONOGRAPH

OF

BRITISH FOSSIL TRIGONIÆ.

GENERAL OBSERVATIONS AND SYSTEMATIC ARRANGEMENTS.

THROUGHOUT the great Mesozoic epoch there are, perhaps, no testaceous forms, with the exception of the Ammonites, of importance superior to the Trigoniæ, or which demand from the Palæontologist more careful discrimination and more extensive acquaintance with their various aspects, to enable him to overcome the difficulties that meet him when he attempts their certain determination. The importance of the genus is based upon its great stratigraphical range, its world-wide occurrence, its great diversity of aspects, locally also by its individual numbers. Its prominence and distinctness as a genus immediately arrests the attention of the observer, and, notwithstanding the diversity of its aspects, a Trigonia is a form recognised without difficulty even by a tyro. The generic characters are so well known that to give a minute description of them would amount to a mere useless repetition of that which has been so fully accomplished by Agassiz and by other Palæontologists. I prefer, therefore, to allude to them with the greatest possible conciseness, and will adopt the terse and brief definition given by my late lamented friend, Dr. S. P. Woodward, in his well-known 'Manual of the Mollusca:'—" Shell thick, tuberculated, or ornamented with radiating or concentric ribs; posterior side angular, ligament small and prominent; hinge-teeth 2—3, diverging, transversely striated; centre tooth of the left valve divided; pedal impressions in front of the posterior adductor, and one in the umbo of the left valve; anterior adductor impression close to the umbo. Shell almost entirely nacreous, and usually wanting or metamorphic in limestone strata."

Moulds which represent the internal cavity of the shell may readily be obtained whenever the investing matrix is in its structure more coherent or compact than the fossilized test, and it usually happens that a rock or matrix of great hardness is associated with a test more than usually friable. The value of these internal moulds to the Palæontologist who wishes to discriminate species is, however, only limited; it rarely happens that any of the external ornaments are visible upon them; and even under the most favourable conditions the impressions of these ornaments are only faintly and insufficiently shown, so that by means of these alone the external aspect even of a single species could never be fully ascertained, and even when both the mould and test have been obtained it is not in every instance that the mould can with certainty be discriminated from those of other allied species. The practice of authors, therefore, who have described supposed new species even partially, and have named them from internal moulds alone, is objectionable, as tending to create doubt and hesitation in the minds of students, and encumbering the list of species with things which for all practical purposes are little more than mere names. Roemer, and afterwards Agassiz, sometimes unfortunately acted in this manner, and D'Orbigny went even further in the same direction; in his 'Prodrome' are new names, together with a few words to each, which are intended to indicate moulds only, that have not even been figured. For the most part, therefore, internal moulds afford but little more than so many proofs that Trigoniæ were buried in the stratum to which they refer, and also that the species had apparently certain peculiarities of figure more or less conspicuous.

The figures of Trigonia given in the 'Petrefacta' of Goldfuss are not very numerous, but are almost invariably founded upon good specimens, and the execution of the plates is equally satisfactory; but the descriptions are only remarkable for a general vagueness and brevity that so often is found to prevail in descriptions of fossils at the period when they were written; it is also certain that the geological positions mentioned are sometimes erroneous. The well-known memoir by Agassiz on the genus Trigonia contains eleven plates, illustrative of forty-eight species; it is much to be regretted that in many instances these figures represent very indifferent specimens or fragments only, and that these deficiencies are not compensated for by sufficiently copious or minute descriptions; they have been felt as a frequently recurring source of embarrassment, and more especially when an author has adopted the name of one of these doubtful forms in a list or description of species. It is, however, only just to remark that the work, upon the whole, exhibits much of that discriminative faculty of mind which we should expect to find in the production of so eminent a naturalist; the sectional divisions which he proposed to establish are so far in accordance with nature, and are so well indicated by his typical examples, that it would, perhaps, have been quite possible to follow out these groups with little error throughout the entire genus, even if the author had given to each of them simply a name without description; in fact, the sectional descriptions are both meagre and imperfect, and appear to have been intended as mere outline sketches, susceptible of future enlargement or modification. D'Orbigny, in his 'Prodrome de Paléontologie' and in his 'Paléontologie Française,' has in some instances endeavoured to correct errors in the identification of species by Agassiz, but not uniformly with success, and apparently with only an imperfect knowledge of British Trigoniæ. Our thanks are, however, due to him for having greatly extended our knowledge of the genus so far as regards the species of the Cretaceous System; but in the Jurassic System his stratigraphical distribution of them is full of errors, and is, perhaps, more calculated to mislead than to instruct. These critical, and it is hoped not unfair, remarks are not however, intended to apply to the descriptions of Trigoniæ to be found in various memoirs and monographs which date within the last twelve years, referred to in the following pages; in these the requirements of modern science are more fully complied with, and by authors whose names are a guarantee for the faithfulness and value of their contributions. The descriptions by Dujardin, Hébert, Munier-Chalmas, de Loriol, Credner, Coquand, and Pictet, are more especially prominent and satisfactory to the student in Palæontology. For the most part, however, each such contribution refers only to the species which pertain to a single formation and locality.

The British Jurassic Trigoniæ are remarkable both for their number of species and the variety in their ornamentation; no other country has produced so considerable an assemblage. The British Cretaceous species, on the other hand, although eminently characteristic, represent only a portion, perhaps scarcely a moiety, of the foreign Cretaceous forms, if we include those of South America and Southern Asia.

The woodcuts illustrative of various foreign forms allied to British Trigoniæ, together with the concise notes which refer to them, will, it is hoped, add somewhat to the value of the Monograph, and be found more satisfactory than any unaided descriptions, however copious they may be. In the order of arrangement each species is placed in its sectional position, adjacent to its allied forms, and at the end of the Monograph a general table will be given indicating the stratigraphical position of each species. The general description of sectional characters which follows conduces to some abbreviation, and obviates the necessity for repetition in the description of each separate species, so far as these sectional characters apply. There are also a few terms employed by which a single word or two suffices to indicate the portion of the shell referred to, and, therefore, materially aids in the same result. Throughout all the fossil species of Trigoniæ there are certain features connected with the general figure and ornamentation which, although only of subordinate importance when considered in relation to the generic characters, nevertheless furnish valuable aids both in the separation of the several groups into which the genus may be divided, but also afford guides to the determination of the stratigraphical position which they occupy. The living Australian species are in their external characters to a great extent unconnected with these distinctive features; they assimilate more nearly to other generic forms of the Conchifera, and therefore constitute a group apart and disconnected from the chain of fossil Trigoniæ of the various Secondary Formations, between which and the living species the missing connecting links will probably only be discovered when other examples of the genus shall be procured in some one of the Lower or Middle Tertiary Formations.

The surface of every fossil Trigonia is clearly divided into two portions; one of these occupies the anteal, the other the posteal portion of the valve; the two parts have their surfaces upon different planes, so that at their junction there is a divisional angle which passes obliquely from the apex of the valve or umbo to the lower and posterior extremity; the divisional angle commonly takes the form of a ridge; this is the *marginal carina*, and the portion of surface posteal to it is the *area*. At the superior border of the area is another ridge or carina, which, like the other, originates at the apex, and forms a curvature posteally to meet the corresponding carina of the other valve. This is the *inner carina*, and the space enclosed between the inner carinæ of the opposing valves is usually heart-shaped, or more lengthened and lanceolate. This is the *escutcheon*, which in its turn embraces at the upper and anteal portion of its border the nymphal or ligamental plates and the ligament itself, which is also not unfrequently preserved. The area is usually divided into two portions longitudinally by a slight furrow, and the superior portion is frequently more depressed than the other; a slight ridge or row of tubercles may border this *median furrow*, and, when present, it constitutes the *median carina*. The anteal portion of the valve, or that adjacent to the pallial border, is invariably more or less ornamented with ribs, which are either plain or tuberculated; the area also usually has some kind of ridges; but, in lieu of designating the former as pallial costæ and those upon the area as cardinal costæ, I prefer, for brevity, to speak of the former as *costæ* and the latter, which are usually smaller, as *costellæ*. One other feature common to the surface of all, and therefore constituting a portion of the attributes of the genus, may also be adverted to, more especially as it has usually remained altogether unnoticed. I allude to the epidermal granulated tegument; in common with the whole group of the Anatinidæ, and with certain other genera of Conchifera, including the allied genus Opis, the granules are arranged in lines perpendicular to the pallial border. They are more minute and closely arranged than is usual in other genera; in the *Trigoniæ clavellatæ* more especially the lines of granules can only be discovered by the aid of considerable magnifying power; in this particular the distinction between the *Clavellatæ* and *Costatæ* is strongly marked, the latter having the lines of granules so much larger and more widely separated that they may frequently be distinguished by the unaided vision.

The seven sections into which Agassiz divided the fossil Trigoniæ are all exemplified by British species, and to those I propose to add an additional section, the *Byssiferæ*. The sectional distinctions are founded upon the figure of the shell and of its proportions, of the ornamentation upon the pallial or costated portion of the surface, of that upon the area, and of that upon the escutcheon; lastly, upon the consideration conjunctively of all these parts in any particular group in comparison with the corresponding parts pertaining to other groups of the same genus. An attentive examination will, it is believed, exhibit fully the natural affinity of the various species comprising these separate groups or sections, and also the changes that the genus underwent in geological time during the great epoch of the Secondary Formations.

INTRODUCTION.

1st.—SCAPHOIDEÆ.

This section, of which the type is the *Trigonia navis* of Lamarck, was constituted by Agassiz to include species in which the form is usually triangular, having a remarkable straightness or truncation of the anterior border; the other two borders converge and terminate posteriorly in an extremity which is more or less truncated or obtuse. The area is nearly smooth, bounded above by an inner, and beneath by a marginal carina; these are always small and sometimes evanescent. The costæ form two series, the larger or posteal series pass almost perpendicularly downwards from the marginal carina to the pallial border, the smaller or anteal series have their general direction horizontal or at right angles to the other series; they are short and occasionally, as in *T. navis*, terminate posteally in a large varix. Agassiz described and figured five species, three of which are Jurassic and two Cretaceous; the present Monograph adds four species, all of which are from the Lower Oolites, and one is new. These are the following :—*T. duplicata*, Sow., *T. gemmata*, Lyc., *T. Bathonica*, Lyc., and *T. recticosta*, Lyc.

2nd.—CLAVELLATÆ.

An essentially and almost exclusively Jurassic section have their sides ornamented with tuberculated costæ in rows, which are either concentric or oblique; the tubercles in the rows usually become indistinct or cord-like as they approach the pallial border, and the last-formed rows of costæ are often interrupted or broken and obscured by numerous plications of growth. The area, which is well separated, is bounded by two tuberculated carinæ, and has usually also a smaller mesial oblique carina bordering upon a furrow, thus dividing the area into two parts. The escutcheon is always plain, as in the *Undulatæ*; it is depressed and is usually well circumscribed by the inner tuberculated carina. Few species have the rows of costæ regular and symmetrical over the whole of the side, or adjacent to the pallial border; most commonly there is some irregularity in the rows near to their anteal terminations; some individuals of certain species have this irregularity so considerable that the tubercles become altogether confused and crowded, in other instances they become absorbed in the longitudinal plications, which form squamous elevations. It rarely happens that this section has the granulated tegument preserved; and it may be remarked that the same clays and shales which have preserved this feature in the *Costatæ* have failed to do so in the *Clavellatæ*. The species of the *Clavellatæ* are very numerous, and in their habits they were usually gregarious, but it seldom happens that any one species can be traced over any considerable geographical area. The interior of the valves have near to their posteal slope a prominent lengthened ridge, which indicates the position and separation of the siphonal currents. The following species, twenty-eight in number, are British examples of this section :—*T. clavellata*,

Sow., *T. perlata*, Ag., *T. corallina*, D'Orb., *T. Rupellensis*, D'Orb., *T. impressa*, Sow., *T. signata*, Ag., *T. spinulosa*, Y. & B., *T. striata*, Miller, *T. Bronnii*, Ag., *T. irregularis*, Seebach, *T. triquetra*, Seeb., *T. Pellati*, Mun.-Chal., *T. Phillipsi*, Mor. and Lyc., *T. Moretoni*, Mor. and Lyc., *T. imbricata*, Sow., *T. muricata*, Goldf., *T. parcinoda*, Lyc., *T. Woodwardi*, Lyc., *T. radiata*, Benett, *T. Ramsayi*, Wright, *T. incurva*, Benett, *T. complanata*, Lyc., *T. Scarburgensis*, Lyc., *T. Juddiana*, Lyc., *T. formosa*, Lyc., *T. tuberculosa*, Lyc., *T. Griesbachi*, Lyc., *T. ingens*, Lyc. Of these, the only species that occurs in the Cretaceous Rocks is *T. ingens*.

3rd.—UNDULATÆ.

This section was constituted by Agassiz to receive species whose general form approximates to the *Clavellatæ*, but whose costæ, whether with or without distinct tubercles, have an undulation or an angle towards the middle or the posteal portions of the costæ; it also not unfrequently happens that the costæ are broken into two distinct series of rows, of which the anteal series are the smaller and more numerous; the rows are usually ridge-like, sometimes nearly plain, or in other species with a few tubercles or varices; the area is narrow and the bounding carinæ inconspicuous; there is also a mesial furrow, bordering which a line of small tubercles may occasionally be traced in immature forms; the escutcheon is always plain. The boundaries between this section and the *Clavellatæ* are by no means clearly defined, as some species of the latter have not uncommonly a kind of angle or undulation in their costæ either mesially or posteally; this feature, therefore, appears more appropriate to a species than to a section, and it is not probably of any further value than as a convenience in the arrangement of the species. Like the *Clavellatæ*, it is a Jurassic section. It has afforded nineteen British species, as follows:—*T. angulata*, Sow., *T. litterata*, Young and Bird, *T. Clytia*, D'Orbigny, *T. tripartita*, Forbes, *T. conjungens*, Phillips, *T. Carrei*, Mun.-Chal., *T. subglobosa*, Mor. and Lyc., *T. Leckenbyi*, Lyc., *T. v-costata*, Lyc., *T. producta*, Lyc., *T. Sharpiana*, Lyc., *T. compta*, Lyc., *T. Painei*, Lyc., *T. paucicosta*, Lyc., *T. flecta*, Lyc., *T. minor*, Lyc., *T. costatula*, Lyc., *T. composita*, Lyc., *T. arata*, Lyc.

4th.—GLABRÆ.

The *Glabræ* or *Læves* was a section founded by Agassiz upon very insufficient materials, which led to errors in his definition of the section; he described it as without ornamentation, without tubercles or costæ, and resembling a large Unio. With some modification in the sectional characters it will be found to constitute a group sufficiently distinctive, not, indeed, devoid of ornamentation, for this appears to be a feature essential to the entire genus. The usual figure, as remarked by Agassiz, is somewhat inflated and

INTRODUCTION.

ovate or ovately oblong, and the area is only slightly separated from the other portion of the valve, thus resembling Unio. Mesially, or immediately anterior to the position of the marginal carina, is a smooth space, which, commencing at the apex or near to it, gradually widens downwards to the lower border; most commonly this smooth space is more depressed than the other portions of the valve, it is also only very slightly impressed by the lines of growth; its breadth in two of the species (*T. Micheloti* and *T. Beesleyana*) is equal to all the remaining surface of the valve. The area is sufficiently defined and has usually the bi-partite character of the preceding sections, but for the most part it is destitute of carinæ or has only indications of them near to the umbones. The anteal portion of the valve has always costæ more or less prominent; usually they are much smaller and more closely arranged or less strongly defined than in other sections; they are either plain or tuberculated, and not unfrequently both kinds of costæ occur in the same specimen. The lines of growth are very conspicuous near to the lower border, and not uncommonly there are two or more arrests of development or transverse sulcations separating the surface into as many zones, and influencing the direction of the rows of costæ longitudinally. The interiors of the valves have the dividing siphonal ridge unusually prominent and forming a considerable indentation upon the internal moulds, the 'Horse Heads' of the Portland Oolite present good examples. Of the following seven species the last only belongs to the Cretaceous Rocks:—*T. gibbosa*, Sow., *T. Damoniana*, de Lor., *T. Micheloti*, de Lor., *T. Manselli*, Lyc., *T. tenui-texta*, Lyc., *T. Beesleyana*, Lyc., *T. excentrica*, Park.

The preceding species, although they possess certain sectional features in common, nevertheless appear to arrange themselves naturally into two groups, which may be termed respectively the *gibbosa* group and the *excentrica* group. If we place those two well-known species each at the head of a group, each of these series has in common the plain wide mesial space which passes downwards from the apex to the lower border, but in other important features the two groups appear to be sufficiently separated; thus, the *gibbosa* group, which is here illustrated by six species, has in the area and escutcheon a repetition of the characters which are seen also in the *Scaphoideæ*, the *Clavellatæ*, the *Undulatæ*, and the *Quadratæ*; both of these parts are well defined, and the area has the usual bi-partite form, with bounding carinæ more or less clearly developed. In the *excentrica* group, on the contrary, there is no distinct escutcheon, and the space representing the area is destitute both of ridges and depressions. Of this latter series we possess only two British species, excluding *T. sinuata*, Park., and *T. affinis*, Sow., which are only synonyms of *T. excentrica*, Park. *T. longa*, Ag., must also be referred to the same group. It is, therefore, not without some hesitation that I refrain from proposing to add another to the already numerous sectional forms of this genus, embracing *T. excentrica* and *T. Beesleyana*, and content myself with looking forward to the probability that the acquisition of additional forms will at some future period induce changes in the sectional arrangement in the direction now indicated.

5th.—QUADRATÆ.

The *Quadratæ* constitute a small section approximating to the *Clavellatæ*, but distinguished by the shorter figure, by the more quadrate outline, by the very large flattened and only slightly separated area, by the ornamented escutcheon, and by the great irregularity and excentricity in the arrangement of the rows of tubercles or varices upon the sides of the valves. *T. rudis*, Parkinson, may be taken for the best known British type of this section, which also includes *T. nodosa*, Sow., *T. dædalea*, Parkinson, *T. Orbigniana*, Lyc.; the latter, which has generally been mistaken for *T. dædalea*, presents in its sub-ovate figure, bi-partite area, and three nodose carinæ, an approximation to or connecting link with the *Clavellatæ*. The interiors of the valves have, as in the *Clavellatæ* and the *Glabræ*, a divisional siphonal ridge. All the British *Quadratæ* are Cretaceous.

6th.—SCABRÆ.

Unlike the Trigoniæ generally, the form is usually lunulate or crescentric rather than trigonal, but much inflated anteally; the umbones are produced and recurved more than usual; the superior or hinge-border is much excavated, the posterior extremity is produced, rostrated, and attenuated; the area has almost disappeared, excepting in the adult condition, which has indications of bounding carinæ towards the posteal portions of the valves; the large upper surface in these crescentric forms is occupied almost solely by a great concavity, which represents the escutcheon, and which is ornamented by transverse costellæ, similar in character to the costæ upon the sides of the valves. The costated portion has the rows for the most part ridge-like and imperfectly tuberculated, or they are scabrous or serrated. The interiors of the valves in the *Scabræ* have, towards their attenuated posteal portions, a lengthened divisional ridge, which separated the excurrent from the incurrent respiratory canal; there is also a lengthened series of small, regular, transverse, dental processes and alternate pits upon a narrow flattened plate that borders the escutcheon, its entire length in both the valves supporting an internal ligament or auxiliary portion of the hinge apparatus, and appears to be special to the present section in the Trigoniæ, reminding us of a similar feature in the genus Leda. Examples will be given in figures of the interiors of *T. aliformis*. All the British *Scabræ*, twelve in number, belong to the Cretaceous Rocks; they are *T. crenulata*, Lam., *T. aliformis*, Park., *T. caudata*, Ag., *T. Fittoni*, Desh., *T. Etheridgei*, Lyc., *T. spinosa*, Park., *T. ornata*, D'Orb., *T. Archiaciana*, D'Orb., *T. Picteti*, Coq., *T. tenui-sulcata*, Duj., *T. Pyrrha*, D'Orb., *T. Constantii*, D'Orb.

Of the foregoing, the first five only possess that remarkable elongation and attenuation of the posteal portion of the valves which tends to separate this section so prominently from all others of the genus. Agassiz placed *T. duplicata*, Sow., with the *Scabræ*; it is here placed with the *Scaphoideæ* for reasons which are given with the description of that species. The *Scabræ*, including the two sub-groups above indicated, constitute, perhaps, the most prominent and characteristic fossils of the Cretaceous formations; from whatever part of the world they are obtained both their natural history and geological position admit of no dispute, so nearly do the American and Asiatic species approach to the more well known of the European forms.

7th.—COSTATÆ.

In the preceding sections the opposite valves present no permanent or systematic differences either in their figure or ornamentation; in the present section, on the contrary, the difference of the valves in both particulars is universal and nearly uniform in their kind, varying chiefly in their degree of prominence or otherwise; separate descriptions of the valves, therefore, becomes necessary.

The Costatæ constitute a numerous and almost entirely Jurassic section recognized by longitudinal elevated plain costæ upon the sides of the valves; it also possesses other not less persistent and distinctive sectional features. The area is well separated from the costæ; it is bounded in each valve by two well marked dentated carinæ, and is divided into two nearly equal portions longitudinally. The superior portion is more depressed than the other; it is divided longitudinally by a small furrow, bordering upon which and placed upon the lower portion is usually a small indented median carina; these two inter-carinal spaces have small longitudinal indented costellæ, which take the same oblique direction as the carina. In the condition of advanced growth the entire ornamentation of the area is usually replaced by transverse irregular plications of growth. The escutcheon varies greatly in its relative size and figure in the outline of its upper border, and not less so in its ornamentation, so that it constitutes an important feature in the *Costatæ*, without which the definition of any of its species is incomplete and insufficient. The area in the two valves presents some well marked differences; the lower or carinal half of the *right* valve has the inter-carinal costellæ fewer and larger than those upon the other valve; they are also irregular and unequal, so that commonly the median carina only exists as one of these larger costellæ, of which there are usually from two to four. There is also a deeply excavated groove upon the area immediately *posterior* to the marginal carina and partly overlapped and concealed by it; this may be termed the *post-carinal sulcus;* in the other valve a similar sulcus exists immediately *anterior* to the marginal carina, which it separates from the extremities of the costæ. It is more conspicuous than the sulcus in the other valve, and is the *ante-carinal sulcus;* its use corresponds to that of the *post-carinal sulcus* in the other valve; the open extremities of the marginal carinæ when the valves were in opposition formed an aperture or incurrent orifice of the gills which admitted of being perfectly closed in the following manner :—The hinge allowed of a slight vertical or sliding motion to the valves by which the mollusk was enabled to bring the produced and internally convex lip, forming the extremity of the sulcus in the left valve, opposite to and in contiguity with the open extremity of the carina of the right valve, which it exactly fitted; in the same manner the sulcus of the right valve closed the extremity of the carina of the left valve, as seen in our engraving. By this exertion of muscular power, therefore,

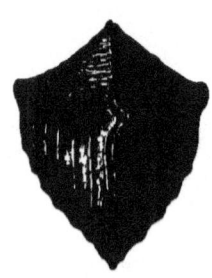

the valves were firmly locked at the will of the mollusk. Another result also followed this arrangement: the orifice for the excurrent aperture existed at the border of the depressed or superior half of the area; and, as that of the right valve is slightly lower than the corresponding portion of the other valve, the border forms a narrow orifice or undulation which became close-fitting when the incurrent aperture was also closed. Specimens, as in *T. elongata*, frequently occur with the apertures closed and locked; in other instances the extremities of the carinæ are rendered open by the relaxation of muscular force at the instant of death. In the *Clavellatæ* and other sections generally there were no open extremities of carinæ, and the excurrent and incurrent orifices were closed by simple muscular effort, but as the respiratory orifices of the *Costatæ* were not closed by simply shutting the valves, it became necessary to protect those organs by a special contrivance which is no less remarkable for its simplicity than its efficiency; it is, in fact, an exact reproduction of the same design as exhibited in the more ancient and allied genus Myophoria. From the foregoing statement it might be concluded that the *Costatæ* represent the most ancient or primordial portion of the genus Trigonia; but, judging from the present state of our knowledge of the organic contents of the Liassic and Triassic strata, such an inference is scarcely justifiable; certain it is that other sectional forms of Trigoniæ occur in Upper Liassic deposits, and that we are altogether unacquainted with the genus in the middle and lower subdivisions of that great formation.

In the left valve the extremities of costæ terminate abruptly postcally, and are separated from the carina by the *ante-carinal sulcus*. As there is no sulcus in that position in the right valve, the costæ touch the marginal carina, and sometimes pass over it as so many plications. The aperture formed by the extremities of the two marginal carinæ does not form a lengthened canal internally as in the genus Myophoria, the deposition of nacreous deposit went on simultaneously with the growth of the valves, so that whatever may be the stage of growth, the figure of an inner canal is preserved even in large species only for a length of about five or six lines.

The cardinal processes are unusually large and massive. In *T. sculpta* they occupy nearly a third part of the interior of the valves; thus it follows that the valves at their apical portions have always much convexity.

INTRODUCTION. 11

There is one other feature common to all the sections, but so much more strongly defined in the *Costatæ*, that it may here be adverted to, more especially as it has not previously been noticed. The oblique median furrow upon the area, forming the lower border of the superior or more depressed portion of the area, indicates the position of an internal rib; in the living mollusk this rib constituted a short process only, which, near to the outer border, served to separate the incurrent from the excurrent respiratory orifice; with progress of growth the process advanced continually, and its former position was obliterated by the deposition of new-formed shell-substance; the external furrow remained, and indicates the former position in growth of the internal divisional ridge.

With the exception of *Trigonia peninsularis*, Coquand, which occurs in the Cretaceous rocks of Spain, all the species of the *Costatæ* are Jurassic. The following thirteen species occur in British strata: *T. costata*, Sow., *T. denticulata*, Ag., *T. pulla*, Sow., *T. Meriani*, Ag., *T. elongata*, Sow., *T. angustata*, Lyc., *T. sculpta*, Lyc., *T. monilifera*, Ag., *T. Cassiope*, D'Orb., *T. Crucis*, Lyc., *T. gregarea*, Lyc., *T. hemisphærica*, Lyc., *T. tenuicosta*, Lyc.

8th.—BYSSIFERÆ.

I propose to constitute this section to include *T. carinata*, Ag., from the Neocomian beds of France and England, a lengthened sub-cylindrical shell, which in general ornamentation resembles the *costatæ*; but in addition it acquired at the period of adult growth a byssal aperture, formed by a slight excavation of the anterior border of each valve; the condition of the area and of the shell generally at that stage of growth indicates that it became fixed or stationary. The elongated and almost cylindrical figure is very abnormal as regards the genus Trigonia, but we perceive a strong resemblance to Bysso-arca and to the relation which the latter holds to the genus Arca; the resemblance is much enhanced by a general worn appearance of the upper surface in adult forms in both these mollusks, which doubtless was the result of similar conditions of existence. The condition of the hinge has not been ascertained.

General Sketch of the Distribution of the Genus Trigonia *throughout the Geological Formations of Britain.*

The lowest geological position in which the genus occurs in Britain is in a single stratum about the middle of the Upper Lias, exposed upon the coast scars of Yorkshire, at the Peak, Robin Hood's Bay; here *T. litterata* occurs in all conditions of growth; it is one of the *Undulatæ*.

The Supra-liassic Sands of the Cotteswold Hills have produced four species very sparingly, and a fifth occurs in a similar position at the Peak; even thus early in the history of the genus these few species exemplify three of the sections into which the genus is divided. Three of these (*T. denticulata*, *T. spinulosa*, and *T. formosa*) pass upwards into the Inferior Oolite, in which they are more abundant. *T. Ramsayi*, and *T. Leckenbyi* are special to this stage.

In Britain the Inferior Oolite appears to have been the very metropolis of the Trigoniæ. Upwards of twenty-seven species have rewarded the industry of collectors; from its geographical position it forms four areas, severally from which the following species have been obtained:

1. *Area to the southward of the Mendip Hills, including the counties of Somerset and Dorset.*

T. costata, T. striata, T. formosa, T. duplicata, T. angulata, T. tenuicosta, T. signata.

2. *Area of the Cotteswold Hills.*

T. costata, T. sculpta, T. denticulata, T. hemisphærica, T. costatula, T. angulata, T. sub-globosa, T. producta, T. signata, T. Phillipsi, T. V-costata, T. pulla, T. tuberculosa, T. formosa, T. duplicata, T. gemmata, T. spinulosa, T. tenuicosta.

3. *Area of the Midland Counties and Lincoln.*

T. sculpta, T. denticulata, T. gregarea, T. hemisphærica, T. pulla, T. formosa, T. signata, T. Phillipsi, T. V-costata, T. Sharpiana, T. compta, T. minor, T. Beesleyana, T. producta.

4. *Area of Yorkshire.*

T. denticulata, T. gregarea, T. conjungens, T. spinulosa, T. V-costata, T. recticosta, T. signata.

From the foregoing lists it will be perceived that the areas of the Cotteswold Hills and of the Midland Counties have many species in common, but that the Somerset and Dorset area and that of Yorkshire have only a single species identical. Of these twenty-seven species two only (*T. denticulata* and *T. pulla*) have been found to pass into a higher formation.

The Bathonian formation has also a greatly varied but less numerous series, seventeen in number, as follows:

T. pulla, T. denticulata, T. Bathonica, T. arata, T. Moretoni, T. impressa, T. Painei, T. Crucis, T. Griesbachi, T. flecta, T. Clytia, T. imbricata; also the following, which appear to be special to the Cornbrash: T. Scarburgensis, T. Cassiope, T. tripartita, T. angustata, T. bivirgata. Of these I have been able to trace only *T. denticulata* into a higher formation.

INTRODUCTION.

Passing upwards we at once take leave of the numerous Trigoniæ that characterise the Lower Oolites; for, although some species occur in immense abundance at certain localities, the number of Jurassic species in any one stage are comparatively few, and in the Middle and Upper Oolites, from the Kelloway Rock to the Portland Oolite inclusive, only twenty-five additional species have been procured. The Oxfordian Trigoniæ are the following: *T. paucicosta, T. denticulata, T. complanata, T. Rupellensis, T. elongata, T. clavellata, T. irregularis, T. triquetra, T. corallina, T. composita, T. perlata, T. monilifera.*

The Portlandian series, including Kimmeridge Clay, Portland Sand, and Portland Oolite, have *T. irregularis, T. Juddiana, T. Pellati, T. muricata T. Carrei, T. gibbosa, T. Damoniana, T. Micheloti, T. Manselli, T. incurva, T. tenui-texta, T. Woodwardi, T. Voltzii.*

In the Cretaceous system of rocks, the species of the Neocomian formation are *T. nodosa, T. caudata, T. carinata, T. ornata, T. spinosa, T. Picteti, T. Pyrrha, T. Orbignyana, T. Etheridgei, T. ingens.*

The Gault has *T. Fittoni.*

In the Upper Green Sand are *T. tenuisulcata, T. Archiaciana, T. excentrica, T. spectabilis, T. crenulata,? T. aliformis, T. dædalea, T. abrupta, T. ornata, T. Cunningtoni.*

The author has much pleasure in recording his grateful acknowledgments for the valuable and varied assistance which he has received in the execution of his task. To Sir Roderick Murchison, Director General of the National Geological Survey, his thanks are due for the liberality which made the resources of the Museum of Practical Geology available for his use. The permission thus accorded has been rendered more especially valuable by the cordial co-operation of Mr. Etheridge in carrying out the wishes of the author, and also in procuring the loan of other valuable specimens. Mr. J. W. Judd, lately an officer of the Survey, has also freely and repeatedly communicated information respecting the range and distribution of the species throughout the district of the Midland Counties, which he is so well qualified to afford. Similar kindness and co-operation on the part of Mr. Henry Woodward, of the British Museum, has enabled the author to utilize to the fullest extent the great resources of that collection, and more especially in the comparison and study of the numerous and splendid foreign examples of Trigonia which form so conspicuous a feature in its fossil conchology.

Mr. Cunnington, of Devizes, whose collection so well exemplifies the fossil faunas of the Upper Jurassic and Cretaceous rocks of that part of England, kindly forwarded everything calculated to illustrate that portion of the subject. The following gentlemen have also afforded material assistance either by the gift or the loan of specimens:—Rev. P. B. Brodie, of Rowington; Rev. J. E. Cross, of Appleby, Lincolnshire; Professor J. Buckman; Mr. J. Leckenby, of Scarborough; Mr. P. Hawkridge, of the same place; Mr. Beesley, of Banbury; Rev. C. L. Smith, of Canfield, Essex; Mr. J. Walker, of York; Mr. E. Witchell, of Stroud; Dr. T. Wright, of Cheltenham; Mr. R. Tate, of Camberwell; Mr. Samuel Sharp, of Northampton; Mr. Mansell, of Blandford, Dorset; also Dr. J. Lowe, of Lynn.

§ I. SCAPHOIDEÆ.

TRIGONIA DUPLICATA, *Sow.* Plate I, figs. 8, 9, 10.

TRIGONIA DUPLICATA, *Sow.*	Min. Con., pl. 237, fig. 4.
— PROSERPINA, *D'Orb.*	Prodr., 1, 10 Et., No. 315, p. 278.
— DUPLICATA.	Ib., No. 317, p. 279.
— — *Morris.*	Catal., 1854, p. 228.
— — *Morr.* and *Lyc.*	Pal. Soc. Mon., 1854, pl. vi, fig. 2.

Shell ovately trigonal, moderately convex; umbones antero-mesial, elevated, obtuse, and somewhat recurved; anterior side much produced, its border curved elliptically with the lower border, posterior extremity more produced, attenuated, and rostrated; hinge-border concave, sloping obliquely downwards; area narrow, flattened, finely striated transversely, and bounded by two very small carinæ, each of which is minutely tuberculated; escutcheon narrow, depressed, and lengthened, forming an excavation at the superior border; it has delicate oblique costellæ, and at the apical extremity are a few transverse ridges, which also pass over the inner carinæ and the area. Costæ numerous (from 12 to 14), narrow, raised, delicately fringed with closely [placed, minute, obtuse, rounded, or ovate tubercles; the first-formed three costæ are concentric, the fourth costa is directed downwards, and has anteally to it four or five short irregular horizontal or supplementary costæ, which are nearly at right angles with the fourth costa; the succeeding costæ—nine, ten, or more in number—pass from the carina downwards in a straight or slightly waved course perpendicularly to the pallial border; two, three, or four of the more anteal of these costæ divide near to the border or near to the middle of their course, each into several smaller costæ, but the few more posteal costæ are undivided. These small supplementary pallial costæ are never precisely alike in any two specimens, but are always small and numerous. The inner borders of the valves are crenulated.

Few Trigoniæ have so much variability as *T. duplicata*, both in the general figure and the arrangement of the rows of costæ; the number of the rows, their closeness or separation, and more especially the number of the small supplementary costæ, all are characterized by this diversity; usually the smaller specimens have the greater convexity, and have their apices more produced and recurved. D'Orbigny was misled by these differences, as exemplified by certain specimens, to separate his *T. Proserpina;* and for some time the larger and more depressed forms, with their closely placed costæ, induced me also to regard them as distinct; the figures now given will sufficiently illustrate this variability.

Agassiz placed *T. duplicata* with the *Scabræ*, influenced, probably, by the examination of insufficient specimens; it is only necessary to direct attention to the characters of the area, with its tuberculated bounding carinæ and transverse striations, to perceive that it cannot be allowed to remain in that section.

Stratigraphical position and localities. This is a delicately ornamented species occurring not uncommonly in the Upper Trigonia-grit of the Inferior Oolite in the Cotteswolds at numerous localities ; impressions in the hard ragstones are common, but it is difficult to separate a specimen in good preservation from the hard matrix. It also occurs in the Inferior Oolite of the Half-way House Quarry near to Yeovil. In France, Guéret is the locality for *T. Proserpina*, D'Orbigny.

TRIGONIA GEMMATA, *Lyc.* Plate I, fig. 7.

TRIGONIA GEMMATA, *Lycett.* Ann. and Mag. Nat. Hist., 1853, pl. ix, fig. 8, p. 425.
— — *Morris.* Catal., 1854, p. 228.

Shell ovately trigonal, moderately convex ; umbones elevated, pointed, and slightly recurved ; anterior border moderately produced, both it and the inferior border elliptically curved ; hinge-border straight, lengthened, sloping obliquely. Area narrow, flattened, transversely delicately striated, but near to the apex costellated, having two distinct, delicately knotted, or cord-like carinæ which circumscribe it. The escutcheon is narrow, lengthened, and depressed, rendering that portion of the slope slightly concave. The costated surface has a dense and salient ornamentation ; the first-formed series of costæ, about seven in number, occupy more than half of the valve. They have the rows very closely arranged, narrow, ridge-like, and concentric, each bearing a row of small, closely arranged tubercles ; the succeeding rows of costæ, eight or nine in number, are similar, but they descend almost perpendicularly, or inclined somewhat forwards from the marginal carina to the pallial border. There is also a third series of short or supplementary costæ, which, originating at the anterior border, pass obliquely upwards to be united to the side of the last-formed costa of the first series, or concentric costæ, producing a singular unsymmetrical but not inelegant aspect to the anteal side of the valve. There are eight of these short costæ. The perpendicular costæ are undivided ; one specimen only has a single intercalated rib ; the tubercles upon the rows are irregular and unequal.

Length, 18 lines upon the carina ; the diameter at right angles to it is 15 lines.

Trigonia duplicata, Sow., is allied to our species, and differs from it in the following particulars :—The species of Sowerby is less convex, and has the ornamentation smaller and more irregular ; it commences with only two or three concentric or horizontal costæ, all the others are directed perpendicularly downwards from the carina ; they are irregularly knotted, are frequently and sometimes alternately dichotomous ; they are also fewer and more widely separated ; the anterior side of the valve has also a few curved and very irregular and narrow costæ. The differences here indicated are very persistent.

T. gemmata is also allied to *Lyrodon sulcatum*, Goldf., from the White Chalk of Havre; but in that species the concentric costæ occupy the whole of the anterior border.

Stratigraphical position and localities. The bed called Upper Trigonia-grit of the Inferior Oolite in the vicinity of Stroud and of Cheltenham; it appears to be one of the most rare productions of that stratum.

TRIGONIA RECTICOSTA, *Lycett*, sp. nov. Plate I, figs. 4, 5, 6.

Shell ovately trigonal, moderately convex; umbones small, antero-mesial, scarcely recurved; anterior side short, truncated, its border slightly curved, lower border nearly straight; hinge-border straight, sloping obliquely downwards; area flattened, very wide, its breadth is equal to two fifths the entire breadth of the valve; it has a mesial furrow, is transversely irregularly plicated, and is bounded by two small tuberculated carinæ; the tubercles upon the marginal carina are regular, and nearly equal in size to those upon the side costæ; the inner carina consists of a row of minute, unequal, transverse, nodose varices. The costæ, nineteen or twenty in number, are moderately elevated, with regular, small, rounded tubercles; the first-formed eight or nine rows are slightly curved or oblique, and are directed towards the anterior border, where they are united to a series of short, ridge-like, narrow, sub-tuberculated, horizontal costæ, of which about nine occupy the anteal portion of the valve; the other costæ pass almost perpendicularly to the pallial border, increasing in size downwards without division or supplementary costæ; about twelve tubercles are in each row. The lines of growth are strongly defined; they form plications where they cross the perpendicular costæ. The general figure of the shell and the direction of the rows of costæ resemble *T. duplicata*, but the costæ are but little elevated or ridge-like; and, unlike that species, the tubercles are regular and symmetrical, the rows are more distantly arranged, and are not divided near to the pallial border into several smaller costæ.

From *T. gemmata*, the second of this group of Lower Oolite forms, it differs both in its figure and in the arrangement of the costæ, which are much fewer and more perpendicular, with wide interstitial spaces, and it has not the numerous first-formed large concentric rows of that species. It is also not without some general resemblance to a young *Trigonia navis*, both in its figure and the design of its ornamentation; for distinctive difference it is only necessary to refer to the numerous costæ and their minute tubercles in the British species.

Stratigraphical position and locality. The Inferior Oolite at Cloughton cliffs, to the northward of Scarborough. The Millepore bed (so called from the prevalence of *Cricopora straminea*) is there somewhat ferruginous; it has produced a few well-preserved specimens of our little Trigonia, and numerous others in an imperfect condition associated with

Trigonia conjungens, Phil., *Pygaster semisulcatus*, Phil., and a considerable series of Inferior Oolite Conchifera, some of the more common forms of which occur also in the Dogger and in the Grey Limestone upon the same coast and vicinity. The position of this marine deposit (from fifteen to twenty feet in thickness), about the middle of the great mass of Estuarine Sandstones and Shales, and between the Dogger and the Grey Limestone, is important as tending to connect the fauna of those two widely separated marine deposits, and as proving that the conditions of sea-bottom and the succession of molluscan life underwent no considerable change during the whole of the northern Inferior Oolite period as exemplified upon the coast of Yorkshire, undoubtedly less than is exhibited by beds of the same geological period at the southern localities. Two other less important marine beds also occur intercalated with the estuary deposits, one between the Dogger and the Millepore bed, the other between the latter deposit and the Grey Limestone; but as their marine testacea are few and ill-preserved, little interest attaches to their presence.

Specimens of *T. recticosta* from Cloughton are in the collection of Mr. Leckenby at Scarborough and in the cabinet of the author; others in a less perfect condition are in the Museum of the Yorkshire Philosophical Society at York. The British Museum has also a fine specimen in the Bean Collection.

TRIGONIA BATHONICA, *Lyc.* Plate I, fig. 3.

TRIGONIA BATHONICA, *Lycett.* Pal. Soc. Suppl. Monog., 1863, p. 52, pl. xl, fig. 3.

Shell sub-trigonal, short, depressed; umbones elevated, mesial and not recurved; anterior and posterior borders nearly straight, sloping obliquely downwards, the surface with numerous (about twenty-four) narrow, elevated, spinose, somewhat undulated and slightly radiating costæ, which are directed from the marginal carina anteally downwards, and all reach the pallial margin; the area is narrow and transversely striated; the marginal carina is very small and rather indistinct.

The narrow ridge-like costæ are very closely arranged, and have numerous minute obtuse spines, which impart roughness to the surface; the general aspect resembles *T. duplicata*, Sow., but it has no bifurcating or interstitial costæ near to the lower border, it is also without concentric costæ near to the apex; the absence of this latter feature will also distinguish it from *T. gemmata*, Lyc. The sole specimen at my disposal is imperfect at the posterior extremity; it has twenty costæ, and would require about four others to complete its surface. The figure is nearly an equilateral triangle, each of the sides having a length of about an inch.

Oppel, in his elaborate work 'Juraformation,' p. 486, makes incidental mention of a Trigonia which is regarded by Messrs. Rigaux and Sauvage as identical with our species.

It is from the Cornbrash of Marquise, and is named by him *T. Bouchardi*; unfortunately no figures of this and other new species incidentally and briefly mentioned in that work have been given, and, in the absence of any allusion to the difference of the costæ near to the apex, I consider that the species in question is not sufficiently characterised to constitute it an authority.

Stratigraphical position and locality. The only example at our disposal was obtained by Mr. Walton in the Great Oolite near to the Box Tunnel, Bath.

§ II. CLAVELLATÆ.

TRIGONIA CLAVELLATA, *Sow.* Plate I, figs. 1, 2.

CURVIROSTRA RUGOSA CLAVELLATA MAJOR, *Luid.* Litho., 1699, p. 36, pl. ix, fig. 700.
TRIGONIA CLAVELLATA, *Sow.* Min. Conch., 1815, pl. lxxxvii.
LYRODON CLAVELLATUM, *Goldf.* Petref., 1834—1840, p. 200, pl. cxxxvi, fig. 6, *c, d, e,*
    and *f,* excl. fig. 6 *b.*
TRIGONIA CLAVELLATA, *Morris.* Catal., 1854, p. 229.
—   —   *Damon.* Geol. Weymouth, Suppl., 1860, pl. iv, fig. 2.

(Exclude the figures of *T. clavellata* in the works of Parkinson, Young and Bird, Bronn, Zeithen, Agassiz, Zwingen, Goldf., Petref., pl. cxxxvi, fig. 6, *b*; also of Hébert, "Trigonées clavellées," Jour. de Conchyl., 1861, pl. vii, fig. 1.)

Shell ovately trigonal, moderately elongated, convex; umbones large, obtuse and incurved, but rarely recurved; anterior side rounded, but not much produced, its lower extremity curved with the lower border; superior border straight, lengthened, sloping obliquely downwards; escutcheon flattened, its length is nearly equal to half the length of the marginal carina; the area is narrow, flattened or slightly convex, transversely irregularly plicated, having three carinæ, of which the mesial carina consists of a row of delicate, regular, small tubercles; the two bounding carinæ have the tubercles much larger, but depressed and closely arranged, those upon the inner carina form lengthened transverse varices; a well marked furrow borders upon the median carina, and in some specimens the furrow is bounded upon each side by a row of tubercles; more frequently, however, the second or upper row is very imperfectly indicated; the superior half of the area is more depressed than the other portion. The sides of the valves have the rows of tuberculated costæ at first oblique, but the later-formed few become more horizontal or more nearly accord with the direction of the lines of growth, so that the greater number of the rows reach the anterior border in the form of small attenuated or sub-tuberculated varices; the postcal extremities of the rows approach the carina at an angle which is somewhat greater than a right angle; the tubercles in the rows are large, not prominent, imperfectly rounded, closely arranged, and are rather unequal both in size and figure; eight tubercles are usually distinct in each row before they degenerate anteally into small varices; the last-formed one or two rows in specimens of advanced growth have the tubercles compressed into lamellar varices, or form almost continuous costæ. The shell in the Lower Calcareous Grit, which I regard as the type, has sixteen or seventeen rows of costæ in adult specimens; this form is special to the Lower Calcareous Grit and is the prevailing clavellated species of that rock both in Dorsetshire and Yorkshire. A clavellated Trigonia special to the Kimmeridge Clay, and very much resembling the typical *T. clavellata* of the Lower Calcareous Grit will, without due care, be placed with the latter species. For a comparison of these forms see the species next following.

Of the mistakes in the identification of species *T. clavellata* is a remarkable instance, the errors respecting it having been chiefly those of Continental authors, who have not had the advantage of comparing authentic English specimens, and have been misled or confused by the figure of Parkinson, which unfortunately has priority; this drawing is in every respect execrably bad, and undeserving of trust, so that after many unsatisfactory attempts at comparison with English Trigonias I have felt compelled to discard it altogether, and to regard the figure in the 'Mineral Conchology' as the typical example, as it is altogether free from doubt, and is readily identified with numerous Weymouth examples obtained in the Coralline Oolite formation or Lower Calcareous Grit of that vicinity. Nevertheless, this variety admits of some variability of figure, and we may regret that one or two additional specimens were not figured in the 'Mineral Conchology;' as an instance, refer to the figure in Mr. Damon's supplement to his 'Geology of Weymouth,' which accurately represents a specimen of abnormal form, with the anterior side very short, and the tubercles in the rows so large as to be partially confluent. Upon the whole, the larger figure of Goldfuss (' Petr.,' pl. 136, fig. *b*, *c*) is the best hitherto given of the adult form, but the locality (Inferior Oolite of Gundershofen) is unquestionably an error.

The *T. clavellata*, Agassiz, from the Oxford Clay of Dives, is remarkable for the great elevation and recurvature of the umbones, and the horizontal direction of the rows of costæ. D'Orbigny has justly separated this species under the name of *T. major*; it has not been recognized in Britain.

The *T. clavellata* of Zeithen is so very unlike the English species that we may be confident no true example of *T. clavellata* had come under his notice; it appears to coincide with *T. signata*, Ag. The *T. clavellata* figured by M. Hébert, from the Oxford Clay of Tronville, is also a different species, remarkable for the great breadth of the area, together with the shortness and prominence of the escutcheon. The *Trigonia Bronnii*, Ag., of which a single defective specimen was figured in the memoir by that author, has since been fully illustrated and described by M. Hébert in an interesting paper on the

"Trigoniæ of the Coral Rag;" he also states that he has seen specimens from the Calcareous Grit of Weymouth, and even expresses doubts as to the real type of *T. clavellata*. The specimens of *T. Bronnii* from Glos (Calvados) are much smaller than the *T. clavellata* of Weymouth; in all the specimens which I have examined the escutcheon has greater breadth and is shorter; the posterior extremity of the valve is broader and less pointed; adult specimens have the rows of tubercles less numerous, and their general direction is more horizontal; their irregularity in size and arrangement is also very conspicuous when compared with *T. clavellata*.

Another very large clavellated species from the Lower Calcareous Grit of Yorkshire and Oxfordshire has sometimes been mistaken for it; this is the *T. triquetra*, Seebach, for which the reader is referred to the description.

A near ally to *T. clavellata* is *T. perlata*, Ag., of which a great profusion of specimens have been obtained in the Coralline Oolite of Pickering, for which also see the description.

In Britain *T. clavellata* has occurred very abundantly in layers of the Lower Calcareous Grit formation in scars, and in the cliff at Sandyfoot Castle, near Weymouth; also in the same formation in Wiltshire and near to Filey Point, Yorkshire. Our figures represent its usual dimensions.

TRIGONIA VOLTZII, *Ag.* Plate X, figs. 1, 2.

   TRIGONIA VOLTZII, *Agassiz.* Trigonées, 1840, p. 23, pl. ix, figs. 10—12.
      —   —   Oppel. Juraformation, 1856—1858, p. 719, No. 88.
      — CLAVELLATA, *Morris.* Catal., 1854, p. 228 (pro parte).

Few Trigoniæ have been the cause of so much perplexity and doubt as the present form, the result of the very unsatisfactory figure given by Agassiz, and also of the insufficiency of his description; so obscure, in fact, has this species appeared to be, that the greater number of authors have been contented to ignore it altogether. The figure in the memoir of Agassiz represents the internal mould of a clavellated Trigonia, some of the tubercles of which are impressed upon its surface, and the only distinctive character that can be ascertained from it is that the form is more elongated than *T. clavellata*.

D'Orbigny ('Prodrome,' 2, p. 51, makes *T. Voltzii* a synonym of *T. muricata*, Goldfuss.

Oppel ('Juraformation', p. 719) has a lengthened note upon *T. Voltzii*, which proves that he was well acquainted with our species, of which he had himself collected several specimens from the Kimmeridge Clay of Boulogne; but he does not appear to have been so certain with regard to the true type of *T. clavellata*, and he therefore gave no further comparison of the two forms than to state that the figure of *T. Voltzii* will readily be distinguished from the other by its greater length.

Having, after great delay, obtained perfect specimens of the Kimmeridge Clay *T. clavellata* (so called), I am enabled to affirm its specific distinctiveness from the typical or Calcareous Grit examples of *T. clavellata*, and to give its distinctive features with sufficient precision, figures of which will be given upon Plate X. Compared with that form, *T. Voltzii* is a larger, and also, in proportion, a more lengthened form, the umbones are somewhat more elevated and attenuated; the anterior side is short, but the posterior side is more produced; the test is also unusually thick; the convexity of the valves is somewhat less, consequently the surface of the area is more nearly upon the same plane with the other portion of the valve; the area in its other features offers but little that is distinctive, excepting that the transverse plications are unusually large, irregular, rugose, and wrinkled; they are united to the tubercles of the carinæ. The rows of tuberculated costæ upon the other portion of the valve are invariably less numerous, and much more widely separated, than in *T. clavellata*; adult examples of the latter shell have sixteen or seventeen rows; the larger, *T. Voltzii*, has only eleven or twelve rows; the tubercles also have some differences, their number in each row is nearly similar, but the Kimmeridge Clay shell has its tubercles compressed, or cuspidated and pointed, and, unlike the other, they are much impressed by the lines of growth, which are unusually large and conspicuous over the whole of the valve. The largest specimens have the anteal portions of the costæ attenuated and cord-like; the largest tubercles are near to the posteal extremities of the rows.

It may be a question of dispute how far the above-stated distinctions are of specific value, or what are the limits of variability possessed by each of these two clavellated Trigoniæ; without discussing the arguments which may be adduced for their distinctiveness or otherwise, it will be sufficient to remark that the peculiarities of each form are observable upon all the specimens in the geological formation where they occur, and are therefore of value in a stratigraphical point of view.

Separate valves of *T. Voltzii* occur in the Kimmeridge Clay of the coast of Dorsetshire, in the same formation at Wootton Basset, Wiltshire, and in Lincolnshire; examples with the valves united are rare. The localities given by Agassiz are Argentenay (Yonne), also Besançon; Oppel gives Boulogne: all in the Kimmeridge Clay.

TRIGONIA PERLATA, *Ag.* Plate III, figs. 1, 2, 3.

TRIGONIA CLAVELLATA, *Young* and *Bird.* Geol. Survey, 1828, pl. viii, fig. 18.
— PERLATA, *Agassiz.* Trigonies, 1840, p. 19. pl. iii, figs. 9—11.
— — *Hébert.* Trigonies clavellées, Jour. de Conchyliologie, 1861, pl. vii, fig. 2.

T. perlata, Ag., has not unfrequently been mistaken for *T. clavellata*, Sow., with which, indeed, it possesses some strong affinities. These errors are for the most part to be referred to the very imperfect single figure of the adult form given by Agassiz, and to his having mistaken the species of Sowerby, and figured another and very different form for his *T. clavellata ;* fortunately M. Hébert has given a good figure and precise description of *T. perlata* in the memoir above cited on some clavellated Trigonias of the Oxfordian Rocks. Although the features which distinguish the *T. clavellata* of Weymouth from the *T. perlata* of Pickering had long been present to my mind, it was the memoir of Hébert that enabled me to identify the latter with the species of Agassiz. Adult specimens of *T. perlata* agree in size with those of the other species. Young specimens are smaller than those of *T. clavellata*, and have their ornamentation much more minute, as exemplified in the rows of tuberculated costæ, and in the tubercles upon the carinæ ; the form also is much more pointed and produced, both at the apex and the opposite extremity of the valves, so that, even when comparing young specimens of both species, their distinctness is evident. Adult specimens of *T. perlata* have much variability in the number of costæ and in the relative size of tubercles ; occasionally the costæ form narrow sub-tuberculated ridges, and the angle at which they approach the carina differs, but the angle always exceeds that which occurs in *T. clavellata.* The apices are more produced, narrow, and more distinctly recurved. After making allowance for occasional variability, this latter feature is very persistent ; the opposite extremity is as constantly more attenuated and even rostrated. This figure is produced, not by an actual difference in the measurement across that part of the valve, but by the greater angle which that portion of the area forms with the costated portion of the valve, so that when the valve is viewed laterally, the posteal portion of the area is but little seen, and is not elevated as in *T. clavellata ;* an appearance of greater breadth and roundness is thus imparted to the posteal extremity of the latter species. The upper portion of the area is more depressed ; but there is never any distinct furrow, and never any indications of a second row of tubercles, as in *T. clavellata.* The three distinctive features, therefore, which are immediately evident are the smaller and more pointed tubercles, the more narrow and recurved apex, and the more produced and narrow posteal extremity in *T. perlata.* The more distinctly ridged specimens have the interior borders of the valves scalloped. The valves of Trigonia in the Coralline Oolite of Pickering occupy about a foot in thickness ; the specimens are of all periods of growth, and the valves are invariably disunited. Their surfaces have frequently been worn by attrition, but there is never any appearance of compression or distortion; many valves are broken, so that only a minority have the surface ornaments well preserved.[1]

TRIGONIA BRONNII, *Ag.* Plate IV, fig. 8.

> LYRODON CLAVELLATUM, *Bronn.* Lethæa Geognostica, 1834—1838, pl. xx, fig. 3.
> TRIGONIA BRONNII, *Agassiz.* Trigonées, 1840, p. 18, pl. v, fig. 19.
> — — *D'Orbigny.* Prodr. de Paléont., ii, 1850, p. 16, No. 259.
> — — *Hébert.* Jour. de Conchyl., 1861, pl. vii, figs. 4, 6, et pl. viii, figs. 1, 2, 3; Note sur les Trigonies clavellées de l'Oxford Clay et du Coral Rag.

Compared with *T. clavellata*, Sow., and *T. perlata*, Ag., the chief distinguishing feature consists in the straightness or horizontal directions of the rows of costæ which approach the marginal carina nearly at right angles, and the costæ last formed take nearly the direction of the lower border. The irregularity and inequality in the tubercles is also remarkable; usually the second and third tubercles from the carina are larger than the others, and the last-formed one or two rows are smaller, irregularly knotted, and cord-like; their general direction is nearly horizontal, so that all the rows have their extremities upon the anterior border. Near to the apices the rows are nearly plain, or only slightly crenulated. The escutcheon nearly resembles that of *T. clavellata*, excepting that it is somewhat larger, and has also greater length; the area and its ornamentation do not offer anything remarkable, its posteal truncation is usually greater than in our figure. The lines of growth are less conspicuous than in the allied species. Measurements of the dimensions are of little utility in a species whose figure varies considerably. More commonly the tubercles upon the costæ are fewer and larger, and sometimes more scabrous, than in either *T. clavellata* or *T. perlata*. The usual figure of the shell is also less elevated or more oblong; in a multitude of examples which I have examined from the Coral Rag of Glos these distinguishing features are persistent.

Stratigraphical position and localities. In the vicinity of Weymouth it has occurred

---

[1] The following section in descending order shows the position of the Trigonia bed at Pickering in the Coralline Oolite:
1. Rubbly coarse limestone with large coralline masses.
2. Thick bed of oolitic building-stone.
3. Hard band, one foot thick, full of shelly fragments, and of disunited valves of *Trigonia perlata*.
4. Thick bed of oolitic building-stone.
5. Flaggy, thin-bedded, hard oolite, full of small mollusca, Cerithium and Nerinæa (basement bed of the Coral Rag.)
6. Yellow, hard, subsiliceous sandstones of the Lower Calcareous Grit.

rarely, and does not usually much exceed the size of our figured specimen; it is found in the Calcareous Grit at Osmington Hill; the same formation at Filey Point, Yorkshire, has also produced it rarely, and not well preserved. In Normandy it is very abundant in the Coral Rag of Glos and of Hennequeville.

TRIGONIA INGENS, *Lycett*, sp. nov.   Plate VIII, figs. 1, 2, 3.

Shell sub-ovate or ovately oblong, convex anteally; umbones obtuse, moderately produced; anterior border short, curved arcuately with the lower border; posterior border nearly straight, sloping obliquely, and terminating in a rounded, wide, posteal extremity; escutcheon large, lengthened, concave, its superior border raised; area moderately large, slightly convex, with a mesial divisional furrow, bordering a small medium tuberculated carina; the area is also bounded by two small minutely tuberculated carinæ; its surface has small transverse striations, which over its posteal half become irregular rugose plications, the carinæ at the same position also disappear. The other portion of the shell has about fourteen rows of large, oblique, tuberculated costæ; the tubercles, six or seven to each row, are large and rounded, but sometimes compressed at their upper sides; they are nearly equal in size, but become suddenly small at the anteal curvature of the valve, where the costæ become cord-like and bend upwards. The last formed two or three costæ are smaller, more depressed, and cord-like, or without distinct tubercles, and in this degenerated condition they proceed anteally in the direction of the lines of growth, or nearly parallel to the lower border. Specimens of adult growth have the lengthened anteal slope occupied by a series of short, narrow, ridge-like, sub-tuberculated costæ, which pass upwards almost perpendicularly to the extremities of the larger costæ; there are about twelve of these supplementary costæ, they gradually disappear at the curvature which unites the anterior and lower borders.

The lines of growth are strongly defined over the whole of the valves.

This is the only British species of the *Clavellatæ* known in the Cretaceous Rocks. Compared with the numerous Jurassic clavellated species, it does not appear to possess any *sectional* distinctive features; its nearest ally is *T. Voltzii*, which it closely resembles in the characters of the tuberculated costæ, excepting that the rows are somewhat more elevated or ridge-like, and that the largest tubercles are those nearest to the marginal carina; the general figure also is less lengthened; the umbones are more obtuse, or less produced, less attenuated, and have not the curvature of the Kimmeridge Clay species; the posterior side is of greater breadth, and is without attenuation or flattening; the short anteal supplementary series of costæ is also another distinctive feature.

The internal mould is inflated anteally, compressed posteally; the apices are widely separated; the posteal muscular scar is unusually large. The height is equal to four fifths of the length; the diameter through the united valves is equal to half the length.

Stratigraphical position and locality. The Neocomian formation of Downham, Norfolk; the rock is a coarse, brownish, or sometimes greyish-brown, incoherent sandstone, locally called Carstone; various specimens, for the most part ill-preserved, and also external casts, have been liberally forwarded to me from the Museum of the Lynn Philosophical Society, through the kindness of Dr. Lowe of that place. Our figures are taken from moulds of gutta-percha pressed into the external casts, and also an indifferently preserved internal cast.

TRIGONIA JUDDIANA, *Lycett*, sp. nov. Plate II, figs. 6, *a*, *b*, *c*; Plate IV, figs. 5, 7.

Shell gibbose, ovately oblong, short and truncated anteally, posteally flattened and angulated; umbones antero-mesial, elevated, much incurved but scarcely recurved; lower border curved elliptically; superior border of moderate length, slightly concave, terminating posteally in a considerable angle with the wide posterior border of the area. Escutcheon large and slightly depressed; its superior border is not raised. Area wide, mesial furrow conspicuous; the superior half of the area is depressed concave; marginal carina small, but well marked, with a row of regular, small, distinct tubercles; transverse plications upon the area very irregular, often wrinkled; they frequently unite to form varices at the median and inner carinæ; near to the apex they become regular, plain, narrow, transverse costellæ. The other portion of the valve has about twelve or thirteen rows of clavellated costæ, which curve obliquely downwards and forwards from the marginal carina, and form short, abruptly attenuated varices upon the curvature of the anteal smooth, flattened space; the rows terminate posteally at a smooth, slightly depressed space, which widens downwards and separates the rows from the marginal carina. The tubercles, from six to eight in each row, are prominent, pointed, and somewhat ovate; they are nearly of equal size, and the rows are symmetrical, excepting the two last formed, which are rendered squamous by the large plications of growth near to the lower border.

The diameter through the united valves is equal to half the length of the marginal carina.

This is one of the most convex, and also short or sub-quadrate, forms of the *Clavellatæ*; it will readily be distinguished by the general shortness of the figure, and the truncated outline, both anteally and posteally, the smooth post-costal space, the general gibbosity, and the short oblique rows of prominent pointed tubercles. It is allied to, but I believe distinct from, a clavellated species found in the Kimmeridge Clay of Boulogne

(*T. Rigauxiana*, Mun.-Ch.[1]); in the latter species the comparatively narrow area and the long oblique slope of the hinge-border are essentially different, and also the absence of the wide, smooth, ante-carinal space. Another allied species from the same formation in North-Western Germany is *T. verrucosa*, Credner,[2] but the latter is more erect, its convexity is much greater; the rows of tubercles are much more concentric, smaller, and more numerous; they become very small and attenuated as they approach to the position of the carina; this latter is also apparently destitute of tubercles; it is, therefore, clearly distinct. A clavellated species still shorter has also been figured in the same work under the name of *T. clivosa*; it has the rows of costæ almost horizontal or sub-concentric, and appears to be destitute both of the smooth ante-carinal space and of tubercles upon the marginal carina; it is, therefore, more remotely allied to our species.

Some specimens obtained from the same bed in Lincolnshire, and at the same locality, are more gibbose, with more numerous rows of costæ, each of which has smaller and more numerous tubercles; the posteal extremities of the costæ curve upwards, and form small faintly defined varices upon the smooth ante-carinal space; these are exemplified by Plate IV, figs. 6, and 7, which afford a marked contrast to fig. 5. In both varieties the lines of growth are strongly defined over the entire surface. As the test has undergone considerable change, no portion of the granulated tegument remains.

Stratigraphical position and locality. The Kimmeridge Clay of Market Rasen, Lincolnshire. The name is intended as a slight recognition of services to Jurassic Geology rendered by Mr. John W. Judd in Lincolnshire and the adjacent counties, during his labours as an officer of the National Geological Survey.

TRIGONIA TRIQUETRA, Seeb. Plate VI, figs. 1, *a*, *b*, 2.

TRIGONIA TRIQUETRA, *Seebach*. Der Hannoversche Jura, 1864, p. 117, pl. ii, fig. 5.

Shell sub-trigonal, depressed; umbones elevated, pointed, and slightly recurved; anterior side very short, its border abruptly truncated; lower and posterior borders slightly curved, giving to the general form, with its pointed posterior and apical extremities, an unusual trigonal appearance.

The escutcheon is large, slightly depressed, flattened; its length is equal to half of that of the marginal carina and to more than twice its breadth; the area is narrow, slightly convex; the posteal half is more depressed than the other; the sides of the valves are flattened; they have rows of large varices, or in other instances tubercles, which pass

---

[1] " Note sur quelques espèces nouvelles du genre Trigonia," par M. Munier-Chalmas, ' Bull. de la Société Linnéenne du Normandie,' vol. ix, 1863–4, (Caen, 1865), pl. iv, fig. 2.

[2] ' Ueber die Gliederung der obern Juraformation, &c.,' Heinrich Credner, Prag, 1863, pl. viii, figs. 23, *a*, *b*, *c*.

downwards to the lower border at right angles to the carina; they are straight or occasionally waved. The other specific features may be conveniently given under two separate descriptions of individual specimens, the originals of our figures; of these the Yorkshire shell appears to be of more advanced growth than the other.

From Lower Calcareous Grit of Cumnor, Oxfordshire.

The median and inner carinæ are represented each by a row of regular rounded tubercles the marginal carina is distinctly elevated, and for two thirds of its length has a row of well separated rounded tubercles; towards the posteal extremity these become large plications. The costæ consist of fifteen rows of large and moderately elevated rounded tubercles, which, towards the border, become continuous rope-like varices; the last three rows are altogether continuous and rope-like; the lines of growth are distinct upon the lower portion of the valve, and are more conspicuous where they cross the area.

From Lower Calcareous Grit of Filey Point, Yorkshire.

The median and inner carinæ are slightly elevated; each consists of an irregular series of unequal transverse varices, which are continuations of the large irregular plications that cross the area; the marginal carina is elevated, consisting of squamous varices or large plications, which are continued across the area. The costæ consist of about fourteen or fifteen rows of oblique, but straight or somewhat waved, broad, depressed ridges, each of which has about thirteen narrow, oblong or slightly rounded varices, which are much impressed by the large, irregular, longitudinal plications upon the sides of the valves; the varices are compressed obliquely from the direction of the carina, and therefore not in the direction of the lines of growth.

Apparently these specimens exhibit the extremes of variability to which the species is liable in the surface-ornaments, and also in the figure; the Cumnor shell is unusually convex, and is also very short compared with the height; the other has the length greater than usual in proportion to the height.

From *T. clavellata*, and not less so from other of the *Clavellatæ*, it is readily distinguished by the sub-trigonal depressed figure, and large, nearly perpendicular, nodulous varices upon the sides of the valves. It has some resemblance to *T. Suevica*, Quenst., but is much shorter, and the apex more elevated, with fewer oblique varices.

Large blocks of stone, detached by marine action from adjacent beds of Lower Calcareous Grit, at Filey Point and at the Castle of Scarborough, contain rough and usually ill-preserved specimens of this Trigonia, of a size comparable to the largest known examples of the genus. In common with other insufficiently known clavellated forms it has been assigned to *T. clavellata*, Sow. Upon the coast of Yorkshire the valves are seldom found disunited; they are in contact, or spread open; originally held together by the ligament, or in the worst-preserved specimens the calcareous spar into which the test was transmuted has disappeared, and the rough, brown, grit-stones still show the ornamentation of the valves more or less imperfectly. The foregoing remarks are founded upon eight examples of various dimensions, two of which are from the same formation at Cumnor, Oxfordshire; three have the valves in contact, the others are in a less satisfactory condition; the largest has a length of six inches upon the marginal carina; the opposite measurement is four and a half inches; but, judging from other imperfect specimens, these are not the largest dimensions attained by the species.

M. Seebach states that *T. triquetra* occurs in the Coralline Oolite of Malton and Pickering, but of this I know of no example; in Hanover he gives as its locality the Hersum beds of the Tönnjesberg, in true Coral Rag. Specimens are in the Museum of Practical Geology, in the Museum of the Philosophical Society, Scarborough, and in the cabinet of the author.

For the larger of the specimens figured (Plate VI, fig. 2) I am indebted to the generosity of Mr. Hawkridge, of Scarborough.

TRIGONIA RUPELLENSIS, *D'Orb.* Plate VIII, fig. 4.

<blockquote>
TRIGONIA RUPELLENSIS, *D'Orbigny.* Prodrome de Paléont., 1850, tome ii, p. 17, No. 261.
— CLAVELLATA, var. *Leckenby.* Quart. Jour. Geol. Soc., 1859, vol. xv, p. 8.
</blockquote>

Shell ovately trigonal, moderately convex; umbones elevated, pointed, and slightly recurved, placed within the anterior third of the valves; anterior side short, both it and the inferior border elliptically curved; posterior side moderately produced, its border somewhat concave. Escutcheon lengthened, narrow, depressed, and flattened; its length is equal to two thirds of that of the marginal carina; the area is narrow and flattened, delicately transversely plicated, with three very small minutely knotted carinæ, which become evanescent posteally; the posteal extremity of the area forms an obtuse angle, both with the escutcheon and with the inferior border; the apical portion of the area has a few transverse costellæ. The costated portion of the shell has about thirteen rows of tuberculated costæ, of which the four or five rows first formed are simply sub-concentric and ridge-like or sub-tuberculated; all the remaining costæ have distinct rounded tubercles, but their direction is very irregular; posteally they are curved upwards to the carina at a considerable angle, and the last three or four costæ have their superior extremities with the tubercles confluent or forming small depressed varices; anteally there are two or three short additional or supplementary rows of tubercles, which form a very irregular and confused ornamentation over a large portion of the valve; nevertheless, the anteal extremities of the rows pass to the border in regular order and attenuated form.

The plications of growth are strongly defined; they impress the three last-formed rows of costæ. The length of the marginal carina is about one fourth greater than the height; the diameter through the united valves is equal to half the height.

Only a few specimens have been procured, and these vary much from each other in the number, character of, and description of the rows of costæ; with a single exception, also, the condition of preservation assumed by these Kelloway Rock Trigonias is very indifferent, and renders the task of description difficult and deficient in definition.

The description of *T. Rupellensis* in the 'Prodrome' of D'Orbigny is very brief, but appears to be sufficient to characterize the species. It was also briefly alluded to, and its more prominent features indicated, by Mr. Leckenby, in his 'Memoir on the Kelloway Rock of the Yorkshire Coast.' It has occurred very rarely; the original of our figure, from Mr. Leckenby's cabinet, is the only perfect example with which I am acquainted.

Geological position and locality. The Kelloway Rock of Red Cliff, near Scarborough, associated with *T. paucicosta* and numerous other fossils characteristic of that formation.

The French specimens are from the Coral Rag of La Rochelle and Nantua.

TRIGONIA SIGNATA, *Ag.* Plate II, figs. 1, 2, 3.

TRIGONIA CLAVELLATA, *Ziethen.* Petref. Würtemburg, 1830, pl. lviii, fig. 3.
— SIGNATA, *Agassiz.* Trigonées, 1840, pl. iii, fig. 8; pl. ix, fig. 5; p. 18.
— — *D'Orbigny.* Prodrome, 1850, tome i, p. 278.
— DECORATA, *Lyc.* Ann. & Mag. Nat. Hist., 1850, vol. xii, pl. xi, fig. 1.
— — *Morris* and *Lyc.* Gr. Ool. Monog. Pal. Soc., 1853, pl. xv, fig. 1.
— — *Morris.* Catal., 1854, p. 228.
— CLAVO-COSTATA, *Lyc.* Ann. Nat. Hist., 1850, pl. xi, fig. 6 (variety).
— SIGNATA, *Oppel.* Juraformation, 1856, p. 408.
— CLAVELLATA, *Quenstedt.* Der Jura, 1856, pl. lx, fig. 13.
— SIGNATA, *Dewalque* and *Chapuis.* Pal. Luxemb., 1857, p. 172, pl. xxvi, fig. 1.

Shell ovately elongated, sub-trigonal, depressed; umbones antero-mesial, small, and not prominent nor recurved, but rarely they are more erect and recurved; the anterior side is moderately produced and rounded; both this and the lengthened lower border are curved elliptically; superior border straight and lengthened, or, more rarely, somewhat concave; area wide, flattened; its posterior extremity is compressed and somewhat truncated, bounded by two delicate, minutely tuberculated carinæ, and traversed longitudinally by a mesial furrow, and sometimes by a minutely tuberculated carina for about the half of its length; it is also transversely plicated, either coarsely or delicately; in the former case the tubercles of the inner carina forms varices or continuations of the transverse plications; the whole surface of the area measured transversely is upon the same plane.

The escutcheon is depressed, lengthened, and narrow; its superior border is somewhat raised.

The costated portion of the shell has a numerous series (about twenty) of oblique rows of tuberculated costæ, of which the first-formed four or five are slightly curved, but are nearly horizontal, delicate, and sub-tuberculated; the rows which succeed are also raised;

they have the tubercles small, separate, rounded, regular, and nearly of equal size, excepting near to the anterior border, where the costæ are attenuated and their tubercles small and cord-like or indistinct; the last-formed six or seven rows pass upwards nearly perpendicularly to the marginal carina, with which they form a considerable angle; these portions of the last-formed costæ are cord-like or imperfectly tuberculated; there are thirteen or fourteen tubercles in each row.

Most commonly the rows of tubercles are symmetrical and continuous across the entire valve, but occasionally the anteal portions of the valves have the tubercles confused and irregular; in such instances the tubercles continue rounded and separate.

The examples upon our plate, which are only of medium size, indicate that the species possessed variability both in the general figure and in the ornamentation. Some specimens obtained near Chipping Norton are remarkable for the raised ridge-like figure of all the costæ, the indistinctness of their tubercles, the coarseness of the plications across the area, the great convexity of the valves, and the recurvature of the umbones. A specimen with the valves spread open, now in the National Museum, Jermyn Street, is remarkable for these peculiarities. The largest example I have seen is in the collection of Professor Buckman, of Bradford Abbas; its length is four and three quarter inches, and is from the Upper Trigonia-grit of Rodborough Hill; the area has coarse plications which render the carinæ obscure. The figure given by Messrs. Dewalque and Chapuis is very distinctive in the characters of its costæ, but the three large cord-like carinæ upon the area differ altogether from the numerous examples that have been brought under my notice.

The very indifferent examples figured by Agassiz, especially that of his Plate III, fig. 8, induced me at first to regard the British forms as a distinct species. The attenuations of the carinal extremities of the costæ, and their increase of size towards the anterior and the lower border, as depicted in the example above quoted, are altogether unlike British specimens, from whatever locality they may be obtained; it is, therefore, just possible that the first figure of Agassiz may really represent another species, even if we allow some latitude for variability in the ornamentation.

Additional examples of *T. clavo-costata*, Lyc., indicate that it is the immature condition of a large variety of *T. signata*, in which both the tubercles upon the costæ and those also upon the marginal carina participate in general increase of the dimensions; the figure is also less lengthened posteally than in the typical form. The few specimens which I have examined are from white limestone in the vicinity of Stroud.

Affinities and differences. From *T. clavellata*, Sow., and *T. perlata*, Ag., *T. signata* is sufficiently separated by the more depressed form, the more numerous rows of costæ, the smaller and more numerous tubercles, together with the considerable angle at which they approach the marginal carina: others of the *Clavellatæ* are more remotely allied.

Stratigraphical position and localities. *T. signata* appears to be limited to the Inferior Oolite, in which it has occurred at numerous localities, both British and Continental, but it is not a common species. In Dorsetshire it appears to be present, judging from the matrix of two specimens which have come under my notice. In the Cotteswold Hills the upper hard ragstones, or Upper Trigonia-grit, yield many impressions of its outer surface; but examples with the test preserved are more rare. Rodborough Hill, near Stroud, and other localities of the same vicinity, have produced good examples; similar conditions apply to the uppermost bed of the same rock in Oxfordshire, near Chipping Norton. In the same county the ferruginous Inferior Oolite Sands at Rollwright Heath and Hook Norton have yielded numerous specimens in a beautiful condition of preservation as regards the test, both externally and internally. Examples of these are in the collection of Mr. Stuttard, of Banbury; and also in the Museum of the National Geological Survey, Jermyn Street.

Following the course of the Inferior Oolite northwards, Mr. Sharp has failed to discover our shell in the Sands of Northamptonshire; and it appears to be equally absent in Rutlandshire and in Lincolnshire, although the fossils of the Inferior Oolite throughout its long course in the latter county have received considerable attention. In the North Riding of Yorkshire, at Cloughton, near Scarborough, the hard grey limestone has yielded it rarely; specimens are in the collection of Mr. Leckenby of that place, and in my own cabinet. Foreign localities are Longwy and St. Pancre, Luxembourg; Guéret and Moutiers, France; also various localities in the Cantons of Soleure and Basle, Switzerland: all in the Inferior Oolite.

TRIGONIA SCARBURGENSIS, *Lyc.* Plate IV, figs. 1, 2, 3, 4.

TRIGONIA SCARBURGENSIS, *Lycett.* Mon. Pal. Soc., 1863, p. 48, pl. xxxvii, fig. 1.

Shell ovately oblong, elongated, somewhat depressed; umbones antero-mesial, pointed, but not conspicuous, much incurved, and somewhat recurved; anterior side moderately produced, but with little convexity; its border curved elliptically with the lower border; posterior border lengthened and straight, or sometimes slightly concave; its extremity attenuated and rounded. The escutcheon is very large, and but little depressed; its length is considerable, or equal to the height of the valves and to nearly three fourths of the length of the marginal carina; its superior border is raised, which renders the superior border of the valve nearly straight. The area is narrow, lengthened, and flattened, delicately transversely plicated, divided by a faintly traced tuberculated median carina and slight furrow; it is bounded by two small carinæ, which, in the young state, are minutely tuberculated; subsequently they form small elevated plications, and in the most advanced stage of growth even these disappear, and the flattened surface of the area has only the usual folds of growth. The costated portion of the shell has the rows, for the first five or six, regular, sub-concentric, and delicately sub-tuberculated; those which succeed are in proportion much more widely separated, very irregular and oblique; they approach the marginal carina at a right angle. Anteally the costæ become attenuated and sub-tuberculated, their direction is more irregular and variable; not unfrequently they form a kind of undulation, and have the tubercles indistinct or cord-like; in other instances their direction anteally is nearly straight or horizontal, and invariably there is a supplementary rib formed upon that side. The posteal extremities of the costæ never reach to the marginal carina; it is separated from them by a smooth diagonal space for the lower three fourths of its length, but this space is neither considerable nor altogether uniform upon each valve in all specimens. The number of costæ are usually about thirteen, but occasionally sixteen; the tubercles upon the few later-formed costæ are large and obtuse posteally, but their number upon each row and their figure are very variable, some costæ having only eight and others about thirteen tubercles.

This is one of the most elongated and irregular of the *Clavellatæ*; it is the Cornbrash shell attributed to *T. clavellata* in the lists of Cornbrash fossils given by Phillips, Williamson, and Bean. In irregularity of the costæ it quite equals *T. irregularis*, Seebach, that beautiful Oxford Clay shell so long and well known at Weymouth; but a comparison of adult forms in the two species will at once show their distinctness. It approaches in figure more nearly to *T. Voltzii*; but in commencing our comparison with the umbones we find that in *T. Scarburgensis* they are less produced and recurved; the anteal side is more produced, and has much less convexity; the superior border is much straighter, resulting from the more raised superior border of the escutcheon; the rows of costæ are much more irregular, the tubercles are smaller and less raised; they do not terminate abruptly anteally, but become gradually attenuated and sub-tuberculated.

Young specimens having only eight or nine rows of costæ have not any strongly defined specific characters: they sometimes have portions of the granulated tegument preserved. The left valve is not unfrequently found with its ornamentation imperfectly developed, as in our specimen Plate IV, figs. 2, 3. The latter figure, which has been exceeded in its irregularity, appears to have resulted from an atrophized condition of the mantle upon that side; a defect which is equally conspicuous upon the left valve of the young specimen, fig. 4 upon the same plate, and is not, therefore, a concomitant of advanced growth.

Stratigraphical position and localities. It is not uncommon in the Cornbrash upon the northern side of Scarborough Castle Hill and in Cayton Bay. The late Dr. Porter obtained it in the same formation near to Peterborough. The officers of the National Geological Survey state that it is an abundant fossil at several localities in the South Lincolnshire district.

CLAVELLATÆ. 33

TRIGONIA TUBERCULOSA, *Lyc.* Plate V, figs. 9, 10.

  TRIGONIA TUBERCULOSA, *Lycett.* Ann. and Mag. Nat. Hist., 1850, pl. xi, f. 9.
   —   —   *Morris.* Catal., 1854, p. 228.
   —   —   *Lycett.* Cotteswold Hills Handbook, 1857, pl. iii, f. 4.
   — CLAVELLATA, *Quenstedt.* Der Jura., 1857, t. 60, f. 14 ?.
   — TUBERCULOSA, *Rigaux* and *Sauvage.* Descr. de esp. nouv. de L'Etage Bathon du Bas-Boulonnois, 1868, p. 19.

Shell ovately trigonal, depressed; umbones small, mesial, and recurved; anterior border produced, curved elliptically with the lower border; superior border sloping and nearly straight; area narrow, with two small delicately tuberculated bounding carinæ, traversed transversely by plain costellæ, which become posteally somewhat rugose and less conspicuous. The sides of the valves have a numerous series (from eighteen to twenty) of rows of curved and delicately tuberculated costæ. The tubercles are regular, very closely arranged, slightly compressed laterally, obtuse and produced downwards, so that their bases almost touch the next succeeding row; they are of equal size, excepting near to the carina, when the rows become smaller.

It is allied to *T. Griesbachi*, Lyc., to which the reader is referred, and also to *T. clavulosa*, Rigaux, and Sauvage, Mem. de la Soc. Acad. de Boulogne, 1867, vol. 3. The latter species appears to differ from it solely in having delicate transverse striations upon the area in lieu of the costellæ upon our *T. tuberculosa.*

Geological position and locality. The Inferior Oolite shelly freestone at Leckhampton Hill, near Cheltenham, where it has occurred rarely. Specimens are in the National Museum, Jermyn Street, in the cabinet of Dr. Wright, of Cheltenham, and of the Rev. P. B. Brodie, of Rowington; for the smaller specimen figured I am indebted to the kindness of the latter gentleman. Our figures do not clearly show the downward prolongation of the little tubercles in each row.

TRIGONIA IMBRICATA, *Sow.* Plate VI, fig. 5 *a, b.*

  TRIGONIA IMBRICATA, *Sowerby.* Mineral Conchology, 1826, t. 507, figs. 2, 3.
   —   —   *Morris* and *Lycett.* Gr. Ool. Monog. Pal. Soc., 1853, p. 63, t. 6, figs. 8, 8A.
   —   —   *Morris.* Catal., 1854, p. 228.
   —   —   *Oppel.* Juraformation, 1857, p. 485.

Under the above name Mr. Sowerby figured a minute Trigonia, which appears to be in an immature or young condition, and of which adult specimens have not been recognised; the peculiar imbrication of the costæ noticed by Mr. Sowerby appears to arise from the erosion of their rounded tubercles.

In the 'Mineral Conchology' it is described as "transversely oblong, depressed, with five or six concentric, dentated, sub-imbricated keels upon the rounded anterior side; the posterior side obliquely truncated, ribbed. The carinæ upon the surface of this little shell resemble terraces, one above the other; each is divided into four or five angular lobes."

The little specimen herewith figured is larger than the type in the 'Mineral Conchology;' it has seven rows of regular, concentric, tuberculated costæ, each of which has five or six distinct tubercles; anteally the rows become attenuated and only slightly tuberculated; posteally they are well separated from the marginal carina, which consists of a row of somewhat smaller tubercles, corresponding in number to the rows of costæ; the area has transverse, plain costellæ, each of which is united to one of the carinal tubercles. Our specimen is slightly broken posteally.

The few minute specimens hitherto examined differ from the young condition of all the known Clavellated Trigonias of the Lower Oolites, and are believed to constitute a distinct species.

Geological position and localities. Ancliff and Bath in the shelly Great Oolite.

TRIGONIA GRIESBACHII, *Lyc.* Plate III, fig. 10, *a*, *b*.

TRIGONIA TUBERCULOSA, *Lycett.* Pal. Soc. Suppl. Monog., 1863, p. 47, pl. xl, fig. 6; not *T. tuberculosa*, Lyc., Ann. and Mag. Nat. Hist., 1850, t. ii, fig. 9.

The little *T. Griesbachi* is only known by a single specimen, which fortunately is in so excellent a condition of preservation that its entire specific characters are fully exposed, and have been faithfully delineated in the magnified figure above cited, and published by the Palæontographical Society in 1863. At that period a single specimen of *T. tuberculosa* was all that remained at my disposal for comparison, and its condition was by no means in so satisfactory a state; it was, without doubt, nearly allied to the Cornbrash Shell, and making some allowance for difference of mineral character, and of geological position, I was induced to regard the two as not specifically distinct or differing only within the limits that might possibly be induced by altered conditions of geological habitat and of fossilisation. The examination of additional specimens of the Inferior Oolite shell have convinced me of the real distinctness of the two little Trigonias, and that their distinctive characters are as follows:

T. Griesbachi has nearly the general outline of its ally, but is more depressed, so that its area is much more nearly upon the same plane as the other portion of the surface; its marginal carina is, therefore, more remote from the superior border of the valve. The rows of tuberculated costæ are much more numerous, and their tubercles are also smaller, so that when viewed in certain directions the rows appear to take a different direction, and to be nearly vertical. The tubercles in their figure accord with those of the Inferior Oolite species; their bases are compressed laterally, and touch the row of tubercles next in succession; but it is only in *T. Griesbachi* that their close proximity produces this deceptive appearance of a vertical arrangement in the rows.

The area is flattened, narrow, with two very small tuberculated, bounding carinæ, and with acute, transverse costellæ, every alternate one of which forms a small varix upon the inner carina, and is prolonged somewhat upon the escutcheon.

Stratigraphical position and locality. The late Rev. A. W. Griesbach obtained this remarkable little species in the Cornbrash of Rushden, Northamptonshire. It has also occurred in the upper zone of the Great Oolite, near to Cirencester.

TRIGONIA FORMOSA, *Lyc.* Plate V, figs. 4, 5, 6.

TRIGONIA STRIATA, *Quenstedt.* Jura., 1857, tab. 46, fig. 2.
— FORMOSA, *Lycett.* Jour. Geol. Soc., 1859, note in Memoir of Wright on the Inferior Oolite formation.

Shell ovately trigonal, depressed; umbones elevated, pointed, and recurved; anterior side moderately produced; both it and the lower border elliptically curved; superior border lengthened and concave; area rather narrow, flattened, with closely arranged acute, transverse striations; a faintly marked oblique, mesial furrow, and bounded by two small densely and minutely dentated carinæ; the escutcheon is concave, smooth, and lengthened, sloping obliquely downwards, forming a considerable angle posteally with the posterior extremity of the area. The costated portion of the shell has very numerous narrow, oblique, knotted ridges, which are small at the carina, but increase in size anteally, where they also curve more or less horizontally, even to the anterior border; the last-formed five or six ridges arrive at the pallial border almost without curvature.

The umbonal extremity of the area has costellæ in lieu of transverse striations. This well-characterised species was long confounded with *T. striata*, Miller, owing probably to the bad figures originally given of the latter species; as a contrast to these the *Trigonia Montlierensis* figured by Goldfuss under the name of *T. striata* is excellent. Upon comparing examples of equal size it will at once be observed that the general figure is very different; *T. striata* is by no means depressed like the other; its superior border is short, straight, and nearly horizontal, so that its posteal extremity is at less than half the distance from the umbo to the lower extremity of the shell; the less conspicuous umbones, and great comparative breadth of the area, are also so remarkable that they impart a sub-quadrate aspect to the whole, and a wide truncation to the posterior side; the fringing tubercles of the costæ are also more dense and delicate than in *T. formosa*. Length of an adult specimen of *T. formosa* 29 lines; height 24 lines.

Stratigraphical position and localities. *T. formosa* has occurred in the Inferior Oolite at Dundry Hill, but probably not at any more southward locality in Somersetshire or Dorsetshire, where it is replaced by *T. striata*. In the Cotteswold Hills it has occurred in the Supra-liassic Sands at Frocester Hill, and also in several beds in the Inferior Oolite, beneath the upper Trigonia Grit, at various localities, more especially at Cold Comfort, near Cheltenham, and at Rodborough Hill, near Stroud; it appears to be altogether absent in the Inferior Oolite, in its extension through the counties of Oxford, Northampton, Lincoln, and York. Another and nearly allied species from the red Inferior

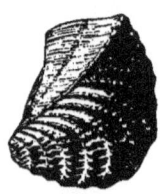

Oolite of Moutiers, Normandy, has the general figure and ornamentation nearly resembling *T. formosa*, excepting that the Moutiers form has greater convexity, and the escutcheon has greater breadth; the rows of costæ increase in size anteally, and the tubercles have each a small pillar, which descends perpendicularly to the costæ next in succession. The space between the anteal extremities of the costæ and the border has a numerous series of small transverse supplementary costæ. All the examples which have come under my notice are smaller than average specimens of *T. striata* or *T. formosa*. The British Museum has a fine series. I propose for it the name *Trigonia Moutierensis*.

TRIGONIA STRIATA, *Miller*. Plate V, figs. 6', 7, 8.

TRIGONIA STRIATA, *Sow*. Min. Chol., 1819, t. 237, figs. 1, 2.
— — *Morris*. Catal., 1854, p. 229.
— — *Oppel*. Juraformation, 1857, p. 407.

Shell subquadrate, short, moderately convex; umbones small, erect, only slightly recurved, antero-mesial; anterior side short, somewhat truncated, lower border curved elliptically; superior border short, horizontal, forming a considerable angle with the wide truncated extremity of the area; the length of this truncated border exceeds that of the superior border. Area very wide and flattened, traversed mesially by an obscure furrow which slightly bends the transverse striations; these are very regular and minute, even to the apex. There is no mesial carina, and almost no inner carina, as the transverse striations are cut off abruptly at that border. The marginal carina is always clearly defined; it is very narrow, ridge-like, minutely tuberculated, and has only a slight curvature.

The escutcheon is narrow, lengthened, and much depressed; its superior border is considerably raised. The other portion of the surface has about twenty-two narrow, obliquely curved, and elevated costæ; they are small towards the carina, which they touch, and descend almost perpendicularly ere they curve towards the anterior border. The last-formed six or seven costæ attain the lower border; each costa is fringed with a densely-arranged row of small, elevated, obtuse tubercles, which are frequently somewhat compressed laterally. The interiors of the valves have the lower border prominently dentated.

The most remarkable features in this species are the short sub-quadrate figure and the large size of the area; the height is equal to, or even slightly exceeds, the length, and the surface of the area is equal to two fifths of the entire valve. Its general aspect is so peculiar that it will not readily be mistaken for any other species; but the only authentic figures are those in the 'Mineral Conchology,' which can only be described as bad specimens badly drawn, and this will account for the fact that both Goldfuss and Agassiz have fallen into error respecting it; each of them has figured for *T. striata* a different species.

Another allied species is *T. spinulosa*, described but not figured by Messrs. Young and Bird in their 'Geology of the Yorkshire Coast,' and soon afterwards figured but not described by Professor Phillips, in his 'Geology of Yorkshire,' as *T. striata*, and subsequently by Agassiz as *T. tuberculata*. In this species the much more lengthened area and the different costæ with their larger tubercles will serve to distinguish it. The name chosen by Miller refers to the transverse striations upon the area. The little shell figured by Agassiz for *T. striata*, Miller ('Trigonies,' tab. 4, fig. 12), is not that species but the single small figure, and insufficient description renders it difficult to be assigned to any one of the allied Inferior Oolite forms. The characters of the rows of tuberculated costæ differ equally from *T. Moutierensis*, *T. formosa*, and *T. Phillipsi*; the general figure has no resemblance to *T. striata*.

Stratigraphical position and localities. The zone of Ammonites Humphriesianus in the Inferior Oolite at various localities in the Counties of Somerset and Dorset. It appears to be altogether absent in the more northern extension of the Inferior Oolite in its course through the Cotteswold Hills, and is also absent in Oxfordshire, Northamptonshire, Lincolnshire, and Yorkshire; it appears, therefore, that the Mendip Hills presented a dividing barrier at the period of the deposition of the Inferior Oolite formation, and thus had an important influence upon the distribution of its testaceous mollusca.

TRIGONIA PHILLIPSI, *Mor.* and *Lyc.*   Plate VI, figs. 3, 4.

TRIGONIA STRIATA, *Agassiz.*   Trigonies, 1840, pl. iv, fig. 12 (not figs. 10, 11).
— PHILLIPSI, *Mor.* and *Lyc.*   Pal. Soc. Gr. Ool. Monog., 1853, tab. 6, fig. 1, p. 62.
—   —   *Morris.*   Catal., 1854, p. 228.

Shell sub-ovate, convex; umbones obtuse, moderately elevated, and scarcely recurved; anterior side produced, its border elliptically curved with the lower border; superior border nearly straight, of moderate length, sloping obliquely; area flattened, transversely lineated, divided by an oblique furrow, and bounded by two very small, minutely tuberculated carinæ; escutcheon excavated, wide, its length is equal to half of that of the marginal carina; the costated portion of the shell has very numerous closely arranged concentric, raised, and minutely tuberculated costæ. About thirty rows may be counted in a specimen fifteen lines in length; the first few costæ appear to be destitute of tubercles; all the rows are nearly of equal size throughout their course, and are very closely arranged, bordering upon the carina; they do not quite reach the anterior border, but form a slight undulation upon the anteal slope; their posteal extremities rise nearly perpendicularly towards the marginal carina.

T. striata, Miller, and *T. formosa,* Lycett, are allied to it. From the first it is separated by the smallness and obliquity of the area, and by the length of the superior border, which offers a marked contrast to the short sub-quadrate figure of Miller's species.

From *T. formosa* it is distinguished by the greater convexity, by the absence of the acute recurved umbones, and by the concentric in lieu of the oblique costæ of that species; the tubercles, also, are much more minute both upon the costæ and the carinæ, and, contrary to *T. formosa,* the few first costæ are plain.

The height is about one sixth less than the length, but specimens differ in their proportions; no example with the valves united has been obtained.

The figure given in the Monograph of the Great Oolite Mollusca (Palæont. Soc., 1853) is unusually short posteally, which may have resulted from the position in which the specimen was placed before the artist.

Stratigraphical positions and localities. It has occurred in several distinct beds of the Inferior Oolite near to Stroud very rarely, in cream-coloured, nearly hard limestone; at Desborough, Northamptonshire, in brown ferruginous oolite; in white oolite at Stamford, Lincolnshire, and in a similar rock at Stoke, near Grantham; at Appleby, North Lincolnshire, in the lowest bed of Inferior Oolite, a hard, brownish oolite. Our specimens figured are from the latter locality, and were presented to me by the Rev. J. E. Cross; the larger one is more lengthened than is usual; at neither of these localities is it common.

It is not certain that the species has been obtained at any foreign locality. Fine specimens are in both of our great national museums, in the collection of the Rev. J. E. Cross, of Appleby, and in that of the author at Scarborough.

CLAVELLATÆ. 39

TRIGONIA IRREGULARIS, *Seeb.* Plate V, figs. 1, *a*, *b*, 2; Pl. VII, fig. 6.

  TRIGONIA, *Damon.* Geol. Weymouth, Suppl., 1860, pl. ii, fig. 3.
  — *Seebach.* Der Hannoversche Jura. See his obs. on *T. triquetra.*

Shell ovately trigonal or oblong; umbones antero-mesial, prominent, and recurved; anterior side short, moderately convex, slightly truncated; its lower portion curved with the lengthened lower border, which is slightly sinuated near to its posteal extremity; the posterior or superior side has its border concave and terminally rostrated.

The escutcheon is very large and depressed; its length exceeds the half of the entire length of the shell; its superior border is only slightly raised. The area is narrow, having three tuberculated carinæ; the inner and median carinæ have each a row of small, transverse, nodose varices rather distantly arranged; these are ultimately lost in the large posteal plications of growth; the marginal carina is small, consisting, for about a third of its length, of a narrow, elevated, finely-indented ridge; subsequently it acquires small transverse nodose varices similar to those of the median and inner carinæ. The transverse plications upon the area are for the most part small excepting near to the apex, where its surface is occupied by about eight regular, narrow, transverse costellæ. The superior half of the area is the more depressed, and has sometimes a minute line of tubercles bordering upon the mesial furrow, and parallel to the median carina. The other portion of the valve has about fourteen rows of slightly elevated costæ decked with distinct, elevated, conical, pointed, and unequal tubercles; the first-formed six or seven rows are regular and concentric; those which succeed are more or less irregular both in their direction and in the size and arrangement of the tubercles; the anteal portions of the rows become broken and confused; the tubercles adjacent to the border are the smaller, and are often compressed laterally. The posteal extremities of the rows are separated from the marginal carina by a smooth and slightly-depressed space which widens downwards, and terminates at the lower border in a well-marked undulation of the border; the number of tubercles in the rows varies from eight to ten. The figure in Mr. Damon's 'Supplement' is an extreme example of that general irregularity in the arrangement of the tubercles which Seebach has adopted as a name for the species. The large imperfect specimen which we have figured (Plate V, fig. 2) exhibits the greatest irregularity observed; the small specimen (Plate VII, fig. 6) is an example of the forms in which the irregularity in the rows of tubercles is so slight as to be quite inconspicuous. The smooth ante-carinal space is always present but varies in its size.

Stratigraphical positions and localities. In the Oxford Clay of Weymouth it is moderately abundant. It has occurred, also, less frequently in the Kimmeridge Clay of Wootton Basset, Wilts.

TRIGONIA WOODWARDI, *Lycett*, sp. nov.

Shell large, ovately trigonal, depressed; umbones elevated, pointed, recurved, placed at about the anterior third of the valves; anterior side produced, its border curved elliptically with the lower border, which is lengthened and nearly straight posteally; the superior border is nearly straight, it slopes downwards obliquely, and forms only a slight angle with the posteal border of the area, the lower extremity of which is pointed. The escutcheon is narrow, lengthened, and concave; its superior border is raised. The area is narrow, its superior or umbonal portion forms a considerable angle with the costated

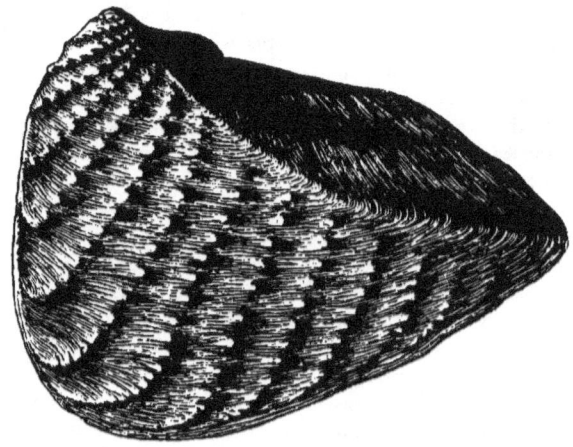

surface of the shell; it has three small tuberculated carinæ, which become evanescent posteally, and transverse irregular plications, which form near to the umbo, acute, regular, small costellæ. The other portion of the shell has the rows of costæ small, widely separated, nearly straight or oblique, sometimes somewhat undulated; the tubercles in the rows are numerous, rounded, closely placed, and unequal; they become smaller, crowded in the rows, cord-like and attenuated near to the anteal and lower borders; the few last-formed rows are smaller, their pallial portions curve much forwards; about twelve distinct rounded tubercles occur in each row. The lines of growth are strongly defined over the whole of the shell.

Length $4\frac{3}{4}$ inches, height $3\frac{1}{4}$ inches, diameter through the united valves $1\frac{3}{4}$ inch.

It is distinguished from *T. clavellata* and also from *T. perlata* by its more depressed figure, by the more produced anterior side, by the straightness of the lower border, by the unusually narrow area, by the straight, equal, or slightly undulated rows of costæ, which also have the tubercles smaller, more numerous, and more equal in size, so that the surface of that portion is remarkably wide, flattened, and uniform in its aspect. The costæ approach the small marginal carina nearly at right angles to it.

Compared with *T. Pellati*, it is less lengthened, and its rows of costæ are more numerous and less curved.

T. Woodwardi has occurred rarely in the Kimmeridge Clay of Dorsetshire and of Wootton Basset, Wilts. Two specimens from the latter locality are in the British Museum, numbered 66, 126, but their state of preservation is only indifferent. Fine examples from the same formation at Villersville, near to Honfleur, are also in the Museum collection.

TRIGONIA PELLATI, *Mun. Chal.* Plate VII, figs. 1, 2, *a, b ;* Pl, XI, fig. 1.

  TRIGONIA PELLATI, *Munier Chalmas.* Bull. Soc. Linn. de Normandie, 1865, tom. ix, pl. iv, fig. 4.
  —   —   *Hebert.* Bull. de la Soc. Geol. de France, 2nd ser., 1865, tom. xxiii, p. 216.
  —   —   *De Loriol* et *Pellat.* Monogr. Paléont. et Geol. de l'Etage Portlandien des envir. de Boulogne-sur-Mer, 1866, tab. viii, fig. 4.

Shell oblong, inordinately elongated, wide upon the superior, depressed or wedge-shaped towards the inferior border; umbones near to the anteal extremity of the valves, obtuse, much incurved and depressed, anterior side very short, truncated, with considerable convexity, its border curved elliptically with the lower border which is very long and straight; the superior border is also very long, its border is slightly concave, its posteal extremity forms an obtuse angle with the posteal border of the area, and terminates with its extremity somewhat pointed and much produced. The area has great length, and slightly convex, with a well-marked mesial furrow bordering a line of minute tubercles and bounded by two delicately traced and minutely tuberculated carinæ, which posteally become transverse, irregular plications. The escutcheon is flattened, of moderate breadth, and unusually lengthened; the ligamental fossa also partakes of the general lengthening of the superior border; the area has transverse, irregular plications, which become large posteally. The sides of the valves are very narrow, and have a few rows of very distantly arranged, oblique, tuberculated costæ; three or four of the tubercles near to the carina are large, rounded, pointed, and much elevated; those upon the lower half of the valve become rapidly smaller and more depressed; their rows curve anteally almost in the direction of the lower border. The first formed five or six costæ are more straight and oblique; each has its upper terminal tubercle unusually large and elevated.

The height is equal to two fifths of the length, and is one fifth greater than the diameter across the superior border.

This is the most elongated of the *clavellatæ*; the great length and horizontal direction of the superior border readily separates it from *T. Voltzii* and from *T. incurva*; others of the section are more remotely allied.

Stratigraphical position and localities. Both the specimens figured upon Plate VII were obtained by Mr. Mansell in the Kimmeridge Clay of Kimmeridge Bay, Dorset. The broken specimen has the area somewhat compressed or spread upwards; it has also occurred in the same formation near to Ely, and in the vicinity of Westbury, Wilts. France, Boulogne-sur-Mer.

TRIGONIA INCURVA, *Benett.* Plate IX, figs. 2, 3, 4, 5, 6.

TRIGONIA INCURVA, *Etherelda Benett.* Catalogue of Organic Remains of Wilts., 1831, pl. xviii, fig. 2.
— — *Sowerby.* In Fitton. Geol. Trans., 2nd ser., 4, 1836, tab. xxii, fig. 14 (internal mould).
— — *Bronn.* Index Palæontologie, 1848, p. 1280.
— — *D'Orbigny.* Prodrome de Paléont., 2, p. 60.
— — *Morris.* Catalogue, 1854, p. 228.
— — *Cotteau.* Etud. sur les Moll. Foss. de la Yonne, fasc. 1, Prodrome, 1953-7, p. 76.
— — *Oppel.* Juraformation, p. 721, 722, No. 145.
— — *Damon.* Geol. of Weymouth Suppl., 1860, pl. vii, fig. 1 (internal mould).
— HÉBERTI, *Munier Chalmas.* Bull. de la Soc. Linn. de Normandie, vol. ix, 1863—4, pl. iv, fig. 5.
— INCURVA, *Hébert.* Note sur la Terr. Jurassique du Boulonnais; Bull. Soc. Geol. Fr., 2nd ser., t. xxiii, p. 214, 1865, p. 220.
— — *Pellat.* Bull. Soc. Geol. de Fr., 3rd ser., 1866, t. xxiii, p. 226.
— — *P. de Loriol* et *E. Pellat.* Monogr. Paleont. et Geol. de l'Etage Portlandien des environs de Boulogne-sur-Mer, 1866, pl. viii, fig. 3.

Shell elongated, curved at the two extremities or sublunate; anterior side convex; posterior side lengthened, curved, depressed; umbones large, obtuse, elevated, somewhat recurved, and placed near to the anterior border, which is curved elliptically with the lower border. Escutcheon concave, lengthened; its superior border is somewhat raised.

CLAVELLATÆ. 43

Area narrow, distinctly bipartite with three delicate tuberculated carinæ and irregular transverse plications; there are also some irregular varices near to the posteal portions of the median and inner carinæ; those of the latter extend a little upon the escutcheon. The ornamentation upon the other portion of the valve varies much in accordance with the development in the growth of the shell. A specimen two and a half inches in length has twelve rows of tuberculated varices, which rise nearly perpendicularly to the carina; anteally they curve much forwards, and are continued in an attenuated condition almost to the anterior border; the superior portions of the varices have the tubercles large, rounded, but somewhat unequal and irregular; anteally they rapidly become small and cord-like; the lines of growth are strongly marked over the whole of the shell. A specimen three and a half inches in length figured by Messieurs de Loriol and Pellat has the costæ more broken near to the anterior border; they lose all distinctness and form an irregular assemblage of small tubercles. The original figure given by Miss Benett represents a condition of growth still more advanced, the anteal gibbosity has much increased, and its border is almost destitute of tubercles; the plications of growth have also become more conspicuous. A specimen greatly larger and still more inflated from Niangle, Boulogne, is No. 36913 in the British Museum; the anteal folds of growth have here replaced all ornamentation.

In the Portland Oolite at Swindon and at the Isle of Portland internal moulds are common, but it rarely happens that any considerable portion of the test is adherent; the characters of the surface may, however, be ascertained upon the adherent portions. The examples figured 5 and 6, Plate IX, represent the more frequent condition of such specimens. Figs. 4 and 5, upon the same plate, may be regarded as a distinct variety from the Kimmeridge Clay, with the anteal portions of the varices curved and unbroken. No. 2 has suffered somewhat from vertical pressure, its umbo is more than usually produced and pointed.

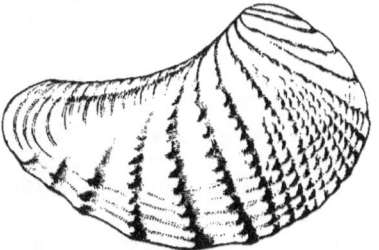

In the Portland Oolite of Boulogne it has occurred in a better condition of preservation and of much larger dimensions; the original of our wood engraving is from the latter locality, and represents an example of adult growth, but not of the largest dimensions, the size is reduced one half; the small anteal supplementary varices are unusually numerous and strongly defined.

TRIGONIA SPINULOSA, *Young* and *Bird*. Plate III, figs. 4, 5, 6.

TRIGONIA SPINULOSA, *Young* and *Bird*. Geol. Survey, 1828, p. 225.
— STRIATA. Phil. Geol. York., 1829, 1, pl. xi, fig. 38 (not *T. striata*, Miller).
— TUBERCULATA, *Agassiz*. Trigonies, 1842, p. 20, t. ii, fig. 17, et tab. ix, figs. 6—8.
— — *Oppel*. Juraformation, 1856, p. 407.
— STRIATA, *D'Orbigny*. Prodrome, vol. i, 1850, p. 278, No. 312 (supposed by him to be young examples of *T. striata*).

Shell ovately trigonal, moderately convex, umbones not prominent, much incurved, obtuse but only slightly recurved, antero-mesial; anterior side short, curved elliptically, with the lower border, posterior side, produced and compressed. Area moderately large, transversely irregularly plicated, divided by a delicately tuberculated carina bordering a furrow; the two bounding carinæ are large, with prominent rounded tubercles; the superior half of the area is more depressed than the other. The escutcheon is very large and depressed, but its upper border is moderately raised throughout its length. The other portion of the surface has about fifteen rows of large, raised, tuberculated costæ, of which the first formed seven are simply concentric; all the rows are small, ridge-like, and only subtuberculated anteally; the tubercles in the rows are distinct, but slightly compressed laterally, and increase in size posteally; the general direction of the rows is nearly horizontal anteally, or slightly directed downwards; posteally they curve upwards rather suddenly and are united to the carina at a considerable angle; they therefore most frequently form a kind of slight undulation posteally at the part where the tubercles are the largest.

The three tuberculated carinæ and median groove are usually well defined, but occasionally, as in the second specimen figured by Agassiz, there is no median row of tubercles bordering the groove.

No figure of this species is given in the work of Messrs. Young and Bird, but their description is sufficiently precise and comprehensive to leave no doubt of the species intended. The figure given in the 'Geology of Yorkshire' is very characteristic and must have been named *striata* from mistaken recollection, as it is impossible that it could have resulted from comparing the species of Miller with that from Blue Wyke; the first figure of *T. tuberculata*, Ag. ('Trig.,' t. ii, fig. 17), has been drawn from a specimen distorted by compression. Specimens differ considerably in the proportions of their length and height.

D'Orbigny ('Prodrome,' i, p. 278) has fallen into the same mistake as Phillips in regarding *T. tuberculata* as the young of *T. striata*; the very coarsely drawn and indifferent figures in the 'Mineral Conchology' of *T. striata* will perhaps account for these errors. *T. spinulosa* is more nearly allied to *T. formosa*, Lyc., but comparison will show that the former is more convex and much more produced posteally; the figure of the rows of costæ is different, the greater size of the posteal tubercles and their separation in the rows are also distinctive. The escutcheon in *T. spinulosa* is larger, the three tuberculated carinæ upon the area also distinguishes that portion of the shell.

Geological positions and localities. *T. spinulosa* occu.s not uncommonly in the shelly bed of the Dogger at Blue Wyke, also in the Ironstone of Glaizedale, associated with *T. denticulata*, Ag.,'and *T. V. costata*, Lyc. Specimens more or less imperfect also occur rarely in the Supra-Liassic Sands at Frocester Hill, Gloucestershire. The localities mentioned by Agassiz for his *T. tuberculata* do not define its geological position clearly in Southern Germany, but, from its association with *T. costellata* and *T. pulchella*, we may infer that its position nearly agrees with that at Blue Wyke.

TRIGONIA CORALLINA, *D'Orb.* Plate III, figs. 7, 8, 9, 11; Plate VIII, fig. 3.

TRIGONIA CORALLINA, *D'Orbigny.* Prodrome de Paléont., vol. ii, 14th et., 1850, p. 16, No. 260.
— CLAVELLATA, var. JURENSIS, *Grewingk.* Gest. u. Geolog. Livonia und Courland, Dorpat, 1864.

"*Espèce voisine pour les petits tubercles rapprochés du côtes du T. concentrica, mais avec les côtes bien plus serrées, moins arquées, l'area anale striée en travers.*"— *D'Orbigny.*

Shell ovately trigonal, convex; umbones antero-mesial, not much elevated, incurved, and somewhat obtuse, anterior side short and curved elliptically with the lower border, posterior slope lengthened, its outline rather convex, its junction with the area is obtusely angulated or somewhat rounded. The escutcheon is lengthened and depressed, its upper border raised; the area of moderate breadth, flattened, with three small imperfectly developed tuberculated carinæ; it has irregular rugose plications posteally, which, near to the umbones, become transverse narrow costellæ. The other portion of the surface has numerous rows of narrow, ridged, curved, costæ; their tubercles are small, irregular, and unequal, closely arranged, becoming attenuated and imperfectly developed towards the anteal extremities of the rows, which, at that portion of the valve, have but little regularity and are but slightly curved. The young examples (Plate III, figs. 8, 9, 11) are from the Coralline Oolite of Wiltshire, and were collected by the officers of the National Geological Survey. The larger examples (Plate III, fig. 7, and Plate VIII, fig. 5) are from the Coralline Oolite at Pickering. As a species it is distinguished from *T. concentrica*, Ag. ('Trigon.,' p. 20, pl. 6, fig. 10), by its shorter figure, its ridge-like sub-serrated costæ, which have less curvature; the area also with its anteal ridge-like costellæ and rugose middle and posteal portions is equally distinctive.

Stratigraphical position and localities. It occurs very sparingly in the Coralline Oolite of Pickering associated with *T. perlata*, Ag., from which it is distinguished by its small dimensions, by the much greater number of the rows of costæ, by their more horizontal direction and more close arrangement. The tubercles are also smaller and much more numerous. The three smaller immature examples are from the Coralline Oolite of Wiltshire. A specimen larger than these latter, but also of immature growth, was forwarded by Professor Grewingk, of Dorpat, to Mr. Leckenby, together with a series of Livonian Oxfordian testacea. The Trigonia was named *T. clavellata*, var. *Jurensis*, obtained in the vicinity of Popilacny, Province of Kowno, associated with a young example of *T. monilifera*, Ag. France, Tonnere (Yonne).

TRIGONIA PARCINODA, *Lycett*, sp. nov.

Shell small, moderately convex, ovately subquadrate, the length slightly exceeding the height; umbones small, antero-mesial; area large, flattened, or slightly concave, traversed transversely by large regular costellæ; there is no mesial furrow; the marginal carina is small but distinct; its sub-umbonal portion has several minute tubercles, there is no distinct inner carina and the escutcheon is small and inconspicuous; the posteal border of the area descends abruptly from the extremity of the escutcheon. The other portion of the surface has about ten rows of small horizontal and gently curved costæ, each of which is crossed perpendicularly by about six small regular varices; these are of moderate size and are distantly arranged about the middle of the valve, becoming small and indistinct towards the anteal border; the costellæ upon the area form continuations of the costæ, but near to the posterior border the former become smaller and more numerous.

Magnified twice.

Height, 4 lines; length, 5 lines; the surface of the area is equal to two fifths of the entire valve.

A pretty little sub-quadrate species remarkable for the few small, widely separated perpendicular varices upon the costated portion of the valves; it does not possess any striking affinities to other of the *Clavellatæ*, and is known only by the sole specimen here figured, which is now in the British Museum, numbered 67,272.

Stratigraphical position and locality. It was collected many years since by Mr. Etheridge in the Inferior Oolite of the Halfway House Quarry near Yeovil.

TRIGONIA IMPRESSA, *Sow.* Plate VII, figs. 4, 5.

TRIGONIA IMPRESSA, *Sowerby.* Jour. Zool., vol. 3, tab. xi.
— — *Prevost.* Ann. Scient. Nat., vol. 4, tab. xviii, figs. 22, 23.
— — *Morris* and *Lycett.* Pal. Soc., 1853, p. 61, tab. v, fig. 14.
— — *Morris.* Catal., 1854, p. 228.
— — *Lycett.* Cotteswold Hills Handbook, 1857, pl. vii, fig. 5.

CLAVELLATÆ.

Shell ovately oblong, depressed, umbones antero-mesial, small, pointed and slightly recurved, anterior and posterior borders curved elliptically; posterior border lengthened, its superior portion straight, and slightly sloping; escutcheon lengthened, narrow, slightly excavated, its outer border raised; area moderately wide, flattened, transversely irregularly striated, but usually delicately impressed, with a slight median furrow, and bounded by two small, minutely tuberculated carinæ. The other portion of the shell has a numerous series (14 posteally) of regular curved and nearly horizontal small tuberculated costæ, which are very delicate and minutely knotted anteally, directed from the border slightly downwards to the middle of the valve; posteally the costæ are somewhat larger, and more distinctly tuberculated; the tubercles enlarge towards the carina, which they meet at a right angle in the few last formed costæ; all the tubercles are depressed, and very few are separated in the rows; anteally there are one or two supplementary or intercalated costæ, the posteal terminations of which are about the middle of the valve.

The aspect of this little species is peculiar, and its several features, although minute, are very persistent; few specimens exceed thirteen lines in length, and the greater number do not exceed ten lines; they were very gregarious, the disunited valves are profusely scattered in a thin layer over the slabs of Stonesfield Slate at numerous localities in Oxfordshire and Gloucestershire; the shell substance is rarely preserved, but the impressions in the soft sandy shale exhibit perfectly all the more delicate features of the species. It appears to be entirely absent in the shelly beds of the Great Oolite.

Its figure is more lengthened than *T. Moretoni*, its costa are more numerous, the area has never the coarse rugose aspect of that species, and the entire ornamentation is much more minute and delicate; other Trigoniæ are more remotely allied. A large example has the height 10½ lines, length 14 lines.

TRIGONIA MORETONI, *Mor.* and *Lyc.* Plate II, figs. 4, 5, 7, 8; Plate IV, fig. 6.

  TRIGONIA MORETONI, *Mor.* and *Lyc.* Gr. Ool. Monogr. Pal. Soc., 1853, p. 57, tab. v, figs. 19, 19 *a*.
  —   —   *Morris.* Catal., 1854, p. 228.
  —   —   *Rigaux* et *Sauvage.* Descr. de Quelq. Esp. Nouv. de l'Etage Bathonien, extr. du vol. iii, Mem. de la Soc. Académ. de Boulogne, 1867, p. 19.
  —   —   *Sharp.* Oolites of Northamptonshire, p. 383; Quart. Jour. Geol. Soc., August, 1870.
  —   CLAPENSIS, *O. Terquem* et *E. Jourdy.* Monog. de l'Etage Bathonien dans le Depart. de la Mozelle, Mem. Soc. Geol. de France, 2 ser., 1869, tom. 9, Pl. II, figs. 1, 2, 3, 4.

Shell ovately trigonal, rather depressed, umbones antero-mesial, elevated and only slightly recurved; anterior border moderately produced and curved elliptically with the lower border; hinge border lengthened and nearly straight, sloping obliquely, and forming an obtuse angle with the posteal extremity of the area; escutcheon depressed, moderately lengthened, its upper border raised; area narrow, flattened, transversely plicated; the plications are usually large and irregular, there is a faintly marked, mesial oblique furrow, and a minutely tuberculated median carina which disappears towards the posteal extremity, the marginal and inner carinæ are well marked; the inner carina has elevated squamous tubercles. The marginal carina is at first a narrow elevated plain ridge, which soon changes to transversely compressed elevated tubercles, from which originate the transverse plications upon the area; the carina ultimately degenerates into these plications. The other portion of the valve has about fourteen or fifteen rows of clavellated but always elevated and sometimes subtuberculated costæ; the first three of these are plain, acute, and concentric, they are interrupted by the marginal carina, but pass unchanged in character across the area; the succeeding costæ are subtuberculated, and pass obliquely downwards with little curvature, both anteally and posteally, the posteal portions are much the larger and more distinctly clavellated, their junction with the anteal portions form so many undulations or angles about the middle of the valve, and the anteal series has one, two, or even more rows than the other series, so that the rows are sometimes rather crowded anteally; all are, however, tuberculated or sub-tuberculated.

The Trigonia which most nearly approximates to our species is *T. arata*, one of the *undulatæ* from the upper zone of the Great Oolite, to which the reader is referred. *T. Moretoni* varies much in its length, as will be seen from our illustrations; occasionally it exhibits a crowding of the narrow anteal portions of the costæ, which imparts an apparent confusion to the ornamentation of the shell.

Our figures exhibit much of the variability which occurs both in the proportions of the general figure and the ornamentation; not uncommonly the costæ are nearly plain, or are only very slightly knotted. Neither of these several variable features appears to be special to any locality or stratum.

Stratigraphical positions and localities. *T. Moretoni* appears to range throughout the whole of the Great Oolite formation, it occurs commonly in all the shelly beds of the Minchinhampton district, and attains to its full dimensions, but the far greater number are immature forms from ten to fifteen lines in length. Mr. Whiteaves has obtained it in Oxfordshire, Mr. S. Sharp in Northamptonshire, the officers of the Geological Survey in the South Lincolnshire district, and in North Lincolnshire the upper zone of the Great Oolite has produced a profusion of beautifully preserved specimens, which are the originals of our figures; for these I am indebted to the liberality of the Rev. J. E. Cross, of Appleby, who has assiduously developed the fossil fauna of a district rich in the testacea of the Lower Oolites.

France, Bas-Boulonnais. See examples in the British Museum.

TRIGONIA COMPLANATA, *Lycett*, sp. nov.   Plate VII, fig. 3.

Shell ovately oblong, depressed; umbones small, pointed, anterior; anterior side very short, curved elliptically with the lower border; posteal extremity produced and pointed; area very large, flattened, its surface is equal to two fifths of the entire valve; hinge-border lengthened, and nearly horizontal; bounding carinæ small, impressed by irregular, small, transverse plications, which also extend across the area; there is also a slight mesial furrow, but the two portions of the area are upon the same plane. Escutcheon large and flattened, its superior border is raised; the other portion of the surface has about thirteen rows of small, tuberculated, oblique, and nearly straight costæ, each of which has about seven small depressed tubercles.

The length is one third greater than the height; it is also nearly equal to three times the diameter through the united valves.

The lines of growth are strongly marked, they modify both the figures of the tubercles and the surface of the area.

This clavellated species is remarkable for the unusual depression of the valves, for the very short, wedge-like, anterior side, for the lengthened figure posteally, for the unusually large size of the flattened area, and for the small depressed tubercles upon the straight rows of costæ which approach the carina at a right angle. These several characteristic features also separate it from *T. clavellata*, *T. perlata*, *T. Voltzii*, and *T. corallina*; other examples of the section are more remotely allied.

Stratigraphical positions and localities.   A few specimens, for the most part deprived of the test, have been obtained in the Kelloway Rock of Scarborough. The British Museum has finely preserved examples from the Oxford Clay of Normandy.

TRIGONIA RAMSAYI, *Wright*.   Plate VI, fig. 6.

TRIGONIA RAMSAYI, *Wright*.   On Upper Lias Sands, Quart. Jour. Geol. Soc., 1856, vol. xii, p. 323.
— — *Lycett*.   Cotteswold Hills Handbook, 1857, p. 26, pl. i, fig. 8.

Shell ovately oblong, convex, short anteally, lengthened, and somewhat attenuated posteally; umbones small, obtuse, erect, placed nearly upon the line of the anterior border, so that the superior border represents almost the entire length of the shell; the lengthened lower border has only a slight curvature, and the superior border is slightly concave. The area is narrow, and slightly convex; it is bounded by two very small, but distinct and minutely tuberculated carinæ, and traversed longitudinally by an inconspicuous mesial furrow; the transverse plications of the area are small and closely arranged, excepting upon its postcal third, where they become large, rugose, irregular, forming some varices at the inner carina. The escutcheon is depressed, of moderate breadth and great length; its upper or inner border is raised.

The other portion of the shell is occupied by a series of numerous and closely arranged costæ, twenty-two in number; they are nearly of equal size, rounded, more or less knobbed or transversely plicated, and are only slightly attenuated near to the marginal carina; their general direction is oblique, or at right angles to the marginal carina, but the seven last formed or postcal costæ are nearly perpendicular; for the most part they are somewhat undulated, or even wrinkled, and there is one supplementary or intercalated costa upon the anterior side. The lines of growth are distinct over the greater portion of the valve. The small tubercles upon the costæ are more distinctly traced upon the few umbonal or more concentric ones, upon the succeeding costæ they are very unequal and frequently indistinct.

The length of the marginal carina is equal to nearly twice the height and to three times the diameter through both the valves; the general aspect is sufficiently distinctive, and does not approximate very nearly to any other species of the Lower Oolites. *T. signata* is more produced anteally and more truncated posteally; the upper portions of its costæ are more attenuated, they also approach the carina at a much greater angle.

Geological position and locality. This rare *Trigonia* has only been obtained in the Ammonite-bed of the Supra-Liassic Sands[1] at Frocester Hill. The specimen in the National Museum, Jermyn Street, and another in the cabinet of Dr. Wright at Cheltenham, are the only examples with which I am acquainted.

TRIGONIA MURICATA, *Goldf.* Plate IX, fig. 1.

LYRODON MURICATUM, *Goldfuss.* Petref. German., 1836, tab. 137, fig. 1, p. 201.
TRIGONIA MURICATA, *Roemer.* Nordd. Ool. Nachtrag., 1839, p. 75.
— — *Agassiz.* Trigonies, 1840, pp. 7 and 51.
— — *D'Orbigny.* Prodrome de Paléont., 1850, vol. ii, p. 51, No. 120.
— — *Oppel.* Juraformation, 1857, p. 719, No. 89.

---

[1] I prefer provisionally to employ this term to designate the sands or marly sandstones which separate the Inferior Oolite from the Upper Lias (*Am. communis* Zone) over such distant areas in England, France, and Germany. After collecting for many years and comparing the faunas of these several zones, the conclusion has long been present to my mind that the fossils of these sands, viewed as a whole, are clearly separable from those of the beds both anterior and posterior to them in chronological order: the boundary lines of this stage, both lower and upper, are neither abrupt lithologically nor palæontologically, each has its species of testacea whose limits vertically may extend to one or both sides of our convenient, but somewhat arbitrary, divisional lines; nevertheless its fauna, viewed as a whole, is well marked and characteristic of the stage.

Several examples of this species have been placed at my disposal from the Portland strata of Wilts and Dorset, collected by the officers of the National Geological Survey, and also by Mr. Cunnington; neither of these, unfortunately, are altogether satisfactory. The Survey specimens are deprived of their tests, as that of Plate IX, fig. 1, which does not exhibit the characters of the area. The specimens of Mr. Cunnington consist of portions only with the test preserved; fortunately the latter gentleman has also made a good gutta-percha pressing, which exhibits the characters of the area as in the adjoining woodcut; together they afford sufficient materials to characterise the species.

The present figure exhibits the entire area and escutcheon, together with the posteal portions of the tuberculated costæ. *T. muricata* has a lengthened oblong figure, the length measuring 48 lines, the height 28 lines; the anterior side is very short; the posterior side is much lengthened and attenuated; the anterior and lower borders are curved elliptically; the convexity of the valves is only inconsiderable; the umbones have but little elevation, but are distinctly recurved; the area is large and flattened, or is slightly convex posteally; it has three tuberculated carinæ, of which the tubercles upon the marginal carina are regular, rounded, and rather distantly arranged; the transverse lineations upon the area are delicate and obscure, excepting the sub-umbonal portion, where they become regular, ridge-like, and closely arranged, but are also small and delicate; the lateral costæ have but little elevation; they are very numerous (about twenty-four), obliquely curved, and are nearly of equal size; their anteal portions curve nearly in the direction of the plications of growth, and become evanescent near to the pallial border; their tubercles are small, numerous, regular, and slightly compressed laterally; the larger tubercles occupy the middle and posteal portions of the rows; they are everywhere well separated.

The more prominent features, as exemplified in the depressed, lengthened, oblong figure, the very numerous rows of curved costæ with their inconspicuous tubercles, together with the delicate features of the area, separate it readily from other of the British *Clavellatæ*. In Portugal it also occurs in the Upper Jurassic Limestone at Torres Vedras.

TRIGONIA CONCENTRICA, *Ag.* (See figures upon the last plate of this Monograph.)

> TRIGONIA CONCENTRICA, *Agassiz.* Mem. sur les Trigonies, 1840, p. 20, tab. 6, fig. 10.
> LYRIODON CONCENTRICUM, *Bronn.* Index, Paléont., 1848, p. 685.
> TRIGONIA CONCENTRICA, *D'Orbigny.* Prodrome, 1850, vol. ii, p. 50, No. 121.
> — — *Pictet.* Traité de Paléontologie, 1855, p. 539.
> — SUBCONCENTRICA, *Etallon.* Thurmann, 1862, Lethea Bruntrutana, p. 203, pl. 25, fig. 6.
> — CONCENTRICA, *de Loriol et Pellat.* Paléont. et Geol. de Etage Portlandien, 1866, pl. 8, fig. 2.

Shell ovately trigonal, moderately convex; umbones small, antero-mesial, not recurved, and but little produced, borders of the valves rounded; area narrow, flattened, distinctly bipartite, with irregular and unequal transverse plications, and bounded by two rows of regular, rounded, small tubercles; the escutcheon is of moderate size, excavated, its superior border is raised. The other portion of the valve is characterised by numerous rows (thirteen in small specimens) of regular concentric costæ, with small, closely arranged, rounded or ovate tubercles. These become somewhat cord-like, and attenuated anteally; both extremities of the costæ curve gently upwards; their posteal tubercles are conformable in size with those upon the marginal carina, which they meet at right angles.

Our specimens, which are dwarfed and not of adult growth, were procured in the Coralline Oolite of Steeple Ashton, Wilts; they agree in their general aspect more nearly with the figure of de Loriol than with that of Agassiz. The minutely tuberculated, closely arranged, concentric costæ distinguish them readily from *T. corallina.*

Length, 10 lines; opposite diameter, $6\frac{1}{2}$ lines; diameter through the united valves, 5 lines.

The foreign localities cited are Laufon (Bale), Switzerland; St. Jean-d'Angély, Auxerre (Yonne), Sennantes (Oise), France; all in the lower Portlandian or Kimmeridge strata.

TRIGONIA WILLIAMSONI, *Lycett*, sp. nov. Plate XVI, fig. 8.

Shell ovately oblong, lengthened, depressed; umbones sub-anterior, not much elevated, obtuse, not recurved; anterior side short, its border somewhat truncated, curved elliptically at its base with the lengthened lower border; superior border nearly straight, or slightly convex, sloping downwards, its extremity rounded with the posteal termination of the lower border. Area narrow and flattened, bounded by two faintly traced, minutely tuberculated carinæ; there is also a similar, rather obscure, line of tubercles indicating the position of a median carina; there are also delicately marked, transverse, plications of growth, which become more prominent posteally. The escutcheon is flattened or slightly excavated, and has great length; it is narrow, in conformity with the area. The other portion of the valve has about ten or eleven oblique, or slightly curved rows of large depressed, nodose varices; the first-formed three or four rows form separate nodes, but with some irregularity and inequality in their arrangement; the succeeding rows have the nodes very large, confluent, and depressed near to the angle of the valve, becoming rapidly small, attenuated, and irregular, near to the pallial border. The lines of growth are strongly defined over the whole of the valve.

This is one of the most depressed forms of the *Clavellatæ*; it possesses some general resemblance to *T. triquetra*, but is more depressed, and more lengthened; the posteal portion is wider and more rounded; its rows of nodose varices are also much more oblique. From *T. clavellata* it is distinguished by the very short anterior side, by the narrower area and escutcheon, by the general depression of the valves, and by the few broad, irregular, confluent nodes at the carinal extremities of the rows.

The name is intended as a trifling tribute to reminiscences of the earlier geological researches of Professor W. C. Williamson, F.R.S., of Owen's College, Manchester, and of his description of the locality whence the examples of this Trigonia have been obtained.[1]

Stratigraphical position and Localities. The Kelloway Rock, of Cayton Bay, near Scarborough. The matrix is a very hard, variable, grey, or sometimes whitish, siliceous rock in the lower portion of that stage; the valves of conchifera occur in abundance, but owing to the very intractable kind of rock, few are separated in good condition; associated with it are numerous valves of *Trigonia Rupellensis*. *T. Williamsoni* appears to be rare, only two examples have come under my notice; it is intended to figure a more perfect specimen upon the last plate of this Monograph.

---

[1] "On the Distribution of Organic Remains upon the Yorkshire Coast." 'Trans. Geol. Soc.,' 2 ser., vol. vi, p. 143; 1838.

§ III. UNDULATÆ.

TRIGONIA ANGULATA, *Sow.* Plate XIV, figs. 5, 6.

TRIGONIA ANGULATA, *Sowerby*. Mineral Conchology, 1826, pl. 508, fig. 1.
— ANGULOSA, *Agassiz*. Trigonies, 1840, p. 9.
— ANGULATA, *Ib.* Trigonies, 1840, p. 50.
— — *Roemer*, Versteinerungen, Oolith, 1836, p. 96.
— — *Morris.* Catal., 1854, p. 228.
— — *D'Orbigny.* Prodrome de Paléont., 1850, i, p. 308, No. 223.
— — *Lycett.* Inf. Ool. Trigonias, Ann. and Mag. Nat. Hist., 1850, p. 427.

Taking the figure in the 'Mineral Conchology' as representing the typical form, obtained from the Upper Ragstones of the Inferior Oolite at Nunney, near Frome, the following will serve for its description:

Shell sub-ovately elongated, convex; umbones obtuse, antero-mesial, much incurved and slightly recurved; anterior side much produced and rounded with considerable convexity; lower border lengthened, nearly straight, but with a slight undulation or excavation posteally; hinge-border lengthened, concave, sloping towards the more produced and narrow posteal extremity of the area, with which it forms a considerable angle. Escutcheon moderately wide, depressed, and concave. Area narrow and flattened, with two very small, but well-defined, bounding carinæ, and in some specimens there is an obscure line of minute tubercles bordering upon the median furrow, which is usually distinct; the bounding carinæ are also minutely tuberculated upon their upper portions, they have always much curvature; the area has transverse irregular plications, near to the apices these become distinct, regular costellæ. The sides of the valves have upon their anteal portions a few narrow, inconspicuous, sub-tuberculated costæ, which are directed obliquely downwards to the middle of the valve, but not in the direction of the lines of growth; they form a curve or undulation, the convex border of which is directed towards the marginal carina; for the most part their posteal extremities are bent suddenly upwards or united to a larger nodulous series of costæ or varices, which approach the carina at a considerable angle, and the few lower costæ approach it almost perpendicularly; these are scarcely so numerous as the anteal series. The first-formed five or six costæ are without undulation, or are nearly concentric; all are slightly nodulous. The larger example figured has some variability in its costæ, which, however, does not entitle it to be considered as a distinct variety. The few last-formed costæ are irregular or confusedly nodulous, and the larger or posteal costæ disappear in their course upwards, leaving a smooth, plain, and depressed space separating them from the carina.

The first of the two figures of *T. undulata*, From., given by Agassiz ('Trigonies,' pl. 6, fig. 1), has been quoted by D'Orbigny and by Oppel as a synonym of *T. angulata;* in this opinion I cannot concur, the large varices upon the marginal and inner carinæ, together with the considerable breadth of the area, appear clearly to separate the Swiss Great Oolite fossil. The second example of *T. undulata* ('Trigonies,' pl. 10, fig. 14) is another equally distinct species, and still more removed from *T. angulata.*

The authors above quoted appear to have had little confidence in the figure of *T. angulata* given in the 'Mineral Conchology,' which, although rudely engraved, is really a good drawing and faithfully renders the characters of the species in a small specimen.

T. angulata has more or less affinities with all the species of the *Undulatæ:* for distinctive differences the reader is referred to the descriptions of the numerous forms depicted upon Plates X to XVI inclusive.

Stratigraphical position and Localities. *T. angulata* has occurred in the Inferior Oolite of Nunney, near Frome, whence the type-specimen, associated with *Astarte elegans*, was obtained; Dundry, in the same county, is another locality. I have obtained it at various Gloucestershire localities in the Oolite-marl and in the Upper Grit-stones of the same formation in the Cotteswold Hills, but the entire number of examples which have come under my notice are inconsiderable, and no evidence has been obtained connecting the species with the southern portion of Somersetshire, or of Dorsetshire; it also appears to be absent throughout the long course of the Inferior Oolite in Oxfordshire, Northamptonshire, Bedfordshire, Lincolnshire, and Yorkshire. Upon the whole, therefore, it may be regarded as a rare species.

In France *T. angulata* has been described by Oppel ('Juraformation,' p. 485) as a species of the Cornbrash at Marquise, near Boulogne; some fine specimens from that formation and locality attributed to our species have been found upon examination to be *T. flecta.* D'Orbigny ('Prodrome,' 1, p. 308) places our species in his Étage 11, Bathonien. Rœmer ('Nordd. Oolith.,' p. 96) records the occurrence of *T. angulata* in the Dogger of Porta Westphalica.

TRIGONIA FLECTA, *Mor.* and *Lyc.* Plate XIV, figs. 7, 8, 9, 10.

  TRIGONIA ANGULATA, *D'Orbigny.* Prodrome de Paléont., 1850, vol. i, p. 308.
  — FLECTA, *Mor.* and *Lyc.* Monogr. Gr. Ool., Pal. Soc., 1853, p. 60, pl. v, fig. 20.
  — — *Morris.* Catal., 1854, p. 228.
  — ANGULATA, *Oppel.* Juraformation, 1857, p. 485, No. 45.

Shell sub-ovate, or ovately oblong, somewhat depressed; umbones antero-mesial, with little prominence, erect, or in other examples slightly recurved; anterior side moderately produced, its border curved elliptically with the lower border; hinge-border lengthened with some concavity, or in other instances nearly straight, sloping obliquely downwards; its length is nearly equal to twice that of the posteal border of the area, with which it forms a considerable angle. Escutcheon lengthened, much excavated, but having its superior border raised. Area of moderate breadth, concave immediately beneath the apex, but expanded and flattened posteally; it has regular, transverse, prominent plications, which become costellæ near to the apex; there is a well-marked mesial furrow; the bounding carinæ are small, but elevated and distinct; they have small, closely arranged, ovate tubercles or varices throughout their entire length; there is no median carina. The other portion of the valve has the rows of costæ rather numerous (about sixteen) anteally; they are plain, narrow, depressed, and horizontal, or are directed slightly downwards to about the middle of the valve, where they enlarge, form two or three nodose varices, and, curving gracefully upwards, become again suddenly attenuated, and meet the marginal carina at a considerable angle; in some instances, as in fig. 9, the costæ become broken mesially, and form an imperfect angle with their posteal portions. In adult specimens two or three of the last-formed anteal costæ coincide in their direction with the lines of growth; they therefore take the direction of the lower border, which is without any undulation, as in *T. angulata;* the anteal costæ are always somewhat more numerous than the others.

Affinities. The nearest ally is *T. angulata,* compared with which it is more depressed both anteally and mesially; its posteal portion is more expanded, and its lower border is destitute of the posteal undulation of that species; its costæ are also more numerous anteally, and do not form a distinct undulation or double curvature upon the middle of the valve, so that their general direction accords more nearly with the lines of growth; their posteal portions are also larger and broader.

T. Painei, another species of the Great Oolite, has also considerable affinities with our species; the latter has the form more lengthened posteally, the umbones are less produced, and are more anteal; the costæ are more numerous; its area more especially differs in having delicately tuberculated carinæ and a rugose plicated surface.

T. paucicosta, of the Kelloway Rock, has greater general convexity, and its area, with the peculiarity of its few, large, widely separated tubercles upon its carinæ, will readily be distinguished.

The young shells of *T. flecta* offer little that is distinctive from specimens of similar size pertaining to *T. angulata* and *T. Painei;* the transverse costellæ upon the area are, however, smaller than in the last-named species; compared with *T. paucicosta* their general ornamentation is much less conspicuous, more especially upon their carinæ.

The largest of our specimens has the length of 23 lines, height $19\frac{1}{2}$ lines, diameter through the united valves 10 lines.

Stratigraphical position and Localities. *T. flecta* appears to be a somewhat rare species;

when figured and described in the 'Monograph of the Great Oolite Mollusca,' published by the Palæontographical Society in 1853, it was only known as a British species from a single specimen indifferently preserved, and not sufficiently exhibiting the characteristic features; it was obtained at Trewsbury Quarry, in Forest Marble, near to the Tetbury Road Station of the Great Western Railway, near to Cirencester. More recently the Rev. J. E. Cross has obtained fine examples in the upper subdivision of the Great Oolite at Thornholm, near to the Village of Appleby, Lincolnshire; to the liberality of that gentleman I am indebted for the specimens now figured. Mr. Cunnington has also kindly forwarded to me four specimens from the Cornbrash of Hilperton, near Trowbridge; Hinton, in the same county, is another locality.

Fine examples of *T. flecta* have also been obtained in the Great Oolite of Marquise, near to Boulogne; they constitute the *T. angulata* of D'Orbigny and of Oppel.

TRIGONIA PAUCICOSTA, *Lycett*, sp. nov. Plate XI, figs. 8, 9; Plate XVI, fig. 7.

Shell ovately oblong, convex; umbones moderately elevated, antero-mesial, and recurved; anterior side rather short, its border curved; lower border much more lengthened, with a lesser curvature; superior border lengthened, somewhat concave, sloping obliquely, forming a conspicuous angle with the posteal border of the area. Escutcheon narrow, slightly depressed, its superior border somewhat raised. Area narrow, flattened, divided by a mesial furrow, bordered by a minute row of tubercles, bounded by two small distinct carinæ, which have each a row of tubercles; those of the marginal carina are large, regular, widely separated, and somewhat compressed by the lines of growth, they become evanescent upon the posteal half of the valve in specimens of advanced growth, upon which portion the surface has conspicuous irregular transverse plications; in the young shell the area has a few plain transverse costellæ. The costæ upon the other portion of the surface consist at first, of three or four narrow, elevated, plain, somewhat angulated ridges, which become, near to the carina, sub-tuberculated; subsequently they form two series, the anteal series are narrow, distinctly ridged, irregularly knotted or sub-tuberculated, they are nearly straight, and pass obliquely downwards to the middle of the valve, when they meet with a much larger less numerous, posteal series of nodose varices; this posteal series turns upwards suddenly at a considerable angle to the anteal series, towards the carina, from which the varices are well separated, and with which they form right angles. In adult forms the lower portion of the valve has the anteal costæ more irregular, and the junction of the two series is less angulated or more obscure; the whole aspect becomes influenced by the folds of growth which are conspicuous. Some specimens which scarcely constitute a distinct variety have the costæ, both anteal and posteal, plain; the posteal series, however, are usually slightly nodulous.

T. paucicosta is nearly allied to *T. angulata*, Sow., but has a much shorter anteal, and more lengthened posteal side; the ornamentation of the area is much larger, especially the tubercles upon the marginal carina, the few larger nodes upon the posteal series of varices are different to the small tubercular rows of *T. angulata;* the angles which the rows form at the middle of the valves are also distinct from the curvature or undulation of *T. angulata*. Another allied form is *T. undulata*, Fromherz (Agassiz, 'Trigonies,' pl. x, fig. 4), also a specimen figured under the same name (pl. vi, fig. 1); the description refers only to the former of the two specimens, which has the varices, both anteal and posteal, equal in size, and united mesially; the marginal carina has unusual prominence. The specimen, pl. vi, fig. 1, may possibly be identical with our *T. paucicosta*, but it appears to have undergone compression, and is therefore scarcely to be relied upon; should its identity with our species be eventually established, the name I have chosen may remain, as it is sufficiently distinct from the typical form figured by Agassiz, which may be regarded as the true *T. undulata* ('Trig.,' pl. 10, fig. 4).

For comparison with *T. flecta*, another allied species, the reader is referred to that shell.

The specimens selected for our figures sufficiently exemplify the general variability of the species, and also the changes of aspect produced by advance of growth; usually examples of very advanced growth are more imperfectly preserved, and are therefore less fitted to exemplify the species.

There is so much variability, both in the figure and ornamentation, that measurements of proportions have but little utility, descriptions also must, to some extent, be subordinate to figures in conveying correct or sufficient ideas of its several aspects.

Stratigraphical position and Localities. Hitherto *T. paucicosta* has been obtained only in the Kelloway Rock, of Cayton Bay, three miles to the southward of Scarborough. Numerous specimens in various stages of growth occur in the higher beds of the formation at that locality, in hard, brownish, sub-siliceous rock, occupying a thickness of eight feet, and associated with a multitude of the characteristic Ammonites of the formation. Other beds, six feet or more, separate it from certain lower hard beds, mostly of pale grey colour, which yield *Trigonia Rupellensis* in considerable numbers, even down to the dark clay which separates it from the Cornbrash. (The discovery of numerous examples of the latter species having occurred since page 28 was printed, I take the present opportunity of mentioning that it is intended to give additional illustrations of it upon the last plate of the present Monograph.) The entire thickness of Kelloway Rock at Cayton Bay, and beneath the adjacent Red Cliff does not exceed twenty-five feet, the section of the same formation to the northward of Scarborough Castle is upwards of three times that thickness, including a portion of the highest beds removed for foundations of houses. These excavations produced the finest

Ammonites of the formation in the collection made by Mr. Leckenby, now in the Woodwardian Museum, Cambridge, including the specimens figured in the plates accompanying his memoir on the Kelloway Rock of Yorkshire ('Quart. Jour. Geol. Soc.,' 1858, vol. xv); but no example of *Trigonia paucicosta* has occurred at that locality.

TRIGONIA PAINEI, *Lyc.* Pl. XII, figs. 2, 3, 4, 5.

>TRIGONIA GOLDFUSSII, *Morris* and *Lycett*. Monog. Moll. Gr. Ool., Pal. Soc., 1853, Bivalves, pl. v, figs. 18, 18 *a*; not *Lyrodon literatum*, Goldfuss.
>
>— — *Morris.* Catalogue, 1854, p. 228.

Shell depressed, ovately trigonal; umbones antero-mesial, large, elevated, and slightly recurved; anterior border moderately produced, and elliptically curved with the lower border; hinge-border nearly straight, sloping obliquely, and terminating in an oblique truncation of the posteal extremity of the area. Escutcheon narrow, depressed, and lengthened; its superior border is much raised: the area is narrow and flattened, divided by an oblique mesial furrow, and bounded by two inconspicuous slightly knotted carinæ, which disappear altogether posteally; near to the apex the area has a few transverse plications, but the surface generally is nearly smooth. The lines of growth are only faintly marked. The other portion of the shell has the first-formed six or seven rows of costæ entire and smooth, they pass obliquely downwards from the anterior border, and curve to the carina at a right angle; the subsequently formed costæ consist of two portions, the anteal series consist of a few narrow and depressed subnodulous costæ, which pass obliquely downwards to the middle of the valve, where their extremities are contiguous to the extremities of a few, much larger, nodose varices which pass upwards almost perpendicularly to the carina.

Much variability in the costæ is observable in different specimens, and not unfrequently the ornamentation over the greater portion of the valve is effaced; the later-formed costæ are usually disunited and very irregular. Our largest example has thirteen costæ. The species was at first mistaken for *T. Goldfussii*, Ag. (*Lyrodon literatum* Goldf.), which is more lengthened, and has a more prominent kind of ornamentation; it occurs in the Upper Oolites, but has not been discovered in Britain.

The smallest of our figures exemplifies a young shell with the costæ prominent, smooth, and ridge-like; they are united to the carina, where they form a projecting angle, and pass undivided across the area as somewhat smaller costellæ. Some doubt may exist whether the minute figure of *T. cuspidata*, Sow., given in the 'Mineral Conchology,' pl. cvii, figs. 4, 5, is intended for a dwarfed example of this form, or for the young state of *T. Moretoni*, as there is no very clear distinction between them; but as the first-formed costæ of *T. Moretoni* are usually smaller and more closely arranged than in *T. Painei*, and as the latter species has not been obtained at Ancliff, which was the locality of Sowerby's little specimen, it is more probable that *T. cuspidata* is the young of *T. Moretoni* in a more dwarfed condition than has been obtained at Minchinhampton.

Adult examples of *T. Painei* are most nearly allied to *T. flecta*, to which the reader is referred.

A shell figured by Messrs. Rigaux and Sauvage as a variety of *T. Arduenna*, Buv., in their interesting Memoir on new species from the Bathonian formation of Boulogne ('Mém. de la Soc. Acad. de Boulogne,' 1868, vol. iii, pl. vi, fig. 4), appears nearly allied to, and perhaps is not really distinct from, *T. Painei*. Our Plate XII, fig. 3, which presents an approximation to the Boulogne shell, represents the most common aspect of the species in the Great Oolite of Minchinhampton, or perhaps with the costæ more than usually prominent; not unfrequently, however, the posteal varices are more imperfectly developed, or more nearly resembling the figure of Messrs. Rigaux and Sauvage.

T. Arduenna, Buv., a much smaller species, is less short in its general figure, with much more numerous and more closely placed costæ and varices; the rugose area is another distinctive feature.

Dimensions.—Our largest specimen has the length upon the marginal carina of $2\frac{1}{4}$ inches; the opposite measurement is 2 inches; the convexity of a single valve 7 lines; the length of the escutcheon 18 lines.

The name is intended as a slight recognition of the success which has attended the exertions of Dr. Paine, of Stroud, as Honorary Secretary of the Cotteswold Naturalists, Field-Club during a long period, and also of his acquirements in the cultitivation of the natural Sciences.

Stratigraphical position and Localities. In the Great Oolite of Minchinhampton Common, in the beds called "planking," where the species occurs of every stage of growth, and is not infrequent, the valves are always disunited, and are often abraded, or have the ornamentation scarcely perceptible. It has also been obtained in the Great Oolite of South Lincolnshire. The specimen figured by Messrs. Rigaux and Sauvage, from Boulogne, is stated to have been procured in the zone with *Clypeus Plotii*. The British Museum has fine examples from the Great Oolite of Normandy.

TRIGONIA PRODUCTA, *Lyc.* Pl. XIII, figs. 1, 2, 3, 4.

TRIGONIA PRODUCTA, *Lycett.* Note in Wright's Memoir on the Inferior Oolite, Quart. Jour. Geol. Soc., 1859, vol. xvi, p. 45.

Shell ovately trigonal, somewhat depressed, elongated posteally, short but curved anteally; superior border lengthened, straight, or slightly concave; lower border slightly curved elliptically; umbones antero-mesial, elevated, obtuse, and but little recurved. Area narrow, flattened, somewhat raised, with numerous very irregular, rugose, transverse plications; it has a well-defined median furrow, bordering upon which, on each side, is a row of minute tubercles; the inner and marginal carinæ are each represented by a small row of inconspicuous tubercles or knotted terminations of the transverse plications of the area. The escutcheon is depressed and flattened, its length is considerable, or nearly equal to the measurement across the valves; its posteal extremity forms an obtuse angle with that of the area. The costæ upon the sides of the valves have but little prominence, the first-formed three or four rows are horizontal or slightly oblique, with small, regular, cord-like tubercles, the succeeding rows form two distinct series; the anteal series are few, oblique, and very irregularly sub-tuberculated; for the most part the rows have but little prominence and sometimes become nearly evanescent towards the middle of the valve; the few last-formed or lower ones are commonly more or less confused or imperfect; about the middle of the valve this anteal series is replaced by a more numerous posteal series, whose lower extremities form nearly right angles with the other series; they are regular, narrow, closely arranged, straight, imperfectly tuberculated, and are somewhat more prominent than the anteal series; they pass upwards nearly perpendicularly to the marginal carina; there are about fifteen rows, and their size continues nearly equal even to the posteal extremity of the valve.

The test is thick, the borders of the valves smooth, and the hinge is remarkable for the great breadth and flattening of the central tooth in the left valve. The larger of our specimens has the length, upon the marginal carina, of $3\frac{1}{4}$ inches; the opposite measurement is $2\frac{1}{2}$ inches; the diameter through the united valves is inconsiderable. *T. producta* has occurred only in single valves; the internal mould is unknown: it is one of the largest and least known of the *Undulatæ*; several young examples have been obtained, these are but little distinguished from similar examples of *T. signata*. The costæ supply the distinguishing features of *T. producta*, the few widely separated and rather obscure anteal series, and the more closely arranged, but separate, straight, and narrow posteal rows, serve to remove it from *T. signata*, which has moreover the umbones more mesial and recurved; the anterior side and border is more produced and rounded; this arrangement of the costæ is very distinct from *T. V-costata*, which has the anteal series much more numerous, and of a different figure, and does not form rows of separate tubercles. It has also some affinity with *Lyrodon literatum*, Goldf., in the characters of the costæ; but the latter species has the general figure more lengthened and oblong, and the posteal portion of the area has much greater breadth; the marginal carina also has a row of large rounded tubercles; the umbones are smaller and less elevated; the anterior side is also more produced. In *T. producta* the whole of the ornamentation has but little prominence, and in some examples it is partially obscured by the plications of growth, which become large and rugose over the lower portion of the valve.

Messieurs Terquem and Jourdy have figured, under the name of *T. producta*, a small species of the *Clavellatæ* from the Great Oolite of the Department of the Moselle ('Mém. Soc. Géol. Fr.,' 2 sér., tom. ix, 1869, pl. xi, figs. 29, 30), this is allied to, and perhaps is not really distinct from, *T. impressa*, Sow.

Stratigraphical position and Localities. *T. producta* is one of the rarer fossils of the Trigonia-grit of the Inferior Oolite, near to Cheltenham and to Stroud; fine examples have also been obtained in Northamptonshire by the officers of the National Geological Survey. Specimens from the Inferior Oolite of Normandy are in the British Museum.

TRIGONIA CONJUNGENS, *Phil.* Pl. X, figs. 5, 7, 8 ; Pl. XIII, fig. 6.

TRIGONIA CONJUNGENS, *Phillips.* Geol. Yorks., 1829, vol. i, p. 156.
— — *Morris.* Catal., 1854, p. 228.

Shell ovately oblong, moderately convex mesially, somewhat depressed near to the anterior and posterior borders; umbones elevated, obtuse, erect, or slightly recurved, placed within, or in other specimens upon, the line of the anterior third of the valves; anterior border produced, curved elliptically with the lower border; hinge-border straight, lengthened, sloping obliquely, and terminating posteally in the wide, rounded, posteal border of the area. Escutcheon large, lengthened, depressed, excepting its superior border, which is raised. Area very wide, occupying about one third of the surface of the valve; it is somewhat raised, expanded, and flattened posteally; it has a well-marked mesial oblique furrow, and is traversed transversely by numerous large plications, which increase in size posteally, and become irregular, prominent, and wrinkled (see Plate XIII, fig. 6); the bounding carinæ are small and distinct; the marginal carina is minutely plicated, excepting its posteal portion, which is occupied by the large transverse plications of the area; these form small varices upon the inner carina; there is no median carina. The costated portion of the valve has numerous rows (eighteen or nineteen in adult forms) of tuberculated or sub-tuberculated costæ, the first-formed six or seven rows are very closely arranged, regular, plain, nearly horizontal, and slightly curved at their two extremities: those which succeed form two series; the anteal series are somewhat irregular in their arrangement, but are always small and inconspicuous ; they are directed somewhat obliquely downwards to the middle of the valve, and are occasionally distinctly tuberculated, but commonly are irregularly sub-tuberculated ; their posteal extremities are united about the middle of the valve to another, less numerous, and somewhat larger posteal series of costæ, which are also either distinctly tuberculated or sub-tuberculated ; they approach the carina at a considerable angle, and the last-formed three or four rows pass perpendicularly down to the lower border; their anteal or lower extremities are for the most part united to the extremities of the more numerous anteal series, with which they form a considerable undulation or angle, which is always less than a right angle; not unfrequently, however, the few last-formed anteal costæ are altogether irregular, presenting only small confused tubercles.

This appears to be the shell indicated by the author of the 'Geology of Yorkshire,' who gave a short notice of it at page 156 of the first edition of that work, but without any figure. It is nearly allied to *T. angulata*, Sow., in the general arrangement of its ornamentation; but it differs from that species constantly and materially in the general figure, which is much more broad and expanded posteally,—so different from the narrow, concave, and delicately plicated area of *T. angulata*; it is also without the undulation upon the lower borders of the latter species, and its rows of anteal costæ are more numerous; but usually the outline of the two forms will at once show their distinctness. It is also somewhat allied to *T. Moretoni*, M. and L., but is more oblong, with a much wider and more rugose area; the marginal carina is much smaller, the costæ are also more disunited mesially, and have the tubercles smaller, and the costæ less ridge-like.

The two small lower figures of plate lxxxvii 'Mineral Conchology,' represented as a variety of *T. clavellata*, have some affinities with our species in the general figure, and the characters of the costæ; but the area is destitute of the large transverse rugose costellæ; there are also indications of a small median carina, it is probably therefore distinct.

T. compta, Lyc., of the Collyweston Slate, is also nearly allied to it, both in the general outline and the ornamentation. The flattened condition of the slate species prevents any comparison of the convexity, but the costæ in adult specimens are fewer; the area is smooth, and has three distinct tuberculated carinæ,—characters which are so different from that portion of *T. conjungens* as to compel their separation as species.

Stratigraphical position and Localities. During many years *T. conjungens* remained one of the more obscure and doubtful forms of Trigonia, and was usually omitted, even in the lists of Yorkshire fossils; this arose from the very few words of description allotted to it by Professor Phillips in his 'Geology of Yorkshire,' and not less so to the very intractable stone of the Millepore-bed, a hard, rough, semi-ferruginous stratum at Cloughton Cliffs, to the northward of Scarborough. A considerable number of examples of *T. conjungens* have been obtained from that locality, but only a few have been separated well preserved. Brandsby, Yorkshire, was the locality given by Professor Phillips, at which place the beds are not now accessible: it has also been obtained in the White Oolite of Whitwell, in the same county. The Cloughton specimens are associated with *T. recticosta*, and a considerable number of Inferior Oolite Conchifera. It also occurs in the same stratum exposed at Cayton Bay, to the southward of Scarborough. Specimens are in the Museum of the Yorkshire Philosophical Society at York, also in the collection of Mr. Reed, of the same place; in the collection of Mr. Leckenby, now forming part of the Woodwardian Museum, Cambridge; also in the author's cabinet. It is now for the first time figured.

TRIGONIA LITERATA, *Young* and *Bird.* Pl. XIV, figs. 1, 1a, 2, 3, 4.

<blockquote>
TRIGONIA LITERATA, *Young* and *Bird.* Geological Survey of the Yorkshire Coast, 2nd ed., 1828, p. 225, pl. viii, fig. 23.

— — *Phillips.* Geol. of Yorkshire, 1829—1835, vol. i, pl. xiv, fig. 11.

— — *Williamson.* Distribution of Fossil Remains on the Yorkshire Coast, Tr. Geol. Soc., 2nd series, 1836, vol. v, p. 243.

— — *Pusch.* Polens Palæont., 1837, p. 60.

— LITTERATA, *Agassiz.* Trigonies, 1840, pp. 8 and 50.

— — *Brown.* Foss. Conchol. of Great Britain, 1845.

— LYRATA. *D'Orbigny.* Prodr. de Paléont., 1850, vol. i, p. 218.

— LITTERATA, *Morris.* Catal. 1854, p. 229.

— LITTERATA, *Simpson.* Fossils of Lias, 1855, p. 116.

— LITTERATA, *Quenstedt.* Der Jura., 1856, p. 442.

— — *Oppel.* Juraformation, 1856, p. 260.

— LITTERATA, *Tate.* Geol. Mag., 1872, vol. ix, p. 306.
</blockquote>

Shell sub-ovate or ovately oblong, convex; umbones large, moderately elevated, obtuse, nearly erect, placed within the anterior third of the valves; anterior side moderately produced, its border curved elliptically with the lower border; superior border lengthened, nearly straight, sloping obliquely downwards, and forming posteally nearly a right angle with the posterior border of the area. Escutcheon wide and somewhat concave, its superior border is moderately raised. Area narrow, slightly convex, with a well-defined mesial furrow. Young examples have two distinct bounding carinæ; the inner carina is characterised by transverse narrow irregular varices; the marginal carina has irregular widely separated tubercles; in examples of more advanced growth the carinæ have small varices which are united to the transverse plications upon the area; much variability exists in the prominence of the plications, but usually specimens of full development have the posteal portions of their areas characterised only by delicate lines of growth, and are altogether without ornamentation. The other portion of the surface has two distinct series of tuberculated costæ, this distinctness commences at the apices even of the youngest specimens; the anteal series has the rows very numerous, small, and extremely irregular; in young specimens they approach the anterior borders horizontally, as smooth attenuated lines; with the curvature of the valve they become sub-tuberculated, and are usually deflected slightly downwards, but their direction is scarcely alike in any two specimens; they are always small and unsymmetrical, one row with another, unequal, and either prominent or obscure, sometimes partially united to the extremities of the larger posteal series, or altogether separated from them and excentric. Well-preserved adult forms have the few lower costæ of the anteal series more or less wrinkled and obscure, these take the direction of the lines of growth, and therefore curve upwards to the anterior border; other examples have the whole of the anteal series forming smooth, irregular and unequal, oblique, wrinkled costæ, but in all instances this series occupies only the smaller portion of the costated surface, their junction with the other series is always anteal to the middle of the costated surface; the posteal series are fewer, much larger, and more regular; they form prominent nodose ridges, which descend almost perpendicularly from the carina, enlarging downwards, and forming acute angles with the anteal series; about a moiety of the posteal series attain the lower border: our largest specimen has twelve of these costæ. Few examples of the genus have the lines of growth so strongly marked as in the adult examples of *T. literata*, they impress the costæ very conspicuously.

Young specimens from nine to twelve lines across the valves are remarkable for the prominence, delicacy, and beauty of their ornamentation; they are slightly more lengthened, than the adult form, but are less oblong and quadrate than the little *Trigonia pulchella*, of Agassiz, to which their ornamentation approximates considerably.

Dimensions of a large specimen.

Length . . .	. $2\frac{1}{4}$ inches.
Opposite measurement . .	. $1\frac{6}{10}$,,
Diameter through the united valves .	$1\frac{6}{10}$,,

Several Jurassic species approximate to *T. literata* in the general plan of their ornamentation. *Lyrodon literatum*, Goldf., a larger species, has the general figure more lengthened, and the area much larger in proportion. *T. subglobosa*, Mor. and Lyc., is distinguished by the more globose form, by the large tuberculated carinal, and by the few very large posteal varices. *T. Painei*, Lyc., on the other hand is much more depressed, and the costæ fewer. *T. V-costata*, Lyc., has not the two series of costæ broken and separated as in the present form, it is also much less convex: other of the *Undulatæ* are more remotely allied to it.

History, Stratigraphical position, and Locality. Messrs. Young and Bird, in their Geological Survey of the Coast of Yorkshire (1822), first described *T. literata*: they assigned it to the Lower Lias Shale, but without any locality: the figure of it given upon plate viii of their work is very indifferently executed. In the year 1829, Professor Phillips, in his 'Illustrations of the Geology of Yorkshire,' gave a much better figure: he assigned the species to the Lower Lias Shale of Robin Hood's Bay, and figured it with the fossils of that stage; he also noticed its occurrence in Upper Lias Shale upon the authority of Mr. Williamson (p. 161). In 1850, D'Orbigny in his 'Prodrome,' also erroneously placed it (printed *T. lyrata*) in his Étage Sinemurien, and gave the vicinity of Metz (Moselle) as a locality. In addition to *T. literata*, Agassiz placed the five following species in the

Upper Lias: *T. navis*, Lam., *T. pulchella*, Ag., *T. tuberculata*, Ag. (*T. spinulosa*, Y. and B.), *T. similis*, Bronn, and *T. costellata*, Ag. Subsequent researches have shown that the latter five species are associated more or less one with another in a single geological position, and that two or more of them occur together at several localities in Southern Germany. Professor Quenstedt (' Der Jura.') has established *T. navis* and *T. pulchella* as species of the lower portion of the Inferior Oolite.

In Britain *T. spinulosa* pertains both to the Supra-Liassic Sands and to the lower portion of the Inferior Oolite; and, as the two remaining species are associated in Southern Germany with the three former, it may be inferred that all of them belong to a higher position than *T. literata*, and that the latter therefore occupies the lowest position of any known species of the Upper Lias.

In Britain *T. literata* has occurred only at a single locality; namely, a little higher than the middle of the Upper Lias Shale at the Peak, Robin Hood's Bay, in scars accessible only at low water; there are no Ammonites in this stratum, but it immediately overlies a bed with *Ammonites crassus*, Y. and B.; it is lower than the beds worked for alum upon the same coast. Specimens occur of every stage of growth, with the valves both united and separated; but, as the greater number have the characters of the surface ill-preserved, good specimens are somewhat rare.

TRIGONIA V-COSTATA, *Lyc.* Plate XIII, fig. 5; Plate XV, figs. 1, 2, 3, 4.

TRIGONIA ANGULATA, *Phil.* Geol. York., 1829, vol. i, p. 156 (not Sow.).
— — *Williamson.* On Distribution of Fossils, Yorkshire Coast, Trans. Geol. Soc., 1836, 2 ser., iii, p. 229.
— V-COSTATA, *Lycett.* Ann. and Mag. Nat. Hist., 1850, p. 422.
— — *Morris.* Catalogue, 1854, p. 228.
— — *Lycett.* Cotteswold Hills Handbook, 1857, pl. vi, fig. 5.

Shell ovately trigonal, moderately convex; umbones nearly mesial, produced, obtuse, and usually somewhat recurved; anterior side produced, its border curved elliptically with the lower border; hinge-border slightly concave, sloping obliquely, its extremity forming an obtuse angle with the extremity of the area. Area narrow, concave beneath the apices, but flattened posteally; it is traversed transversely by delicate plications, which near to the apices form a few regular, smooth costellæ; it has a mesial longitudinal furrow, and in the young condition three closely tuberculated carinæ, which become evanescent posteally with advance of growth. The escutcheon is much depressed compared with the inner carina; it is lengthened, of moderate breadth, and perfectly flat. The other portion of the surface has the rows of costæ numerous (twenty to twenty-four) and narrow; they are but little raised, and are rather inconstant in their characters; sometimes they are plain, but more frequently they are subtuberculated; all of them commence at the anterior border and curve obliquely downwards; the first few rows are simply curved upwards at their posteal extremities to the carina; those which succeed are more straight and oblique; their posteal portions form with the anteal portions a more decided angle, which increases with every succeeding costa, until they form acute angles upon the middle of the valve, the posteal portions passing upwards perpendicularly to the carina, but without any increase in their size. In adult forms the last three or four posteal costæ pass downwards perpendicularly to the pallial border.

Three Jurassic Trigoniæ are allied to *T. V-costata*. *T. tripartita*, Forbes, a much smaller species, differs in having a few large posteal, straight, oblique costæ, which are distinct from the far more numerous and smaller anteal costæ. The figure of *T. angulata*, Sow., is much more produced, and attenuated posteally; the hinge-border is more lengthened and concave; the umbones are more prominent and recurved; the costæ are very much fewer, and posteally they form an undulation rather than an angle; it is very correctly represented by the coarse figure in the 'Mineral Conchology.' *T. producta*, Lycett, is somewhat allied to it in the characters of the costæ, but has the anteal series few and distinctly tuberculated; the general figure also is essentially different; the anteal not recurved apices, the lengthened and flattened area, together with the greater general length of the shell, separate them very clearly.

Adult specimens of *T. V-costata* have the length one sixth greater than the height; of small specimens the number is far more considerable; these latter have been obtained at numerous localities in beds of very different mineral character; usually they have a more lengthened figure transversely; they differ materially one from another in the closeness or wider separation of the costæ.

Stratigraphical position and Localities. The figure given in the 'Annals of Natural History' was from a large example obtained in the Upper Trigonia-grit of Rodborough Hill, near Stroud; a few specimens have also occurred in a similar position at several localities near to Cheltenham; it is, however, rare throughout the Cotteswold Hills. In Northamptonshire it occurs more commonly in the ferruginous beds, and is of smaller dimensions. Numerous specimens (apparently dwarfed) also occur in the very fossiliferous bed of the Dogger at Blue Wyke, near Robin Hood's Bay, Yorkshire; they are of various stages of growth; three are depicted upon Plate XV. A considerable number of small, imperfectly preserved Trigoniæ also occur in the layers of oolitic slate at Collyweston; they are deprived of the test, and have probably undergone vertical compression; their condition, therefore, does not admit of a rigid comparison; they also differ much one with another in the general figure, and in the prominence or indistinctiveness of the costæ; all have the appearance of young shells, and occasionally specimens have the figure more lengthened transversely than is observed even in young examples of *T. V-costata*; there does not, however, occur any constant characters which will justify their separation from that species, to which, therefore, they are provisionally united; two examples are figured on Plate XI, figs. 6, 7. In the absence of better illustrative specimens it is therefore necessary to cancel " *T. minor*," as this obscure form was designated in the list of Inferior Oolite Trigoniæ, page 12 of this Monograph.

The internal mould is not known; impressions occur in the Ferruginous Oolite of Glaizedale, North Yorkshire. All the localities are in Inferior Oolite.

TRIGONIA SUBGLOBOSA, *Lyc.* Plate XII, figs. 8, 9, 10.

> TRIGONIA SUBGLOBOSA, *Lycett.* Ann. and Mag. Nat. Hist., 1850, p. 421.
> — — *Morris* and *Lycett.* Gr. Ool. Monog., Pal. Soc., 1853, p. 55, pl. v, fig. 21.
> — — *Morris.* Catal., 1854, p. 229.

An inflated shell with prominent ornamentation and nodose angulated costæ.

Shell ovately globose; umbones antero-mesial, large, produced, much incurved, and slightly recurved; anterior and inferior borders curved elliptically; posterior or superior border short, concave, its posteal extremity forming an obtuse angle with the posteal border of the area. Escutcheon wide, depressed, and rather short. Area wide, flattened, slightly excavated, its surface forming a considerable angle with the other portion of the shell; it is conspicuously bi-partite, the lower constituting the larger portion; it is traversed transversely by irregular rugose plications, which near to the apex are replaced by a few, narrow, regular, plain costellæ; it is bounded by tuberculated carinæ; the tubercles upon the marginal carina are unusually large and nodose; the median carina is also represented by a similar, but smaller row of tubercles. The other portion of the shell has upon its anteal side thirteen or more, narrow, depressed, subtuberculated costæ, which are rather irregular in their course, but, for the most part, curve obliquely downwards to within a short distance of the carina, when they meet with a much larger and more prominent nodose posteal series, fewer in number (about eight or nine), which pass upwards perpendicularly to the carina, their point of junction with each of the anteal series having a large tubercle; the angles thus formed are more considerable than right angles. Upon the lower third of the adult valves the plications of growth become strongly marked anteally, and all other ornamentation then ceases; this change is also accompanied by a sub-concentric sulcation which crosses the valve longitudinally.

Affinities. It is allied to a larger species, *T. Painei*, Lyc., in the general features of its ornamentation, but is greatly more inflated; the area is much wider and more ornamented, and more especially by the presence of the prominent marginal carina with its few large tubercles; this latter feature, together with the absence of large transverse plications upon the area, and its sub-globose figure, serve also to separate it from *T. conjungens*. The valves are separate, or spread open and held in contact by their ligament.

UNDULATÆ.

Stratigraphical position and Localities. Near to Stroud and Nailsworth, Gloucestershire, in the middle portion of the Inferior Oolite; in a bed of pale, tough, cream-coloured limestone (Coralline mud), associated with *Trigonia costatula*, Lyc., *T. angulata*, Sow., *T. Phillipsi*, Mor. and Lyc., a crowd of small, sub-cylindrical Nerinææ, sub-acicular Chemnitziæ, and a numerous group of Molluscan forms, both of Gasteropoda and Conchifera: but all the Trigoniæ are rare.

TRIGONIA GEOGRAPHICA, *Ag.* Plate X, fig. 6.

<blockquote>
TRIGONIA GEOGRAPHICA, <i>Agassiz.</i> Mém. sur les Trigonies, 1840, p. 25, tab. 10, fig. 7 (Excl. tab. 6, figs. 2, 3).

— — *D'Orbigny.* Prodr. de Paléont., 1850, vol. ii, p. 17, No. 267.

— — De Loriol, E. Royer, and H. Tombeck. Descr. Géol. et Paléont. des Étages Jurassiques Supérieurs de la Haut Marne, 1872. Mém. Soc. Linn. de Normandie, tom. 13, pl. 17, fig. 7.
</blockquote>

Shell ovately trigonal, moderately convex; umbones submesial, not much produced, and only slightly recurved; anterior side produced, its border curved elliptically with the lower border; hinge-border straight, sloping obliquely downwards. Area narrow, somewhat concave, divided into two portions by a considerable depression of the upper half; the whole is transversely striated; the bounding carinæ are very small, without any distinct tubercles. Escutcheon narrow, depressed, and moderately lengthened. The other portion of the shell has numerous rows of narrow, plain, closely arranged costæ, all of which originate at the anterior border and pass over the middle of the shell obliquely downwards with a slight curvature; they enlarge somewhat posteally, and curve upwards towards the carina; about twelve or thirteen of the costæ first-formed are plain, those which follow gradually become tuberculated and enlarge at their posteal portions; each succeeding row becomes more tuberculated, so that the rows last-formed have only a short portion plain anteally; they thus gradually form two series of costæ, of which the posteal or tuberculated series is the larger and less numerous in consequence of the intercalation of three or four rows of short anteal costæ, which impart much irregularity to the few last-formed rows; the tubercles are neither regular nor symmetrical, some are distinct and oval, others are united in the rows which all curve upwards to the carina.

Our specimen resembles only the second figure of Agassiz ('Trigonies,' tab. 10, fig. 7), but is a smaller example with less prominent posteal tubercles; the area is also nearly destitute of ornamentation, which is probably due to defective preservation. The figure given by De Loriol, Royer, and Tombeck is more lengthened, and the costæ have less posteal curvature towards the carina; the tubercles also appear small for a specimen of such advanced growth, differences which indicate a variety.

Stratigraphical position and Locality. Our specimen was obtained in the Trigonia-bed of the Coralline Oolite at Pickering; no second specimen has come under my observation.

TRIGONIA COMPTA, *Lyc.* Plate XV, figs. 5, 6, 7.

> TRIGONIA COMPTA, *Lycett.* Suppl. Mon. Gr. Ool. Mollusca, Pal. Soc., 1863, p. 50, pl. xl, fig. 1.

Shell ovately trigonal, somewhat depressed; umbones moderately elevated, antero-mesial, not recurved, obtuse; anterior side produced, curved elliptically with the lower border; posterior border truncated at its extremity. Escutcheon narrow and inconspicuous. Area rather wide, flattened, with a mesial furrow and three small tuberculated carinæ; its general surface is smooth, with faintly traced lines of growth. In other specimens the carinæ are evanescent, and the area is altogether smooth; its entire surface is raised, so that the marginal carina, although so little conspicuous, forms a prominent ridge compared with the more depressed, adjacent costated portion of the valve.

The other portion of the surface has about twelve rows of costæ, which pass from the anterior border obliquely downwards; they are very narrow, elevated, and sub-tuberculated, or occasionally (plain; posteally they increase in size, become partially disunited, and form two distinct, large, depressed nodes or varices, which curve upwards and meet the marginal carina at a right angle; each row of varices corresponds to every alternate row of the narrow anteal costæ.

The general ornamentation has but little prominence, and not unfrequently it is rather obscure.

Young specimens, when only three lines across, have narrow, horizontal, plain costæ, which are slightly curved upwards at their posteal extremities; the marginal carina is then prominent.

T. compta is a small species not uncommon in the slate of Collyweston; the specimens are usually more or less compressed; it also occurs more rarely in the sand of Northampton, in which it retains its original convexity. It differs from *T. Moretoni* in having much fewer costæ, which do not form continuous curves as in that species, but are disunited posteally, forming a less numerous series of short large varices; the area destitute of large transverse rugose folds, is another distinctive feature.

From *T. costalula* it is separated by the more lengthened form, less convexity, larger area, and the presence of the short, curved, large posteal varices. The same general features, together with the larger ornamentation, also separate it from *T. impressa.*

Compared with *T. conjungens*, its ornamentation is much less prominent; its posteal varices are larger and more depressed; the area is less expanded posteally, and is destitute of the large rugose plications which are so conspicuous on that species.

Locality. The finely laminated, slaty sandstone has preserved the ornamentation in a very perfect manner, although the test has wholly disappeared. It has only been recognised in the vicinity of Collyweston, and in an apparently similar geological position in the Inferior Oolite of Northamptonshire.

TRIGONIA LECKENBYI, *Lycett*, sp. nov. Plate XVI, figs. 1, 2.

Shell ovately oblong, lengthened and attenuated posteally, much depressed; umbones sub-anterior, obtuse, not conspicuous; anterior side short and rounded; lower border lengthened, curved elliptically; superior border lengthened, slightly concave, and having a gentle curvature downwards posteally to the lower extremity of the area, which is somewhat pointed. Area narrow and flattened with some obscure transverse plications; the bounding carinæ are scarcely elevated, the marginal carina has a row of small tubercles, which disappear posteally; there is no distinct mesial furrow. The escutcheon is lengthened, very narrow, and depressed. The other portion of the valve has the rows of costæ in two series; the posteal series consists of depressed, curved, rounded, sub-tuberculated costæ, which pass downwards almost perpendicularly from the carina to the middle of the valve; their number is from fourteen to sixteen, they are cord-like, and become attenuated near to the carina; the anteal series is much smaller and more numerous, the rows are sub-tuberculated, and, for the most part, nearly horizontal in their direction; they are not distinctly united to the extremities of the posteal series, but become broken into small, irregular, isolated tubercles, which occupy the middle of the valve, even to the lower border; the general direction of the rows of posteal costæ is therefore not conformable with that of the anteal series. The lines of growth appear to have but little prominence; no portion of the test has been preserved in the larger example, and only partially so in the smaller.

This remarkable example of the *Undulatæ*, conspicuous for its very much depressed, lengthened figure, and the delicate, but composite, character of its ornamentation, possesses some general resemblance to *T. angulata*, from which it will readily be distinguished by its large obtuse umbo and depressed figure; or, when the costæ are preserved, the very numerous and minutely knotted anteal series is a characteristic feature. The lengthened depressed form equally removes it from *T. literata*, Phil., to which it approximates in the general character of its costæ. Other species more remotely allied are *T. v.-costata* and *T. paucicosta*, to the descriptions of which the reader is referred. *T. Ramsayi*, from nearly the same geological horizon in Gloucestershire, with a more lengthened form and cord-like costæ, is without the small anteal series. In the general lengthened outline it is also not without some resemblance to *T. Pellati;* but they differs in all the essential features of ornamentation.

Stratigraphical position and Locality. *T. Leckenbyi* ranks as one of the most rare of the Trigoniæ. I am not aware that more than four entire examples of single valves have been procured, and fragments of a few others; all are more or less flattened from vertical pressure, and only portions of the test remain; but the characters of the surface are sufficiently well preserved. The rock is a dark grey, highly micaceous, thin-bedded, shaley sandstone, a member of the stage which may be provisionally termed Supra-liassic, and according in position with the *Jurensis*-beds of the Cotteswold Hills; it is associated with some of the characteristic fossils of that stage, and more especially with *Terebratula trilineata,* Y. and B. It has been procured only over a small area in shore-beds covered by the tide at high water, between Blue Wyke and the Peak, near Robin Hood's Bay, upon the coast of Yorkshire. The geological position is therefore higher than the Alum Shale or the zone of *Ammonites communis,* and lower than the stratum with *Lingula Beanii.*

TRIGONIA CARREI, *Mun.-Chal.* Pl. XII, fig. 1.

TRIGONIA CARREI, *Munier-Chalmas.* Note sur quel. esp. nouv. du genre Trigonia, Bull. Soc. Linn. de Normandie, 1865, vol. ix.

— — *De Loriol* et *Pellat.* Mon. Paléont. et Géol. de l'étage Portlandien des env. de Boulogne, 1866, pl. viii, fig. 5.

— — *Hébert.* Terr. jurass. du Boulonnais, Bull. de la Soc. Géol. France, 1866, 2 sér., t. xxiii, p. 216.

Shell ovately trigonal, moderately convex; umbones sub-anterior, elevated, obtuse, and erect; anterior side short, its border somewhat truncated, but having its lower portion curved elliptically with the lower border; cardinal border moderately lengthened, straight, sloping obliquely, and curved suddenly with the posterior extremity of the area. Escutcheon excavated, its superior border raised. Area narrow, flattened, conspicuously bi-partite; the plane of its surface forms a considerable angle with the other portion of the shell; it is bounded by two well-defined tuberculated carinæ; the tubercles upon the marginal carina are more especially large and distantly arranged; there is also a line of small tubercles bordering upon the mesial furrow; the transverse plications upon the area are irregular and not conspicuous. The other portion of the surface has a series of large elevated varices, about twelve in number, which curve downwards from the carina, occupying more than half the costated surface; each row has about nine depressed nodes; the lower extremities of these varices meet the posteal extremities of a much smaller anteal series, which are nodose, short, and nearly horizontal in their direction, but with some irregularity. The lines of growth are conspicuous over the greater portion of the shell. Portions of the epidermal granulated tegument are preserved, and have the lines of granules larger than is usual in the *Undulatæ* or the *Clavellatæ*. Our figure exemplifies a specimen, the posteal extremity of which is somewhat defective from disappearance of the test.

Locality and History. Our specimen is from the Portland Oolite of Tisbury; Mr. Cunnington has also obtained the species in the same formation near to Devizes; at both localities it ranks as one of the most rare Testacea in the formation. Allusion may here be made to *T. radiata*, Benett, the absence of which in this Monograph as a recognised species requires some explanation.

In the year 1831, Miss Etherelda Benett, of Warminster, published a small folio volume, intended as an illustrated catalogue of Wiltshire fossils; plate xviii, fig. 3, of that work, represents *Trigonia radiata* from the Portland Limestone of Tisbury; the figures generally in Miss Benett's work are carefully drawn, it may therefore be assumed that the deficiencies in the Trigonia represent the defective condition of the specimen. The general aspect seems to indicate flattening or compression; the area has no clearly defined features; the marginal carina has some indistinct tubercles; the rows of varices which pass downwards from the carina are more clearly expressed, and appear to be obtusely nodose; there are also some partially preserved tubercles upon the pallial portion of the valve near to the lower border; all else is left to the imagination. The type specimen was removed to the Philadelphian Museum, with the whole of Miss Benett's collection, at the decease of that lady; but, judging from the figure, the possession of the specimen would have added but little to our knowledge of the species, and no second British example recognised as *T. radiata* is known.

In the year 1865 M. Munier-Chalmas figured and described under the name of *T. Ferryi* ('Bull. Soc. Linn. de Normandie,' vol. ix), a Trigonia which in the general characters of its varices possesses a considerable resemblance to the figure given by Miss Benett; the anteal portion of the valve is represented as devoid of ornamentation, but as this, in common with the other figures upon the same plate, is reduced in size, we are the less able to judge of the actual condition of the specimen; he also described in the memoir accompanying the plate an allied Trigonia from the same formation, in the vicinity of Boulogne, under the name of *T. Carrei*, but of this no figure was given.

In the following year appeared the almost simultaneous publication of Memoirs on the Portland formation of Boulogne by Professor Hébert, and Messieurs de Loriol and Pellat, each of these palæontologists discovered the apparent identity of *T. Ferryi* with *T. radiata*, and the last-named authors figured a very perfect example of *T. Carrei*, and another of *T. radiata;* the latter form is, however, apparently defective, the anteal half of the valve being separated from the other portion with varices by a line of fracture, which apparently indicates the absence of the test upon the anteal portion of the valve, the varices of which end abruptly at the line of fracture. Several large clavellated

Trigoniæ, or portions of them from the Portland formation of Oxfordshire and Wiltshire, which have come under my notice, are too imperfect in their general condition to admit of any satisfactory comparison with other known forms; at present, therefore, I can only allude to *T. radiata* as a British species with much hesitation, as being represented possibly by examples which are too defective to be submitted to the artist as illustrations of Miss Benett's species. It may also be remarked that an entire absence of ornamentation upon the anteal portion of the shell, such as appears to be indicated by the specimen of Miss Benett, by that of Munier-Chalmas, and also by the defective figure of De Loriol, represents a feature to which we discover nothing analogous throughout the entire series of the genus Trigonia. In the absence of any more satisfactory example it is intended to figure an imperfect specimen in the Oxford University Museum, obtained in the Portland Limestone of Shotover Hill, and kindly brought under my notice by Professor Phillips.

TRIGONIA TRIPARTITA, *Forbes*. Pl. XII, fig. 7.

TRIGONIA TRIPARTITA,	*Forbes*.	Quart. Jour. Geol. Soc., 1851, vol. vii, p. 111, pl v, fig. 11.
—	—	*Morris*.	Catal., 1854, p. 229.
—	—	*Lycett*.	Pal. Soc. Suppl. Gr. Ool. Monogr., 1863, p. 51, p. xl, fig. 4.

Shell ovately trigonal, short, rather depressed; umbones elevated, obtuse, and scarcely recurved; anterior and lower borders curved elliptically; hinge-border short, sloping obliquely. Area of moderate breadth, flattened, distinctly bipartite, the inner or superior half being the more depressed; it is traversed transversely by delicate lines of growth, and near to the umbones by a few small costellæ; the bounding carinæ are distinct, but have little prominence, and are only imperfectly tuberculated; the posteal border of the area forms only a slight angle with the superior border of the escutcheon, which is small and depressed. The costated portion forms a considerable angle with the plane of the area; the costæ form two distinct series; the more numerous or anteal series has the rows very small, plain, and closely arranged; their direction is uniform, passing from the anteal border obliquely downwards to the middle of the valve, where they are cut off by a much larger, less numerous, posteal series of costæ, seven or eight in number, which are slightly tuberculated, and pass downwards from the carina almost perpendicularly, forming right angles with the anteal series. The comparative size and direction of the two species of costæ will serve to distinguish this little shell from *T. v.-costata*, from *T. undulata*, and from *T. detrita*; others of the *Undulatæ* are more remotely allied. The anteal ribbing in Fig. 7 is more minute than in the specimen figured by Professor E. Forbes, and this is the sole difference observable.

Stratigraphical position and Localities. Our specimen was obtained by Mr. Walton, of Bath, in the Cornbrash of Chippenham; the type specimen was found by the late Professor Edmund Forbes in a stratum of soft crumbling yellowish limestone and shale beneath the Oxford Clay at Loch Staffin, Isle of Skye, associated with Testacea, which are for the most part estuarine or fluviatile forms; the geological position is probably not very different from that of the Chippenham Cornbrash, but modified by peculiar local conditions.[1]

TRIGONIA DETRITA, *Terq.* and *Jour.* Pl. X, figs. 3, 3 a, 4.

<blockquote>TRIGOIA DEETTRITA, *Terquem* et *Jourdy.* Monog. de l' Étage Bathonien dans le Département de la Moselle; Mém. Soc. Géol. de France, deux. sér., tom. neuv., 1869, pl. xii, figs. 1 and 2.</blockquote>

Shell ovately trigonal, moderately convex; umbones placed within the anterior third of the valves, moderately elevated, only slightly recurved; anterior side very short, its border curved elliptically with the lower border; superior border straight, sloping obliquely downwards and forming an obtuse angle with the posteal border of the area. Escutcheon lengthened, concave, its superior border somewhat raised. Area of moderate breadth, flattened, having a well-defined mesial furrow; it is transversely rugosely plicated and bounded by two small elevated, tuberculated, or rather closely arranged, plicated carinæ, which are formed by enlarged continuations of the transverse rugæ upon the area. The other portion of the shell has numerous, very closely arranged, plain, rounded costæ; the few first-formed are regular and almost horizontal, succeeded by rows more oblique or directed obliquely downwards from the anterior border; posteally they form a sudden flexure somewhat downwards and then turn upwards perpendicularly to the carina; at first the posteal angle is but small, but it regularly increases with the succeeding costæ, so that the few last formed have their posteal portions somewhat disunited; they become slightly nodulous and, forming each a right angle, pass per-

---

[1] The association of the generic forms of Testacea in the Hebridean deposit described in the Memoir quoted and depicted upon the plate which accompanies it, is curious and instructive, especially when compared with the habitats of two of the living Australian *Trigoniæ. T. tripartita* was found in the Loch Staffin shale with a Perna and an Ostrea, and with ten other species, belonging to the fluviatile genera, Cyrena, Potamomya, Unio, Neritina, and Hydrobia. That this association was not accidental may be inferred from the following analogous facts. The recent *Trigonia Lamarckii* occurs in Sydney Harbour, partially buried in black mud, and also in the Paramatta River, within reach of the tide; in like manner the larger *T. pectinata* is found in Launceston River, Tasmania, in similar black mud, exposed to the alternating influences of fresh and of tidal brackish water. Any section of these Australian deposits may, therefore, be expected to expose the *Trigoniæ* associated, as at Loch Staffin, with the fluviatile or estuarine testacea with which they lived and were buried.

pendicularly upwards; the last-formed two or three anteal costæ become slightly waved or irregular, and their direction accords nearly with the lines of growth, which become conspicuous near the lower border; they also form, near the middle of the valve, an horizontal sulcation or arrest of growth.

Localities. The materials upon which the foregoing description is founded are somewhat scanty, consisting of three British and as many foreign examples. The larger of the specimens figured, with the valves united, is in a very perfect condition, and was obtained by the Rev. P. B. Brodie, in Forest Marble, near to the Tetbury Road Station of the Great Western Railway, North Wilts; the small example is from the Cornbrash of Hilperton, which has also produced other specimens. There are several in the British Museum, of medium size from the Great Oolite of Ranville, Normandy. This appears to have affinity much larger and imperfect fossil figured by Goldfuss ('Petrefac.,' pl. 136, fig. 5 c) from the Lower Oolite of Pegnitz, forming one of a group which he attributed to *T. literata*. Its nearest ally probably is *T. undulata*, which is obtained in a similar geological position; compared with the latter species it has greater convexity, more especially at the umbonal portion of the valves, which has also the costæ much more closely arranged, more horizontal, more numerous, and more rounded; the slight posteal flexure downwards and the considerable angles which they form with the short, nodulous, perpendicular series is also distinctive. In *T. undulata* the angle or curvature is situated nearly at the middle of the valve.

Messrs. Terquem and Jourdy have figured a large example of *T. detrita* in their 'Monograph on the Great Oolite of the Department of the Moselle.'

TRIGONIA CLYTIA, *D'Orb.* Pl. XI, figs. 4, 5; Pl. XVII, fig. 7.

TRIGONIA CLYTIA, *D'Orbigny.* Prodrome de Paléont., 1850, vol. i, p. 309.
— — *Lycett.* Gr. Ool. Suppl. Monog., Pal. Soc., 1863, p. 48, pl. xl, fig. 5.
— — *Rigaux* et *Sauvage.* Descr. de quelq. esp. nouv. de l'Étage Bathonien du Bas-Boulon., 1868, p. 19.

Shell small, sub-trigonal, moderately convex; umbones elevated, sub-mesial, pointed, and recurved; anterior side produced and curved elliptically with the lower border; superior border short, somewhat convex, passing abruptly downwards to the pointed posteal extremity. Escutcheon small and depressed. Area narrow; the plane of its surface forms nearly a right angle with the other portion of the shell; it is traversed transversely by a few large depressed costellæ, which are waved as they pass the mesial furrow; there are also three slightly nodulous carinæ, of which the inner and median carinæ are very small, the marginal carina is much larger. The other portion of the shell has the costæ plain, numerous, small, and very closely arranged, convex upon their lower and concave upon their upper sides; their direction has a slight curvature obliquely downwards towards the carina; the first-formed four costæ are simply horizontal, the succeeding costæ are bent suddenly upwards at their posteal extremities, forming a series of right angles, one of which proceeds from every second costa. The entire number of costæ in adult shells is nineteen or twenty in a specimen eight lines in height. In aged specimens the few last-formed costæ are small, or indistinct, broken, or undulating, and their posteal extremities become somewhat nodulous when the whole of the ornamentation becomes nearly effaced.

The height and lateral diameter are nearly equal; the diameter through both the valves is one fifth less. The size varies from three to ten lines across the valves.

Stratigraphical positions and Localities. The Great Oolite of Box, near Bath, has produced numerous examples of varying dimensions and stages of growth. Mr. Cunnington has also obtained it in the Cornbrash of Westbrook and Trowbridge. Specimens are in the British Museum; in the Museum of Practical Geology, Jermyn Street; the Woodwardian Museum, Cambridge; in the Cabinet of Mr. Walton, of Bath; Mr. Cunnington, of Devizes; and in my own collection. France Luc, Calvados, in the Great Oolite.

Trigonia undulata, *Fromherz.*   Pl. XVI, figs. 9, 10, 11; Pl. XVII, figs. 5, 6.

  Trigonia undulata, *Fromherz.*   Agassiz, Trigonies, 1840, p. 34, tab. x, figs. 14, 16, exclude tab. vi, fig. 1.
  — arata, *Lycett.*   Suppl. Monogr. Gr. Ool. Moll., Pal. Soc., 1863, p. 52, tab. xl, fig. 2 (variety).

Shell sub-ovate or ovately trigonal, moderately convex; umbones little produced, obtuse, nearly erect, antero-mesial; anterior border moderately produced, curved elliptically with the lower border; hinge-border lengthened, nearly straight, sloping obliquely, forming a considerable angle with the posteal extremity of the area. Area flattened but slightly convex, the plane of its surface differing only slightly from that of the other portion of the valve; its breadth is equal to one third of the costated surface; it is traversed by large transverse plications which become prominent costella upon the upper third of its length; the marginal carina has little prominence; the median furrow is well defined, but there is no median carina; the posteal extremities of the costellæ form a tuberculated inner carina. The escutcheon is narrow and much excavated, its superior border is raised; one of our specimens has several oblique varices across its surface, but this appears to be an abnormal feature. The rows of costæ are numerous, small, and not quite regular in their arrangement, the first-formed six rows are plain, slightly curved and acute; the succeeding rows descend obliquely from the anterior side, and about the middle of the valve form a sudden curvature or an imperfect angle; their posteal portions curve upwards towards the marginal carina, which they meet at a considerable angle. The rows are nearly equal in size throughout their length, but become somewhat attenuated towards their two extremities. Some specimens have the costæ almost entirely plain, or only sparingly and slightly knotted (Pl. XVI, fig. 11), others have the tubercles small, numerous, and irregular; usually the anteal portions of such have the tubercles more or less lamellated, their posteal portions more frequently form rounded tubercles; in other instances they are only sub-tuberculated or cord-like (Pl. XVI, fig. 9; Pl. XVII, fig. 5); each row extends to the marginal carina, the tubercles of which form the terminations of the rows of costæ, but towards the middle and lower portion of the carina its tubercles become somewhat compressed, and are more numerous than the rows of costæ. An examination of numerous specimens in various conditions of preservation and in different stages of growth prove that the costæ have great variability, both in their figure and in the presence or absence of tubercles. Occasionally a specimen will occur with the costæ plain and oblique, forming a considerable angle with the posteal portion, which, over the lower half of the valve, rise upwards almost at a right angle with their anteal portions, but even such specimens are never altogether destitute of small depressed nodose elevations upon the costæ; others are irregularly tuberculated or nodose over two thirds of the height of the valve, and are more or less angulated; again, other specimens have the few last-formed costæ altogether variable both in their direction and in the size and arrangement of their nodose tubercles which meander across the valve (Pl. XVI, figs. 9, 10), but their posteal portions are invariably nodose and curve suddenly upwards to the carina. Without exception fully developed forms have one or occasionally two supplementary anteal costæ.

The example figured under the name of *T. arata* in the 'Supplementary Monograph of the Great Oolite Mollusca,' above cited, was one of several very indifferently preserved specimens obtained by Mr. Walton in the Forest Marble of Farleigh, near Bath; the specimen is not of fully developed growth, the costæ are plain or have only slight indications of tubercles, and there is no distinct carina; other examples forwarded to me by the same gentleman were of more advanced growth and tuberculated, but so imperfect as to be unfit for purposes of illustration. The portions of their surface preserved agree with the larger and more irregular of our Lincolnshire examples.

Affinities and differences. From *T. v.-costata* it differs in the figures of the costæ, which are not V-like; the area is also much more rugose, and it is without the median carina.

It is more nearly allied to some forms of that very variable species, *T. Moretoni;* unlike the latter, the posteal portions of the costæ do not enlarge, but become somewhat attenuated near to the carina, to which, in the best preserved examples, they are united, each small tubercle upon the carina forming the terminal tubercle of one of the rows of costæ; the median carina in *T. Moretoni* has no counterpart in *T. undulata;* the latter-formed costæ in our species have also much greater irregularity than is seen in *T. Moretoni*. Upon the whole the enlargement of the posteal portions of the costæ in the latter species appears to be the most reliable distinctive feature, as it is always present. The remarkable variability above described will account for our having failed to identify with the second figure of Agassiz the very imperfect examples from the Forest Marble of Farleigh; and it has only been after comparison with examples from several localities that I have become convinced of the necessity of merging *T. arata* in *T. undulata*. By this latter species is intended only the specimen upon table x of the work of Agassiz, excluding table 6, figure 1, which appears to be a different species with a few large, widely-separated varices upon the angle of the valves or marginal carina.

In no British specimen examined, is the marginal carina and its row of tubercles so large, as in the figure of Agassiz, table x, which appears to constitute the extreme limit of its variability in one direction; other examples with the marginal carina nearly plain constituting the opposite limit of variability.

Stratigraphical position and Localities. *T. undulata* has been obtained only in the upper subdivision of the Great Oolite. Mr. Walton procured specimens in the Forest Marble of Farleigh; Mr. Cunnington in the Cornbrash of Hilperton, near Trowbridge. The officers of the National Geological Survey have also obtained it in Northamptonshire and Southern Lincolnshire; our figured examples are from Edenham, near Bourne, Lincolnshire. The specimen figured by Agassiz was from Piedmont.

TRIGONIA SHARPIANA, *Lycett*, sp. nov.   Pl. XV, fig. 11; Pl. XVI, figs. 3, 4, 5, 6.

.... *Morton.* Natural History of Northamptonshire, 1712, tab. vi, fig. 9.

Shell ovately trigonal, convex; umbones elevated, and slightly recurved; anterior side short, its border curved elliptically with the lower border; hinge-border short, nearly horizontal, terminating posteally in the wide, rounded, and produced posteal border of the area. Escutcheon depressed, wide, and short; its upper border is somewhat raised. Area very wide, flattened, but occasionally with some convexity, bipartite, and bounded by two regular, small, delicately, and closely tuberculated carinæ; it is traversed transversely by narrow, sparingly arranged, regular costellæ, which become evanescent posteally in specimens of advanced growth; each costella has for its carinal termination one of the small carinal varices; the area occupies fully one third the surface of the valve. The other portion of the surface has numerous (16—17) rows of minutely tuberculated costæ; the first-formed four or five rows are nearly concentric, narrow, elevated, and only slightly tuberculated; the succeeding rows are closely arranged posteally; they descend from the carina almost perpendicularly to the middle of the valve, where they are suddenly bent horizontally forwards, and form a slight undulation nearly to the anterior border; the last-formed three or four rows descend perpendicularly to the lower border; all the rows have great regularity in their arrangement, and are nearly of equal size throughout their course.

The internal mould is unknown.

Stratigraphical position.—A specimen unusually large but without name as a species was figured by Morton in his 'Natural History' of Northamptonshire. The figure gives the characteristic features with truthfulness and minuteness, and is the only notice of it which I have discovered. Our smaller figures represent the usual size of specimens preserved in the form of external casts upon slabs of the sandy iron oolite of Northamptonshire. These preserve the more minute features of the Trigonia with great delicacy; for examples I am indebted to the kindness of Mr. S. Sharp, of Dallington Hall, Northampton, who has investigated the geology and palæontology of his district with long-continued perseverance and success. His collection has the only few Northamptonshire specimens with the tests preserved which I am acquainted with; my own cabinet has a single much larger example, but of less mature growth, from the shelly bed of the Dogger at Blue Wyke, near Robin Hood Bay, Yorkshire. The coarse ferruginous Oolite of Glaizedale in the same county also very commonly contains external casts of this Trigonia under circumstances of lithological character closely resembling the beds in Northamptonshire, and in like manner associated with *Astarte elegans*, Sow., *A. rhomboidalis*, Phil., and other well-known testacea of the Inferior Oolite. The Glaizedale casts of the Trigonia are much larger than those of Northamptonshire. It, therefore, appears to be a characteristic shell of the lower portion of the Inferior Oolite in the midland and northern counties of England.

Affinities. It is allied to *T. compta*, from which it is distinguished by having the area more expanded, by the absence of the few large isolated varices at the posteal extremities of the costæ, by their greater upward posteal curvature, and by their anteal undulation. From *T. Phillipsi* it is separated by the great breadth of the area with its distantly arranged transverse costellæ, together with the undulation and occasional angularity in the curvature of the rows of costæ.

T. Sharpiana has also considerable affinities with a still smaller species, viz. the *T. pulchella*, of Agassiz (' Trig.,' pl. xxv, t. 2, figs. 1—7), from strata, which he assigns to Upper Lias, at Urweilar and Mühlhausen (Haut-Rhin). Subsequently Quenstedt figured this little species in his 'Handbuche der Petrefacten' (tab. 43, fig. 14), and assigned it to the lowest zone of the Inferior Oolite. Compared with the British species all these forms are more oblong or sub-quadrate; their umbones are therefore more anterior, and are scarcely raised above the superior border. In *T. Sharpiana* the umbones are much raised, and have so much prominence that the general figure is ovately trigonal; the number of tuberculated costæ are variable, and differ much in their mesial angularity or undulation, but they are always more numerous, and never assume the broken and irregular aspect exhibited in all the specimens figured by Agassiz. The figures of Quenstedt have the ornamentation approximating more nearly to our species, but the general figure coincides with the shell of Agassiz, and cannot be identified with *T. Sharpiana.* Various specimens from Normandy in the British Museum from the Inferior Oolite at St. Vigor and Montiers have an absolute specific identity with the British examples, and serve materially to establish their distinctness from *T. pulchella.*

TRIGONIA COSTATULA, *Lyc.* Pl. XV, figs. 8, 9, 10; Pl. XII, fig. 6, 6*a*.

TRIGONIA COSTATULA, *Lycett.* Ann. and Mag. Nat. Hist., 1850, p. 421, tab. xi, fig. 5.
— EXIGUA, *Lycett.* Ann. and Mag. Nat. Hist., 1850, tab. xi, fig. 3 (young example dwarfed).
— COSTATULA and T. EXIGUA, *Morris.* Catal., 1854, p. 228.

Shell convex, ovately trigonal, or sub-quadrate; umbones elevated, obtuse, sub-mesial, scarcely recurved; anterior side produced, its border, together with the lower border, curved elliptically. Escutcheon narrow, short, and depressed; only slightly more lengthened than the posterior border of the area, with which it forms an obtuse angle. Area wide, flattened, divided by an oblique mesial furrow, and bounded by two inconspicuous, knotted, small carinæ; the knots upon the inner carina assume the form of small varices, which are occasionally somewhat extended upon the escutcheon; the intercarinal space is occupied by small, irregular, transverse plications, which are sometimes only faintly traced. The other portion of the shell has a numerous series (21—22) of smooth, narrow, horizontal, or somewhat concentrically curved costæ, which near to the lower border become less elevated, and have also less regularity; a large specimen of the left valve has the last-formed two or three costæ more or less broken posteally; other specimens have their last-formed four or five costæ attenuated posteally, and bent slightly upwards with some irregularity to the marginal carina.

Trigonia exigua was founded upon very perfect examples of the young shell of the present species, obtained with numerous other dwarfed and immature testacea in the shelly freestone of Leckhampton Hill, near Cheltenham. In this young condition the first-formed costæ, to the number of thirteen or fourteen, are united almost uninterruptedly to the knotted elevations which constitute the marginal carina; they then pass across the area in the form of smaller costellæ, each of which terminates at the slight nodosities which form the inner carina; occasionally an intercalated costella is formed upon the area, which also possesses only slight indications of a median furrow; the uniformity, close arrangement, elevation, and acute edges of these first-formed costæ are remarkable, and differ greatly from the condition of specimens of more-developed growth, obtained in a bed of hard cream-coloured limestone (coralline mud), in which specimens have been cleared with great difficulty with the help of cutting instruments. Due allowance being made for these adverse circumstances, it will readily be understood that the specific identity of the two forms, apparently dissimilar, was not at first discovered, and that they were believed to represent different species; the cream-coloured rock, which also abounds with the genus *Nerinæa* in the vicinity of Stroud and Nailsworth, is little used for economical purposes, and the Trigonia therefore has very rarely been obtained.

Trigonia costatula appears to occupy in its sectional characters a position between the *costatæ* and the *undulatæ*, or to connect these groups, but as the area and escutcheon agree with those parts of the *undulatæ*, and are altogether distinct from the *costatæ*, I have preferred to place it with the former section, notwithstanding the presence of a series of plain horizontal costæ upon the other portion of the shell.

Specimens of this rare species are in the Museum of the Royal School of Mines; in the Woodwardian Museum, Cambridge; also in the collections of Dr. Wright, Cheltenham; Mr. Witchell, Stroud; Rev. P. B. Brodie, Rowington, near Warwick; and in my own cabinet; all from the middle portion of the Inferior Oolite in the vicinity of Cheltenham and of Stroud.

TRIGONIA JOASSI, *Lycett*, sp. nov.   Pl. XX, figs. 2, 3, 4.

Shell ovately oblong, convex; umbones placed within the anterior third of the valves, moderately elevated; anterior side short, curved elliptically with the lower border; posterior side much produced and somewhat depressed, pointed at the extremity of the marginal carina. Area of considerable breadth, flattened, rugosely plicated transversely. Escutcheon lengthened, narrow and flattened; marginal carina slightly elevated, transversely knotted. The other portion of the surface has the rows of costæ very numerous, each row having a double undulation resembling in figure the radii upon the Ammonites of the group of *Falciferi*, but with less regularity. The nodes in the rows are usually small, with much inequality in size and variability in figure, but for the most part they are either rounded or ovate; anteally they become much attenuated or cord-like; they are curved upwards obliquely from the anterior border, and at the end of about two fifths of their course form a sudden flexure directed obliquely downwards and become more distinctly nodulous. The nodes are irregular in the rows, and about the middle of the valve some of the anteal rows terminate; the remainder of the rows form another undulation, curving upwards to the marginal carina at a considerable angle, becoming more ridge-like and attenuated at their extremities. The nodes upon the middle portion of the valve are commonly confused, unequal in size, and irregular in figure; but this does not appear to be an invariable feature, as a specimen in the collection of Mr. Grant, of Lossimouth, has the rows of nodes regularly falciform and nearly equal. This specimen, although imperfect, shows that the rows near to the umbones become simply transverse and lose the falciform aspect. The materials upon which the foregoing description is founded are all more or less imperfect, but in the aggregate they exemplify nearly the whole of the more important features. The specimen with the ornamentation preserved over the greater part of the costated surface is in two portions, exhibiting only a moiety of the marginal carina, and is destitute of the area. The internal mould gives the general outline and proportions; the large striated hinge teeth, the muscular scars, and pallial lines; fortunately also, owing to the thinness of the test, the exterior ornamentation is obscurely visible, and even the posteal portion of the area has delicate, faintly-marked, transverse striations. The mould and also the specimen belonging to Mr. Grant have the few last-formed posteal portions of the costæ almost effaced, a condition which appears to be a concomitant of the last stage of growth. The specimen figured with the surface preserved retains its costæ at that portion of the valve.

The third specimen, which exhibits the greater portion of the area, is slightly defective near to the pallial border, and also at the apex; it is a gutta-percha pressing taken from a well-preserved external cast, kindly forwarded by the Rev. Mr. Joass; the compact siliceous rock has retained the impression of the surface of the shell with minuteness and delicacy. In the aggregate the examples figured upon our Plate appear sufficiently to elucidate the species.

Affinities. No one of the British *Undulata* presents any near approximation to *T. Joassi*; there is, however, one foreign *Trigonia* from the Oxford Clay of Gundershofen, which might possibly be mistaken for it, the *Lyrodon litteratum* of Goldfuss ('Petref.,' vol. ii, p. 200, tab. 136, fig. 5 *b*); the largest figure is the nearest ally; excluding other testacea upon the same plate, also named *litteratum*, which belong to two other species; the broken specimen pertains to the Inferior Oolite, the other to the Neocomian formation. Limiting the *litteratum* of Goldfuss in this manner, it will be found to have the area larger than in our species, which is also without the row of distinct rounded tubercles upon the marginal carina; upon the other portion of the valve the species of Gundershofen possesses in its ornamentation a certain amount of resemblance to *T. Joassi*, differing from the latter in the posteal portions of the costæ, which are larger and for the most part almost perpendicular and disunited from the other portions; the anteal costæ are also larger and less numerous. For examples of this fine species of the *Undulatæ* I am indebted to the kindness of Mr. J. W. Judd, who obtained them at Brora, in the far north of Scotland, during long and persevering researches in the Jurassic rocks of that little-known region. The imperfect specimen in two portions, but having the surface preserved, is from plaster casts of originals in the British Museum, obtained by Mr. Charles Peach in the Brora region. The species is dedicated to the Rev. J. M. Joass, of Golspie, at the suggestion of Mr. Judd, as a fitting recognition of his own obligations to that gentleman, for untiring efforts in his assistance, during the survey of a region heretofore but little explored by geologists.

Stratigraphical position and Localities. The Brora specimens are in a pale whitish argillaceous grit, which Mr. Judd believes to be upon the horizon of the Lower Calcareous Grit of Yorkshire; they are associated with *Pecten vimineus*, Sow.

§ IV. GLABRÆ.

TRIGONIA GIBBOSA, *Sow.* Plate XVIII, figs. 1, 2, 2 *a*, 3, 4, 5, 6; Plate XIX, figs. 1, 1 *a*, 1 *b*, 2.

TRIGONIA GIBBOSA, *Sowerby.* Min. Conch., tabs. 235, 236, p. 61, vol. iii.
— — *Benett.* Catal. Org. Rem. County of Wilts, 1831.
— — *Deshayes.* Coq. Carac., 1831, pl. xi, fig. 8, p. 37, 1831.
— — *De la Beche.* Manual of Geology, 1833.
— — *Fitton.* Géol. Trans., 2 sér., 4, p. 356, 1835.
— — *Pusch.* Potens Paléont., 1837, p. 60.
— — *Fitton.* Bull. Soc. Géol. de France, 1839, 1 sér., tom. x, p. 445.
— — *Agassiz.* Trigoniés, 1840, p. 10.
— — *Bronn.* Index Paléont., 1848, p. 686.
— — *D'Orbigny.* Prodrome de Paléont., 1850, vol. ii, No. 42, p. 60.
— — *Buvigier.* Statist. Minér. et Géol. Meuse, pp. 370, 401, 1852.
— — *Morris.* Catalogue, 1854, p. 228.
— — *Pictet.* Traité de Paléont., t. iii, p. 539, 1855.
— — *Hébert.* Terr. Jurass de la Bassin de Paris, p. 73, 1857.
— — *Oppel.* Juraformation, p. 722, No. 144, 1856.
— — *Contejean.* Étage Kimmeridgien de Mont Belliard, pp. 60 et 217, 1859.
— — *Coquand.* Synopsis des Foss. de la Charent., 1860, p. 36.
— — *Rigeaux.* Notice Stat. sur le Bas Boulonnais, 1865, p. 26.
— — *Pellat.* Bull. Soc. Géol. de Fr., 2 sér., tom. xxiii, pp. 208, 209, 1866.
— — *Hébert.* Note sur le terrain Jurassique du Boulonnais, Bull. Soc. Géol. Fr., 1866, 2 sér., tom. xxiii, p. 21.
— — *De Loriol* and *Pellat.* Monog. Paléont. et Géol. de l'étage Portlandien des env. de Boulogne, 1866, pl. iii, figs. 1, 2, 3.
— — *Wright.* Correlation of Jurassic Rocks, Proceedings of Cotteswold Nat. Club, 1869, p. 88.

Shell somewhat inflated, sub-ovate, or ovately oblong; umbones large, obtuse, elevated, antero-mesial and erect; anterior and inferior borders elliptically curved, hinge-border concave, its posteal extremity curved gently with the posteal border of the area.

The area is narrow, slightly convex, having a mesial oblique furrow; there are no distinct bounding carinæ, but near to the umbo the area forms a distinct angle with the more depressed ante-carinal space. The escutcheon is of moderate breadth, smooth and depressed. The ante-carinal space is much depressed near to the umbo; downwards it becomes more flattened and widens regularly towards the lower border, where its breadth exceeds that of the area. The entire valve in the adult state is divided into four or more zones by large, deeply-indented, transverse sulcations which are always conspicuous; they curve upwards at their extremities in accordance with the lines of growth. The costated portion occupies more than half the valve; the costæ in their prominence, number, and general aspects possess so much variability that, without the possession of numerous connecting specimens, other species may possibly be united with it; whenever the costæ are distinct upon the first or umbonal zone they are moderately numerous, plain, and oblique; upon the next and the succeeding zones they have greater curvature; anteally the extremities of the costæ in each zone curve upwards, external to the extremities of the costæ in the preceding zones, so that anteally the costæ appear to be unsymmetrical; upon the last zone the costæ become smaller and less distinct, or are confused irregularly with the lines of growth. *T. gibbosa* may be arranged under three varieties as follows:

Var. a.—Figure, Unio like or produced posteally; longitudinal sulcations large and deep, irregular and unequal near the pallial border; ante-carinal space narrow and not well defined, excepting near the apex; surface generally destitute of ornamentation, occasionally with some indications of costæ. Plate XVIII, figs. 5, 6.

Var. b.—Costæ prominent and numerous, covering the greater portion of the valve; narrow, ridge-like, small, and plain anteally, forming large, oblong, or subovate nodes posteally; ante-carinal space much larger and more defined than in *Var. a;* the area also more distinctly marked, sometimes with slightly knotted elevations at the positions of the marginal and inner carinæ. Longitudinal sulcations distinct, but smaller than in *Var. a*. Plate XVIII, figs. 1, 2, 2 *a*, 4; Plate XIX, fig. 2.

Var. c.—Ante-carinal space very large and depressed; sulcations only slightly defined; rows of costæ very numerous and irregular, with small, crowded, but prominent nodes, producing a roughened surface; area narrow, strongly defined, transversely coarsely plicated, its bounding carinæ knotted anteally, plicated posteally. Plate XXI, fig. 1.

The figures in Sowerby's 'Mineral Conchology' represent the varieties *a* and *b*.

T. gibbosa in the work of De Loriol and Pellat is exemplified only by the variety *a*.

The *T. gibbosa* of Seebach ('Der Hannoversche Jura,' tab. 2, fig. 6 *a*, 6), from the Pterocera Beds of Tongesberge (a lower zone of the Jura formation), is evidently a distinct species characterised by a more trigonal form; by the absence of the large longitudinal sulcations and of the wide ante-carinal space, the costæ are also different.

Stratigraphical postion and Localities. In Britain *T. gibbosa* is the most well-known Trigonia of the Portland Oolite, and is limited to that rock and the subjacent sands; at Chilmark and at Tisbury all its varieties are exemplified throughout beds, the entire thickness of which is not less than sixty feet. Other reputed localities are the Isle of Portland, Devizes, Brill, Hartwell, and Swindon, but the examination of a multitude of examples of the *Glabræ* from those places demonstrate that *T. Dasmoniana* is their prevailing Trigonia.

Chicksgrove Quarry, Tisbury, has disunited valves of *T. gibbosa* in great profusion, sometimes covering large slabs of the bed called *Troughstone*, to the exclusion of all other testacea. The Portland formation in Britain is well characterised by the presence of its *Trigoniæ*, all of which appear to be special to it; those of the oolites and sands are *T. Damoniana*, *T. Manseli*, *T. Micheloti*, *T. tenuitexta*, *T. muricata*, *T. incurva*, and *T. Carrei*; those of the lower or Kimmeridge Clay are *T. Voltzii*, *T. Juddiana*, *T. Pellati*, *T. Woodwardi*, also another lengthened form, somewhat doubtful, which I have provisionally placed as a variety of *T. incurva* (Pl. IX, fig. 2).

In the vicinity of Boulogne *T. gibbosa* is also an abundant species at numerous localities, and in various beds of the upper portion of the Portland Formation. For ample details the reader is referred to the important memoirs by M. Hébert and by De Loriol and Pellat above cited.

TRIGONIA MANSELI, *Lycett*, sp. nov.   Pl. XIX, figs. 3, 4, 4 *a*, *b*.

Shell subovate or ovately oblong, inflated mesially, compressed near to the pallial border; umbones antero-mesial, prominent, large, and obtuse, much incurved and nearly erect; anterior and lower borders curved elliptically; hinge-border rather convex, curved gently with the posteal extremity of the area and terminating in an extremity which is somewhat produced and pointed. Escutcheon smooth and concave, but having its upper border somewhat raised. Area narrow, convex, and raised, divided conspicuously by a deep mesial furrow, which has bordering upon it, immediately upon either side, a slightly defined row of small tubercles, or in other specimens they are evanescent; there is also a well-defined line of small tubercles or varices which forms an inner carina; these varices are extended somewhat upon the escutcheon. The position of the marginal carina is indicated near to the umbones by a well-marked obtuse ridge, but this soon disappears with advance of growth, and the carina is then represented only by the rounded elevation which the border of the area forms, adjoining to the wide and very depressed smooth or ante-carinal space; the plications of growth are well defined over the whole of the area. The other portion of the surface has a very numerous and well-marked series of obliquely directed tuberculated costæ; upon the umbones the costæ are different; they there form a densely-arranged minute or linear series which pass horizontally across the whole of the valve uninterruptedly; they are plain and the breadth of the series does not exceed three lines (this feature, unfortunately, is not depicted upon our figures 4 *a* and 4 *b*); the costæ then change abruptly to oblique tuberculated rows which continue with only slight irregularity even to the lower border; the costæ (about twenty-four in number) are narrow, closely arranged, curved, and somewhat attenuated near to the pallial border; they pass upwards in a manner sometimes somewhat waved and meet the depressed ante-carinal space at a considerable angle, the tubercles in the rows (about twenty) are rounded or ovate and closely arranged, but upon the anteal attenuated portions of the costæ they are indistinct or cord-like; the costal terminations, posteally, are abrupt but do not form a regular line, so that the anteal boundary of the ante-carinal space is irregular. The arrangement of the rows is so close that it is sometimes difficult to discover the real direction of the lines of tubercles; in such instances the attenuated pallial extremities of the rows of costæ afford the real guide. The ante-carinal space is less wide than in *T. Micheloti* and *T. gibbosa*, but it is always conspicuously depressed longitudinally, which imparts an additional apparent convexity to the area. The valve has three transverse sulcations or arrests of growth; these are not very conspicuous and do not appear materially to have interfered with the direction of the rows of costæ which pass across them.

This pretty species constitutes one of the most clearly defined examples of the *Glabræ*, the direction and arrangements of the rows of tubercles immediately suffice to determine the species; perhaps it may be the shell of which a fragment of the anteal side was figured by Agassiz, 'Trigonies,' tab. 6, fig. 11, under the name of *T. picta*, from the White Coral Crag of Hoggerwald (Canton of Soleure), the general direction of the lines of tubercles and their attenuation towards the border agree with *T. Manseli*, but in the absence of the greater and more important portion of the valve I prefer only to allude to their possible specific identity.

Upon the whole there is much variability in the characters of the tuberculated costæ; occasionally the anteal portions of the umbonal rows are somewhat angulated or are curved concentrically, or the anteal portions of the latter-formed series have an occasional intercalated costa. The varying size of the tubercles in different specimens, and their more close or distant arrangement in the rows, impart much variability to the aspect of the species, but in every instance the tubercles are rounded and have much prominence.

The size is usually smaller than *T. Damoniana*.

Dimensions of a specimen rather smaller than usual:

Length 22 lines, height 18 lines, diameter through the united valves 14½ lines.

The name is intended as a slight recognition of most welcome assistance rendered to the author by the loan of some interesting *Trigoniæ* from the Kimmeridge Clay of the Cliffs of Dorsetshire, one of the results of extensive explorations made by J. C. Mansel-Pleydell, Esq., of Longthorns, near Blandford, Dorset, upon his property in the Kimmeridge Clay, a formation which has been but little exposed in England.

Stratigraphical position and Localities.—*T. Manseli* has occurred somewhat rarely in the Limestone of the Isle of Portland and Tisbury. Specimens are in the Museum of Practical Geology, Jermyn Street; in the collection of Dr. Wright, of Cheltenham; in the collection of Mr. Cunnington, of Devizes; also in my own cabinet.

TRIGONIA DAMONIANA, *de Lor.* Pl. XVIII, fig. 3; Pl. XX, figs. 1, 2, 2 *a*, 2 *b*; Pl. XIX, figs. 1, 1 *a*, 1 *b*.

TRIGONIA GIBBOSA, a new variety, *Etherelda Benett.* Catal. Org. Rem. County of Wilts, 1830, pl. xviii, fig. 1.

— — *Damon.* Geol. Weymouth, Suppl., 1860, pl. vii, fig. 2.

— DAMONIANA, *de Loriol* and *Pellat.* Monog. Paléont. et Géol. de l'étage Portlandien des envir. de Boulogne, 1866, pl. xvii, figs. 4, 5.

Shell sub-ovate, lengthened obliquely, convex, umbones large, erect, very prominent and somewhat pointed, much incurved and rendered bipartite by the narrow deep sulcation produced by the apical termination of the ante-carinal space; borders of the valves elliptically rounded, excepting the hinge-border, which is straight and lengthened, sloping obliquely; the anterior face of the valves has also a large, rounded, depressed space or lunule, which gives a slightly truncated aspect to that portion when viewed laterally. The escutcheon is depressed, cordiform, and strongly marked by the lines of growth; the area is narrow, slightly elevated or convex, traversed transversely by irregular folds of growth; it has a well-marked mesial furrow, and is bounded at its upper or umbonal portion by two rows of minute, sparsely-arranged tubercles; some specimens have also a median line of minute tubercles bordering upon the groove; more frequently these three lines of carinal tubercles cannot be traced or only partially so even upon well-preserved specimens. The anti-carinal space or sulcation is narrow, smooth, and only slightly depressed, excepting near to the umbones, where it forms a deep sulcation; more frequently in adult forms its slight depression is the only feature which separates it from the area. The costated portion is divided into three or four zones by as many elliptical horizontal furrows; these are much less conspicuous than in *T. gibbosa*; like to that species the direction of the rows of costa are not conformable with the sulcations; they are more horizontal, thus rendering their aspect rather excentric; upon the anterior face of the shell they are entire, much attenuated, and form a slight angle or undulation; the costæ upon the umbo form a very numerous, plain, minute, closely-arranged, horizontal series, which pass also across the ante-carinal sulcation and area with slightly diminished prominence; subsequently the rows have their middle and posteal portions closely arranged and very unequal in size; they are usually slightly knotted or nodose, becoming larger posteally, and terminate abruptly at the smooth ante-carinal space, but occasionally over the lowest zone they are confusedly crowded, and minute or even continued across the ante-carinal space. Our example, Pl. XVIII, fig. 3, represents a well-marked variety, with few prominent costæ, each of which has about seven, large, widely separated tubercles; a specimen of more advanced growth has the last zone crowded with minute tubercles, which afford a remarkable contrast to the other rows of costæ, but the umbones have their costæ plain, minute, and dense, as in the typical form.

Compared with *T. gibbosa* the general figure differs considerably. It is shorter transversely, or more obliquely lengthened or ovate; the concentric sulcations are smaller, the umbones are more elevated, the costæ are smaller, more numerous, and more minutely nodose; the very numerous linear series which occupies the first zone of the shell and passes across the whole of the valves transversely is also another remarkable and distinctive feature. As there is much variability in the obliquity of the valves, measurements of proportions would have but little utility.

The first notice of *T. Damoniana* as distinguished from *T. gibbosa* occurs in a thin quarto volume published by Miss Etherelda Benett, of Warminster, in the year 1831; intended as an illustrated catalogue of organic remains in the County of Wilts, it is therein mentioned as a distinct variety of *T. gibbosa;* the drawing is characteristic and satisfactory, but the extreme scarceness of the work in question, together perhaps with some supineness or absence of sufficient investigation by British palæontologists, rendered it altogether neglected as a species.

Stratigraphical position and Localities. *T. Damoniana* is an abundant fossil at several localities, more especially in the white limestone of the Isle of Portland; at Brill, Bucks, and at Swindon, at the latter place the reddish ferruginous sandy beds at Dayhouse Farm have produced very numerous separate valves of *T. Damoniana*, having the ornamentation for the most part very well preserved; the specimens are of every stage of growth, but are frequently distorted by vertical pressure; they are spread out laterally and slightly flattened; specimens are in the Museum of Practical Geology and in the collection of Mr. Cunnington, of Devizes. Internal moulds are abundant at Swindon. They are shorter and more oblique than those of *T. gibbosa.*

A very large majority of examples of *T. Damoniana* obtained in the Isle of Portland coincide with the foregoing description; rarely, however, certain features become prominent, indicative of an additional variety, characterised by the great breadth of the smooth ante-carinal space. Apparently this is the form figured by De Loriol and Pellat for the *T. variegata* of Credner (' Monog. de la Portland. de Boulogne,' 1865, pl. 7, figs. 6, 7); the latter species, which is from the Kimmeridge strata of Fritzow, has large oblong nodes upon the posteal portions of the costæ; these are much fewer, and the ante-carinal space is much smaller than in our variety. In the expectation that examples of *T. variegata* might occur in the Portland formation of England I have attentively examined a multitude of examples of the *Glabræ* from its beds, but have failed to ascertain its presence; perhaps, considering the general very limited stratigraphical range of the *Trigoniæ*, we should scarcely expect to discover one of its Kimmeridge Clay species in the Portland Oolite; our British examples from the upper and lower stages of the formation indicate this distinctness very conclusively. The figures in the little work of Credner (' Ueber die Gliederung der Obern Juraformation und der Wealden Bildung in Nordwestlichen Deutschland, &c.,' Prag, 1863) are, upon the whole, coarsely engraved, and the drawings of the *Trigoniæ* are apparently not very reliable for correctness; in them we observe only a remote resemblance to the two figures of *T. variegata* given by De Loriol and Pellat from Boulogne.

For the most part the size and general aspect of the ante-carinal space affords a good distinctive feature for the *Glabræ*, and aids materially in the separation of its species, but in the present, as in some other instances, it assumes an amount of variability indicative of its subordination to some other specific characters. Other variable examples of this feature are seen in the smooth or typical form of *T. gibbosa*, in which the space is smaller and less distinct than in the other varieties; the same remark will also apply to certain examples of *T. Juddiana* and of *T. irregularis* in the *Clavellatæ* and to *T. angulata* in the *Undulatæ*; but it is only in the *Glabræ* that this feature, from its constancy and prominence, becomes of sectional importance.

TRIGONIA TENUITEXTA, *Lycett*, sp. nov. Plate XX, figs. 1, 1 a.

Shell with the general outline of *T. Damoniana*, but with less convexity; the most striking peculiarity is afforded by the ante-carinal space, which is nearly absent; there is only a narrow slight depression indicating its position; the knotted costæ upon the side of the valve are remarkable for their minuteness, close arrangement, and irregularity or undulations, so that they appear partially confused; they are also continued more or less obscurely even across the ante-carinal space; upon the umbonal portion the minute tubercles with which the costæ are crowded, disappear, and they there form a very numerous, minute, plain, or almost linear series. Upon the specimen figured the escutcheon has a few regular oblique plications; as this feature is one altogether foreign to the *Glabræ*, and occurs only in the *Quadratæ*, the *Scabræ*, and the *Costalæ*, its occurrence in the present instance may be regarded as only an abnormal or individual peculiarity.

Affinities. Compared with the allied form, *T. Damoniana*, the partial disappearance of the ante-carinal space, the smaller, much more numerous and crowded, meandering costæ, serve sufficiently to distinguish it. These characters also separate it from other *Trigoniæ* of the same section.

The magnified figure 1 *a* represents the umbonal costæ constituting a very numerous, minute series, directed horizontally, without interruption, across both the ante-carinal space and the area.

Stratigraphical position. The specimen figured is from the limestone of the Isle of Portland. A series of specimens more or less imperfect, kindly forwarded to me by Mr. Cunnington, proves that the species also occurs in the Portland Oolite at Devizes, Crookwood, and Tisbury.

TRIGONIA BEESLEYANA, *Lycett*, sp. nov. Plate XVII, figs. 2, 3, 4.

Shell ovately oblong, depressed, transverse, thin; umbones antero-mesial, small, depressed, only slightly raised above the superior border; anterior side moderately produced; its border rounded; lower border lengthened and curved elliptically, its posteal extremity pointed; superior border moderately lengthened, nearly horizontal, curved downwards posteally. Each valve is divided into two unequal portions by a plain oblique angle. The escutcheon is of moderate breadth, flattened, but somewhat concave, traversed by very numerous, rounded, regular, depressed, delicate costellæ, which pass across its surface horizontally, and are slightly indented transversely by the lines of growth; a few of the costellæ are bifurcated near to the superior border. There is no marginal carina, but a distinct oblique divisional angle, such as occurs in a portion of the *Scabræ*; to this latter group it is also allied by the entire absence of an area and by the ornamentation of the escutcheon, the costellæ upon which are similar in character or slightly scabrous. The middle portion of the valve comprising nearly the half of the whole surface is plain, and its surface is traversed only by the very delicate lines of growth; it is also without the oblique depression which usually is characteristic of the *Glabræ*. The anteal portion of the valve has numerous delicate, narrow, smooth, small costæ, which are conspicuous at the anteal border; they pass towards the middle of the valve obliquely downwards; they are slightly waved, and become evanescent ere they have traversed little more than one fourth of the length of the valve; towards the umbo they are scarcely perceptible. The lines of growth are numerous, unequal, and delicate.

The internal mould is well preserved, it exhibits very wide-spreading, coarsely striated dental processes, and owing to the general depression of the valves, there is little convexity excepting near to the umbones; the muscular scars have but little prominence, and there are no traces of the external ornamentation. I have also succeeded in exposing the dental characters in a specimen of the right valve; the processes are narrow, widely divergent, and have little prominence, corresponding with the internal mould.

Dimensions.—Height 20 lines, length 27 lines, diameter through the united valves 6 lines.

My attention was first directed to this remarkable form by my friend Mr. J. W. Judd, who during his labours in the geological survey of Oxfordshire saw this Trigonia in the collection of Mr. Beesley, of Banbury, and was much struck with its novel aspect. So singular is the combination of characters which it presents, that the term *paradoxa* might be fitly applied to it were it not already employed in the genus. Its analogues belong altogether to the Cretaceous rocks; the general outline and also the ornamentation of the anterior side much resembles *T. excentrica*, Park. The posteal angle and slope with the characters of that portion of its surface assimilate it to some of the cretaceous *Scabræ*, and more especially to *T. Hondeana*, Coq., *T. tenuisulcata*, Duj., and *T. Archiaciana*, D'Orb., and this remarkable combination of sectional characters, so foreign to the Jurassic species, occurs in a Trigonia from almost the base of the Lower Oolites associated with a numerous series of Conchifera special to that stage.

Stratigraphical position and Locality. The two fine examples herewith figured, several others less well preserved, and a few internal moulds, constitute all the materials known; Mr. Beesley, to whom we owe their discovery, informs me that the locality of the quarry is Tynehill, in the parish of Adderbury, between that place and Great Barford, Oxfordshire. The rock is coarse, brown, shelly Oolite; amongst the Inferior Oolite Testacea found with it are *Cricopora straminea*, Phil., sp., *Serpula socialis*, Goldf., *Natica Leckhamptonensis*, Lyc. *Lima bellula*, Mor. and Lyc., *Nerinæa Jonesi*, Lyc., &c. The specimens figured are from the collection of Mr. Beesley; the University Museum, Oxford, also possesses a specimen.

TRIGONIA MICHELOTTI, *De Lor.* et *Pellat* (*variety*). Plate XX, fig. 7.

    LYRODON EXCENTRICUM, *Goldfuss* and *Munster*. Petrefacta, 1836, vol. ii, page 203, plate 137, fig. 8. (Not *Trigonia excentrica*, Park.)

    TRIGONIA MICHELOTTI, *T. de Loriol* et *E. Pellat*. Monog. Paléont. et Géol. de l'étage Portlandien des env. de Boulogne-sur-Mer, 1866, plate 7, fig. 9.

    — MUNIERI, *Hébert*. Note sur le terr. jurass. der Boulonnois, Bull. de la Soc. Géol. de France, 2 sér., 1866, tom. 23, page 216.

---

[1] I am unable to determine to which of the eminent French geologists above cited priority should be given, as their publications are dated in the same year. Some consideration is perhaps due to the fact that the beautiful Monograph by De Loriol and Pellat gives the only figure of the species which has appeared since the great work of Goldfuss and Munster, in 1836.

Shell subovate, very convex; umbones prominent, obtuse, antero-mesial, much incurved and slighly recurved; anterior side short, curved elliptically with the lower border; hinge-border short, convex, curving downwards posteally with the siphonal border, the length of which nearly equals that of the hinge-border. Area narrow, its surface forming nearly a right angle with the other portion of the valve; it is slightly concave near to the apex, becoming convex posteally; it is divided by a deeply marked mesial furrow, and is traversed transversely in common with the whole shell by delicate lines of growth; it is bounded by two small carinæ, which are conspicuous near to the apex; the marginal carina has a few small distantly arranged tubercles; the inner carina is also slightly knotted. The escutcheon is small, depressed, but becomes somewhat elevated at its upper border. The ante-carinal space is remarkable for its great breadth, which at the pallial border exceeds that of the area, and is equal to one third of the length across the valve, its upper portion forms a considerable concavity. The anteal or costated portion of the shell is comparatively narrow, occupying only half the surface of the valve; the costæ are plain, oblique, and have some irregularity, curving downwards from the anterior border, and terminating abruptly at the smooth and more depressed ante-carinal space; for the most part their posteal extremities become irregularly nodose. An arrest of growth or concentric sulcation occurs beneath the middle of the valve; the costæ subsequently have less obliquity, or are more concentric, curving upwards anteally, and externally to the extremities of the costæ upon the upper half of the valve.

The large proportion which the smooth ante-carinal space bears to the other portions of the surface, together with the few, plain, oblique, and irregular series of short anteal costæ, constitute the most conspicuous distinctive characters.

Our figure is taken from a gutta-percha pressing of an external cast in the collection of Mr. Cunnington, and was obtained by him in the Portland Oolite of the vicinity of Devizes; a second specimen in a condition nearly similar is in the same collection. Specimens of the typical form of *T. Michelotti* from the Kimmeridge strata of Boulogne were figured by Goldfuss under the name of *Lyrodon excentricum*, and by De Loriol under that of *Trigonia Michelotti*; it was also described and stratigraphically determined by Professor Hébert under the name of *T. Munieri*; it has a lower position than that of our Devizes specimens; the Boulogne shell also possesses some conspicuous distinctive features; the figure is more lengthened, the umbones are much less elevated, and less recurved; the short anteal costæ are less prominent, or are somewhat obscure, and are therefore not nodose; the carinæ have less distinctness, and are without tubercles. These differences are of considerable importance, and were they founded upon a sufficient number of specimens, both French and British, there would remain no doubt of the propriety of separating them as species, but with the very limited materials of either form at my disposal or brought under my notice, and knowing the variability exhibited by some of the *Trigoniæ Glabræ* of the Portland formation, more especially by the two more abundant of its forms, *T. gibbosa* and *T. Damoniana*, I prefer to regard (provisionally at least) the two allied forms from Boulogne and Devizes as constituting only well-defined varieties of one species.

Dimensions.—Length 16 lines, height 13 lines, thickness through a single valve 4 lines.

TRIGONIA EXCENTRICA, *Park.* Plate XX, figs. 5, 6; Plate XXI, figs. 6, 7; Plate XXII, figs. 5, 5 a.

TRIGONIA EXCENTRICA, *Parkinson.* Org. Rem., vol. iii, pl. xii, 1811.
— SINUATA, *Ib.* Ibid., fig. 13.
— EXCENTRICA, *Sow.* Min. Conch., vol. iii, p. 11, tab. 208, figs. 1, 2, 1821.
— AFFINIS, *Miller* and *Sow.* Ibid., tab. 253, fig. 3, 1821.
— — *Defrance.* Dict. des Scien. Nat., tab. lv, p. 297, 1828.
— — *Pusch.* Polens, Paléontologie, p. 61, 1837.
— EXCENTRICA, *Ib.* Ibid.
— — *Agassiz.* Trigonies, p. 9, 1840.
— AFFINIS, *Ib.* Ibid., pp. 9 et 52, 1840.
— EXCENTRICA, *D'Orbigny.* Prodrome de Paléont., 1850, vol. ii, p. 160, No. 328.
— SINUATA, *Ib.* Pal. Franç., Terr. Crét., tom. iii, p. 147, pl. 293, 1843.
— — *Ib.* Prodrome de Paléont., vol. ii, p. 161, No. 323, 1850.
— — *Morris.* Catalogue, p. 229, 1854.
— EXCENTRICA. *Ib.* Ibid., p. 228, 1854.

Shell inequilateral, subovate, rather depressed and thin in the very young condition, becoming thick, with a considerable convexity, in an advanced stage of growth; umbones pointed, erect, little produced, situated about two fifths the length of the valve from the anterior border. Anterior side produced, its border curved elliptically with the lower border; hinge-border nearly straight, or in some examples slightly concave, sloping obliquely downwards, lengthened, terminating in a posteal extremity, which is rounded but attenuated. Area narrow, slightly concave near to the umbo, where the valve forms an oblique angle, separating the area from the anteal portion; the angularity soon disappears, the area then acquires some convexity, and has no distinct separation from the other portion of the surface excepting that a space anteal to the area becomes somewhat depressed near to the lower or pallial border. The other portion of the shell is covered by a series of very numerous, inconspicuous, slightly elevated, longitudinal or horizontal costæ, which are indented anteally by oblique intersecting lines of growth; the costæ are regular and distinct, crossing the entire valve near to the umbo, but they soon disappear over the posteal third of the surface, and examples of adult growth have the costæ everywhere evanescent near to the pallial border, where the surface is occupied almost solely by the lines of growth, which are large, irregular, and rather distantly arranged. Several sulcations or arrests of growth are usually visible at irregular intervals. The costæ have their anteal portions horizontal or directed slightly downwards; this excentrical direction has been depicted both by Parkinson and by Sowerby in a manner somewhat exaggerated where the costæ are crossed by oblique lines of growth.

The length compared with the height is as ten to seven, or, as in other examples, as ten to eight.

The hinge-teeth diverge widely, and are larger than is usual in the *Glabræ*; the adductor scars are deeply impressed, more especially the anteal adductor, which forms a deep sinus, passing upwards towards the apex of the valve concealed by the anterior dental process in each valve; the borders of the valves are plain, their inner surfaces exhibit remains of the iridescent nacre in well-preserved specimens.

The figures upon Plate XX illustrate the young condition of the species; fig. 6 closely resembles the *T. sinuata* of Parkinson; fig. 5, a specimen of much more advanced growth, retains the surface ornaments similar to the smaller specimen; Plate XXI, fig. 6, and Plate XXII, figs. 5, 5 *a*, represent common examples of *T. excentrica* in which the umbonal portion of the test does not retain the characters of the surface; the large specimen, Plate XXI, fig. 7, which agrees with the *T. affinis* of the 'Mineral Conchology,' has the valve thickened from advanced growth, and the horizontal costæ are obscure. Much of the variability seen in this species is produced by differences in the general figure which are not dependent upon any one stage of growth; thus, the short and thick example fully developed, depicted upon Plate XXI, which fairly represents the *T. affinis* of the 'Mineral Conchology,' is nearly allied in figure to certain young forms which are only three or four lines in length; these latter also pass gradually into the undoubted young condition of *T. excentrica*, Plate XX, fig. 6, which is more lengthened. The six figures upon our plates, although exhibiting much diversity of aspect, do not sufficiently exemplify the medium-sized and fully developed forms of the more lengthened specimens; through the kindness and discrimination of Mr. Vicary this defect may be rectified, that gentleman having recently forwarded to me so considerable a series of specimens from the Greensand of the Blackdown region as to enable me both to verify the unity of these three supposed species and to select from them specimens exhibiting the surface ornaments of the umbonal portions of *T. excentrica* and their identity with the small specimens commonly referred to *T. sinuata*. These will be given upon a future plate.

Dimensions.—Length of the largest of our specimens $2\frac{3}{4}$ inches, height $2\frac{1}{4}$ inches, convexity of a single valve $\frac{3}{4}$ inch. Occasionally the species attains larger dimensions.

Affinities.—The umbonal portion of the shell in the posteal sinuation of its delicate costæ resembles a similar feature in *T. semiculta*, Forbes, from the Cretaceous rocks of Verdachellum, near Pondicherry, Southern India (see the description of *T. aliformis*); in the Indian species the costæ are much larger, and the posteal slope or area forms a greater angle with the other portion of the valve.

A similar feature is also conspicuous in *T. Sanctæ Crucis*, Valang, (Pictet, Paléon. Suisse,' plate 128, figs. 1—5). The latter shell has the anterior side shorter and its umbones more obtuse; the anteal portions of its costæ are also less distinctly horizontal or excentrical.

For *T. Coquandiana*, D'Orb., which is also an allied form, see the next species.

Owing to the fragility of the test, and the more compact matrix, specimens in Greensand collections are usually very imperfect, and afford no adequate means for testing the distinctness or affinity of other examples of the *Glabræ* from the same formation. These remarks also apply to specimens upon the tablets in the two national metropolitan museums, and enhance the value of the aid which has been afforded by the contribution from the collection of Mr. Vicary.

As both *T. excentrica* and *T. sinuata* are figured upon the same plate in the 'Organic Remains' by Parkinson, neither form possesses priority; I have made *T. sinuata* a synonym, as it exemplifies only the very young condition of the more fully developed *T. excentrica*. The internal mould does not appear to have been identified. The valves are always disunited.

Stratigraphical Positions and Localities.—*T. excentrica* in its different aspects occurs in the Greensand of the Blackdown and Haldon regions at several localities, as at Hembury Fort, at Staple Hill, and near to Collumpton. The Chloritic Marls and Sandstones at Dunscomb Cliffs, to the eastward of Sidmouth, is another locality.

D'Orbigny records *T. sinuata*, including *T. affinis*, in the lower beds of his Terrain Turonien or Chloritic Chalk of the Ligerian and Pyrenean basins; the localities given by him are Mans, Saint-Calais, Coudrecieux (Sarthe), Fouras, and the Isle d'Aix (Charente Inférieure), Ambillon (Marne et Loire). He retained *T. excentrica* as distinct from *T. sinuata*, but the only locality attached to it is Blackdown.

TRIGONIA LÆVIUSCULA, *Lyc.*, sp. nov. Plate XXII, fig. 6.

Shell depressed, lengthened; umbo moderately produced, small; placed upon the boundary line of the anteal third of the valve and slightly recurved; anterior and lower borders rounded elliptically; superior border somewhat concave; posteal extremity of the valve produced, attenuated, and depressed, its outline rounded. The portion of the valve adjacent to the posteal or superior border is slightly convex, and is without any angular division; its surface is smooth, or traversed only by delicate lines of growth. The other portion of the surface has horizontal, broad, depressed, plain, slightly irregular, and unequal ridges, which over the middle and lower portion of the valve become evanescent posteally; the space thus rendered plain has three or four obscure longitudinal sulcations, and is somewhat more depressed than the costated portion; its boundary anteally is nearly perpendicular, and extends somewhat anteal to the posteal third of the valve. The umbonal costæ are very delicate and closely arranged; near to the pallial border the costæ become widely separated, irregular in their directions, and more obscure.

The lines of growth are very delicate, they decussate the horizontal anteal extremities of the costæ.

Internally the borders of the valves are smooth, the test is rather thin, and the hinge dental processes have but little prominence. I have no knowledge of the internal mould.

This fine specimen was obtained by Mr. Vicary, of Exeter, in the Greensand of the Blackdown Hills, near to Collumpton, Devon; it is shorter posteally than *T. longa*, Ag., and its few lower costæ are more widely separated and irregular. Compared with *T. excentrica*, Park., the latter has the general figure shorter, more especially anteally; the convexity is much greater, the umbones are more conspicuous, and the longitudinal ridges are less widely separated. The test generally has greater thickness, and the hinge dental processes project more considerably; the posteal smooth, wide, depressed space in *T. læviuscula* is also distinctive.

Mr. Meÿer has obtained the species ill preserved in chloritic sandy marl at Dunscomb Cliffs between Beer Head and Sidmouth; these imperfect specimens and the single example herewith figured are only materials known to me.

More especially allied to *T. Coquandiana*, D'Orbigny ('Pal. Fran.,' vol. iii, pl. 294,) for which the imperfect specimens first collected were mistaken. It differs from the species of D'Orbigny in the following features: the convexity of the valves is less; the posterior extremity is shorter or more rounded; the costæ disappear altogether over a considerable portion of the surface posteally; there is also no indication of the little intercalated rib between each of the rows, as in *T. Coquandiana*; the latter species has the costæ well defined and passing across the valve continuously its entire length.

TRIGONIA LINGONENSIS, *Dum.* Plate XXII, figs. 1, 1 *a*, 2, 3, 4.

> TRIGONIA LINGONENSIS, *Dumortier.* Études Jurrassiques du Rhône, p. 275, pl. xxii, figs. 6—8, 1861.
> — — *Tate.* Discovery of the oldest known Trigonia in Britain, 'Geol. Mag.,' vol. ix, No. 97, p. 306, 1872.

Shell ovately trigonal, very convex; umbo antero-mesial, pointed, much produced, and slightly recurved; anteal, lower, and posteal borders curved elliptically; hinge-border nearly straight, sloping obliquely downwards, and forming an obtuse angle with the rounded siphonal border of the area. Escutcheon wide, depressed, traversed by transversely oblique, delicate plications, which become conspicuous and more strongly marked upon the obtuse inner carina. Area concave, bounded by two raised, obtusely rounded prominences or carinæ, traversed mesially for about a moiety of its length by a small furrow; both the area and its carinæ are traversed obliquely by very numerous unequal rugose elevations; near to the umbo these form a row of minutely knotted papillæ upon the border of the narrow ridge-like, inner carina. The marginal carina is distinct and elevated only as compared with the area, but has no distinctiveness or separation when compared with the other portion of the shell whose rugæ pass across it without interruption; near to the apex, however, it becomes elevated, narrow, and ridge-like, and the surface anterior to it has densely arranged acute rugæ. The other portion of the surface has a numerous irregular and unequal series of rugæ, which take the direction of the lines of growth; all originate at the pedal border as narrow, densely arranged plications, which become less conspicuous and nearly evanescent upon the middle of the valve; they are continued uninterruptedly across the area and escutcheon. There are also, in some instances, several longitudinal sulcations which are conformable with the rugæ in their direction, and are similarly unequal in their distinctness and distances; other specimens are nearly destitute of these sulcations. The area forms nearly a right angle with the other or pallial portion of the valve, so that, when a specimen is placed in a horizontal position and viewed from above, the area and escutcheon are scarcely visible.

Dimensions.—The largest of the specimens herewith figured has the length, measured from the apex to the posteal extremity, 28 lines; from the upper extremity of the siphonal border across the valve, at right angles to the length, 24 lines; thickness through the single valve 9½ lines, length of the siphonal border 10 lines, length of the superior border of the escutcheon 18 lines.

These dimensions do not refer to the largest specimen obtained, as they are exceeded by one in the museum of the Philosophical Society at York; very rarely also the valves are found in position.

GLABRÆ.

The hinge dental processes are exposed in two of the specimens figured; they agree with those of the *Glabræ* generally, and are less massive than in those of other Jurassic sectional forms. The internal mould has the adductor scars well developed; there are also traces of the external encircling rugæ; the test appears to be thin. The external rugæ differ much in their prominence in different specimens, so that in some instances the general surface is smooth, and has rugæ only near to the pedal border.

M. Dumortier described *T. Lingonensis* from the Marlstone beds of the valley of the Rhone, near to Langres; his figure has a slight depression of the surface anteal to the angle of the valve. None of the British specimens possess this feature, which probably, therefore, is only accidental.

Mr. R. Tate, now a resident at Redcar, to whom we owe its discovery as a British species, states that it occurs in the main seam of ironstone throughout the Cleveland district in the zone of *Ammonites spinatus*, and that it has been obtained at the following localities:

Skinningrone mines; Hobb Hill mine, near Saltburn; Eston mines, near Middlesboro'; Belman Bank and Challoner mines, near Guisborough; that the species is rare, excepting at Eston, but well-preserved specimens are everywhere rare. The two larger of our specimens were obtained by Mr. G. Lee, manager of the Eston mines at that locality, and generously presented by him for the present Monograph.

Examples of this remarkable Trigonia are in the museum of the Royal School of Mines; in the museum of the Yorkshire Philosophical Society; in the collection of Mr. R. Tate at Redcar; in that of Mr. G. Lee at Eston; and in my own cabinet—all from the ironstone of the Cleveland district. The British Museum has a specimen obtained by the late Miss Baker, of Northampton, in the Middle Lias at Preston Capes in that county; no other example from the Lias of the midland or southern counties has come under my observation.

Affinities.—The almost entire absence of ornaments upon the surface associates it with the *Glabræ*, a section which has only a small number of ascertained species, and, unlike other sections of the genus, is not limited to one portion of the Mesozoic period, but occurs at intervals widely separated stratigraphically; thus, the *Clavellatæ*, the *Undulatæ*, and the *Costatæ*, are limited almost exclusively to the Jurassic formations. The *Quadratæ* and the *Scabræ* are not less strictly Cretaceous forms, but the *Glabræ*, although represented by few species, constitute a section which embraces nearly the entire limits of the Mesozoic period.

T. Lingonensis, the oldest known Trigonia, is limited to the Middle Lias. The next known example of the section is our *T. Beesleyana*, which occurs not less rarely in the Inferior Oolite at a single locality. From that position the section appears to be absent until we arrive at the Portland formation, where it becomes the predominating section of the genus, and is represented in Britain by five species, two of which are abundant. Again, after a long stratigraphical interval, the section reappears in the middle portion of the Cretaceous rocks represented by two species in the Devonshire Greensand and Upper Greensands and Chloritic Marls of the South Devon coast. Two other European forms are also recorded. The Cretaceous rocks of India have yielded to the researches of Forbes and of Stoliczka four characteristic examples of the section. It is chiefly in the Cretaceous examples of the *Glabræ*, both European and Asiatic, that we discover affinities with the Liassic *T. Lingonensis*, species placed almost at the opposite stratigraphical limits of the Mesozoic formations; these affinities, however, have only a general or sectional resemblance, and refer chiefly to the characters of the area, so little separated from the pallial portion of the surface.

§ V. QUADRATÆ.

In Britain the *Quadratæ* are represented by four species only, two of these each constitute two varieties; they possess the several features special to the section prominently developed. Externally their escutcheons have small nodose varices; internally the posteal portion of the pallial border has a short row of quadrate pits and elevations; there is also a smaller oblique series of pits or furrows at the posteal extremity of the hinge-border. Plate XXIV, fig. 1 *a*, exemplifies the pallial pits of *T. nodosa*. *T. dædalea* and *T. spectabilis* have less than half the number of these pits upon their inner surfaces.

TRIGONIA DÆDALEA, *Park.* Plate XXII, figs. 7, 8; Plate XXIII, figs. 2, 3. Var. *confusa*, Plate XXIII, fig. 1.

    TRIGONIA DÆDALEA, *Parkinson.* Org. Rem., vol. iii, pl. xii, fig. 6, 1811.
    — ? RUDIS, *Ib.* Ibid., fig. 10.
    — DÆDALEA, *Sowerby.* Min. Conch., vol. i, tab. lxxxviii, 1815.
    — QUADRATA, *Ib.* Geol. Trans., 2nd series, vol. iv, pl. xvii, fig. 10 (young example), 1836.
    — DÆDALEA, *Pusch.* Polens, Paléont., p. 60, 1837.
    — — *Agassiz.* Trigonies, p. 52, 1840.
    — PALMATA, *Deshayes.* Leymerie, Mém. Soc. Géol. Franç., vol. v, pl. viii, fig. 5 (variety), 1842.
    — QUADRATA, *Morris.* Catalogue, p. 229, 1854.
    — DÆDALEA, *Ib.* Ibid., p. 228 (pars). Exclude Lower Greensand.

Exclude the *Trigonia dædalea* of the following authorities:

    De la Beche, Geological Manual, p. 287, 1832.
    Mantell, Geology of the South-east of England, Appendix, p. 388, 1833.

Ibbetson and Forbes, Proc. Geol. Soc., vol. iv, p. 414, 1844.
Fitton, Quart. Journ. Geol. Soc., vol. iii, p. 317, 1847.
D'Orbigny, Paléontologie Française, Terr. Cret., vol. iii, p. 145, pl. 292, 1843.
Ib., Prodrome de Paléont., vol. ii, p. 161, No. 322, 1850.
Pictet and Renevier, Foss. du Terr. Aptien de la Perte du Rhone et des Environs de Ste. Croix, pl. xii, fig. 1, 1857.

The figure of *T. dædalea* is ovately quadrate and depressed; the apices are small, pointed, slightly recurved, anteal and terminal; the hinge-border is short and slightly rounded, forming nearly right angles with the anterior and posterior borders; the latter or siphonal border is equal in length to the hinge-border, its lower extremity is curved; the lower border is curved elliptically with the anterior border. The area constitutes half the surface of the valve; it is flattened, without any distinct mesial division, excepting near to the umbo; its separation from the costated portion of the surface forms an angle which is slightly ridge-like where it is crossed by the first-formed or apical costellæ; there is, therefore, no distinct marginal carina; the positions of the inner and median carinæ are each indicated by an ill-defined row of small, rounded, widely separated tubercles; there are also numerous, small, irregular depressed tubercles scattered confusedly over its surface, and also some transverse rugose plications posteally. The upper or apical portion of the area has about eight rows of narrow, ridge-like, delicately knotted costellæ, which, originating at the anteal portion of the superior border, and passing across the valve obliquely downwards, enlarge rapidly and become rounded after they have passed the divisional angle, forming tuberculated varices about the middle of the costated portion of the valve, where they are bent suddenly upwards to the anteal border, each forming nearly a right angle; these are succeeded by several short supplementary varices, which pass backwards horizontally from the anterior border until they are interrupted by the short bent varices. The remaining portion of the surface is occupied by about eight larger, curved rows of tuberculated varices, each row having about nine distinct rounded tubercles; the rows commence at the angle of the valve and enlarge rapidly downwards towards the pallial border, the smallest tubercle in each row is, therefore, at the angle of the valve or at the usual position of the marginal carina. The escutcheon is narrow, lengthened, and flattened; it has several obscure rugose varices. The fragment figured by Parkinson, which has priority as an example of this species, appears to represent a specimen with the apical portion of the ornamentation unusually small and more than usually irregular anteally. This arrangement differs somewhat from the fine adult specimen figured in the 'Mineral Conchology,' but approximates to some of the Blackdown examples (see Plate XXII, fig. 7). The few first-formed rows of varices form, in some specimens, angulated ridges, which are only slightly tuberculated, resembling the little *T. quadrata*, Sow., but commonly, as in the figure of Parkinson, the first-formed rows are distinctly tuberculated. The interiors of the valves have the cardinal teeth much smaller and less prominent than in *T. nodosa* and *T. spectabilis;* the pallial posteal pits and elevations are small, closely arranged, and only three in number, thus affording a contrast to the same feature in *T. nodosa*, which has them much larger and greatly more numerous, so that in that species they occupy much of the surface near to the pallial border.

The foregoing detailed description will, it is trusted, suffice to rectify an error into which some palæontologists have fallen, who have taken for their guide the figure of *T. dædalea* in the 'Mineral Conchology' of Sowerby, an inaccuracy in the drawing having produced much confusion from its supposed identity with that fine variety of *T. nodosa* of which such considerable numbers have been obtained in the beds called "Crackers" in the Neocomian formation at Atherfield. Sowerby's drawing of *T. dædalea* has the angle of the valve occupied by a row of rounded nodes, which are larger than those in the adjacent rows of varices and have no accordance with them either in number or position. This arrangement differs materially from the original specimen now in the British Museum, but has much resemblance to the corresponding parts in *T. nodosa*. As the figure of *T. dædalea* in the 'Mineral Conchology' represents the only example of **adult growth** to be found in any British work since the fragment figured by Parkinson, our figures are the more deserving of scrutiny and comparison.

The variability so frequent in the ornamentation of the surface in the typical form may now be adverted to. Occasionally the minute tubercles upon the area are arranged with regularity, forming oblique rows which curve upwards and backwards from the angle of the valve. Another arrangement has the row of small nodes upon the angle of the valve, assuming the same comma-like figure which is so conspicuous in the larger carinal nodes of *T. nodosa;* however, this feature is scarcely ever distinct upon the whole of the row. Near to the pallial border, occasionally, are one or several short supplementary rows of varices resembling the same feature which is common in *T. nodosa*. The internal mould appears to be unknown.

A larger and well-marked variety of *T. dædalea*, var. *confusa*, is illustrated by Plate XXIII, fig. 1, which represents a specimen obtained by Mr. Vicary, of Exeter, in the Greensand of Little Haldon; it is distinguished by the generally increased size of the several features which ornament the surface, by the larger pallial varices and their nodes, and more especially by the unusually large, confusedly irregular, and unequal nodes scattered over the area; there is also a series of short anteal supplementary pallial varices, several of which are intercalated with the anteal extremities of the large oblique varices. This variety is rare, no second specimen has come under my notice.

Measurements of two examples of the typical form in my collection :

No. 1.—Length across the valve	29 lines;
Height	25 lines;
Length of the escutcheon	17 lines;
Length of the siphonal border	17 lines;
Across the area	16 lines;
Across the pallial surface	18 lines.
No. 2.—Length across the valve	24 lines;
Height	22 lines;
Length of the escutcheon	15 lines;
Length of the siphonal border	15 lines;
Across the area	11 lines;
Across the pallial surface	12 lines.

Specimens of less mature growth have the length and height equal, the measurement across the area is also equal to that across the pallial surface.

Stratigraphical positions and Localities.—The typical form of *T. dædalea* is not uncommon in the lower beds of Greensand at several localities in the Blackdown region, more especially in the Whetstone Pits near to Lyme Regis, Honiton, and Collumpton.

The south-western outliers of the Greensand at Great and Little Haldon have not produced the typical form, excepting in its very young condition. It appears to be also absent in the coast sections of Upper Greensand and Chloritic Marls upon the coast of South Devon, and equally so at the Isle of Wight, in Wiltshire, and in Kent and Sussex, and as no foreign example is known it may be presumed to have been a species eminently localised and restricted to a very limited area.

The large variety *confusa*, Plate XXIII, fig. 1, was obtained by Mr. Vicary at Little Haldon, near Dawlish, in the second bed of Greensand, which underlies the bed with Orbitulites. As these latter fossils are abundant and do not occur at Great Haldon, which is separated from the other hill only by the narrow valley of Escombe, Mr. Vicary inclines to the opinion that the upper beds at Little Haldon are somewhat higher in the series than the other. The pebbly stratum enclosing Greensand fossils, including *Trigonia pennata*, *T. sulcataria*, and *T. Vicaryana*, with numerous other forms well preserved, caps the hills both of Great and Little Haldon. These *Trigoniæ* do not occur in the Blackdown region.

Affinities and differences.—The only figure of *T. dædalea*, other than those of Parkinson and of Sowerby, which has come under my notice is a variety which constitutes the *T. palmata* of Deshayes (Leymerie, 'Mém. Soc. Géol. Fr.,' sér. 2e, vol. v, plate viii, fig. 5). Compared with British specimens, the small but distinctly raised marginal carina is destitute of nodes; the rows of pallial varices are much fewer; small at the carina, they enlarge rapidly downwards towards the pallial border; the spaces between the rows are plain and very wide. This well-marked variety has not been observed in Britain.

The fine specimen of *T. dædalea* figured in the 'Mineral Conchology,' now in the British Museum, having so frequently been regarded as identical with the variety of *T. nodosa* so abundant in the beds called "Crackers" of the Neocomian formation at Atherfield, a comparison of the two forms becomes necessary. *T. dædalea* has much less convexity; this feature alone will usually be sufficient to separate them; it is also shorter in the general figure, more especially upon the superior border, which has the escutcheon small and inconspicuous. The area is more flattened and its ornamentation is much more minute; it is much less decidedly bipartite, or is without the concavity formed by the upper half of the area in *T nodosa*. Its three rows of carinal nodes upon the apical portion of the valve are minute, very numerous, and closely arranged, so that the transverse ridges which cross that portion of the area and are continued to the anteal border of the valve are also greatly more numerous and closely arranged. The rows of pallial varices have greater curvature, they pass upwards towards the angle of the valve almost perpendicularly, or form a much greater angle with it than is seen in *T. nodosa*; the nodes in the rows are also more elevated and pointed; they lessen in size rapidly and regularly from the border upwards, so different from the irregular and unequal nodes in *T. nodosa*, which, for the most part, have the largest nodes about the middle of the rows. But apart from these details, a first glance at the depressed, short figure of *T. dædalea* will usually suffice to separate it from the Neocomian species.

The foregoing details will also serve sufficiently to separate the Devonshire *T. dædalea* from that different example of the *Quadratæ* afforded by the Greensand of Le Mans, with which, misled by Parkinson's very insufficient figure, Deshayes and D'Orbigny united it. Sowerby ('Geol. Trans.,' 2nd series, vol. iv, pl. xvii, fig. 10) depicted a very young, almost embryotic example of *T. dædalea* under the name of *T. quadrata*. A similar but larger specimen is given, Plate XXII, fig. 8, of the present Monograph; it should be compared with the young example of *T. nodosa*, Plate XXIV, fig. 3. Four years subsequently to the appearance of Sowerby's figure, Agassiz, who had received specimens of *T. dædalea* from England, was therefore fully aware of their distinctness from the large species of Le Mans, figured and described the latter under the name of *T. quadrata* ('Trigonies,' p. 27, tab. vi, figs. 7—9); apparently he was unaware that Sowerby had appropriated that name for his little Greensand specimen. By reuniting the little species of Sowerby to *T. dædalea* the name *quadrata* given by Agassiz to the large species of Le Mans will thus be entitled to remain; it has not been obtained in Britain. Possessing such materials for comparison it is remarkable that Agassiz, upon page 26 of the same work, should have tabulated *T. dædalea*, Park., as identical with a Trigonia from the Portland formation of Besançon, which he figured and described under the name of *T. Parkinsoni* ('Trigonies,' tab. x, fig. 6), but he avows his uncertainty as to the correctness of this identification. Judging from the drawing upon the plate of Agassiz, it is not identical with any known British species; of its several features the general form alone possesses any similarity to the section of the *Quadratæ* in which that author placed it.

Three years after the publication of the memoir of Agassiz D'Orbigny figured and described the large Le Mans shell for the *T. dædalea* of Parkinson ('Paléont. Fran., Ter. Crét.,' vol. iii, p. 45, plate 292).

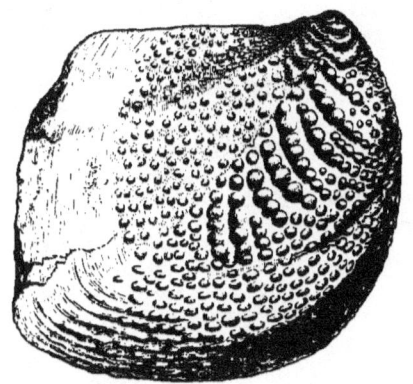

Our wood engraving exemplifies *T. quadrata*, Ag., or *T. dædalea*, D'Orbigny ; the original is one of a fine series in the British Museum, numbered 32,394 ; it is remarkable for the great length of the siphonal border, which exceeds that of the upper border of the valve ; the absence of any row of nodes at the carinal angle of the valve, the more numerous and smaller rows of pallial nodes, their general confusion, or, in other instances, the attenuation and crowded bifurcation of the rows near to the pallial border, the simple curvature of the few first-formed rows, and the general absence of any distinct separation between the pallial and siphonal portions of the valve, supply very evident distinctive differences. The valves of *T. quadrata* are also more depressed, more thin, their surface ornaments are much less prominent than in *T. dædalea*, their condition of preservation is also less satisfactory ; not unfrequently large portions of their surfaces are deprived of the test or have the ornaments imperfectly preserved.

TRIGONIA NODOSA, *Sow.* Plate XXV, figs. 1, 2. *Var.* ORBIGNYANA, Plate XXIV, figs. 1, 1 a, 2, 3.

TRIGONIA CLAVELLATA, *Mantell.* Geol. Sussex, p. 73, No. 10, 1822.
— NODOSA, *Sowerby.* Mineral Conchology, vol. vi, tab. 507, fig. 1, p. 7, 1829.
— DÆDALEA, *De la Beche.* Geol. Manual, p. 287, 1832.
— — *Mantell.* Geol. of South-east of England, p. 179, 1833.
TRIGONIA CINCTA, *Agassiz.* Trigonies, p. 27, tab. vii, figs. 21—23 ; and tab. viii, figs. 2—4, 1840.
— — *Matheron.* Catal. de Corps Org. Foss. du Départ. des Bouches du Rhone, p. 166, 1842.
— RUDIS, *D'Orbigny.* Pal. Fran. Terr. Crét., pl. 289, 1843.
— DÆDALEA, *Ibbetson* and *Forbes.* Proc. Geol. Soc., vol. iv, p. 144, 1844.
— — *Fitton.* Quart. Journ. Geol. Soc., vol. iii, p. 317, 1845.
— RUDIS, *D'Orbigny.* Prodrome de Paléont., vol. ii, p. 78, No. 291, 1850.
— CINCTA, *Buvignier.* Statist. Géol. Minér. et Paléont. du Départ. de la Meuse, p. 473, 1852.
— NODOSA, *Morris.* Catal., p. 229, 1854.
— — *Cotteau.* Moll. Foss. de l'Yonne, p. 76, 1857.
— DÆDALEA, *Pictet* et *Renevier.* Foss. du Terr. Aptien de la Perte du Rhone et des Env. de Ste. Croix, pl. xii, fig. 1, 1857.

Shell ovately oblong, moderately convex anteally and mesially, depressed posteally; umbones small, anterior, pointed, scarcely elevated above the superior border; anterior side very short, its border curved elliptically with the lower border; superior border lengthened, straight, its posteal extremity forming an obtuse angle with the siphonal border of the area, which is sinuated, and its lower extremity curves elliptically with the lower border. Ligament of the hinge occasionally preserved; large, in its absence the ligamental plates are conspicuous in their wide fossa. Escutcheon narrow, horizontal, flattened, its surface with small, oblique, irregular, nodose elevations. Area large, flattened, equal to two fifths of the surface of the valve in the variety *Orbignyana*, or to a moiety of the surface in the typical form; a row of nodes divides the area into two portions, the superior or outer portion is the larger, and is somewhat more depressed than the other, adjacent to the umbo; the area has three conspicuous rows of large depressed nodes, representing the inner, median, and marginal carinæ; the inner carina has its nodes ovately lengthened, and near to the posteal extremity of the escutcheon they become mere plications of growth; the median and marginal carinæ have their nodes more rounded, but often somewhat concave upon their anteal sides, producing a commu-

QUADRATÆ.

like figure. This, however, is not constant, and not unfrequently the nodes are ovate or slightly oblong, or in other instances the marginal carina has its nodes unequal in size and irregular in figure; all the rows become evanescent upon the posteal half of the valve, where the area is occupied by large, irregular, knotted, transverse plications; the portion of the area near to the umbo has regular transverse ridges, which are united to the carinal nodes; the whole of this portion of the area has rows of regular, minute, transverse, papillary elevations, more or less distinct, which impart a highly ornamented aspect to the valve. The rows of nodose varices or costæ upon the other or pallial portion of the valve commence, as in some other species, with about seven elevated biangulated, ridge-like rows, which pass obliquely downwards from the anterior border to the middle of the costated surface, where they are bent suddenly and pass uninterruptedly across the area obliquely, forming small elevated nodes at the marginal and median carinæ, and smaller ridges as they are prolonged to the inner carina; the succeeding rows consist of large, closely arranged, rounded varices, curved obliquely downwards from the marginal carina; their nodes usually increase in size downwards in each row; as they approach the pallial border they curve upwards in the direction of the lines of growth; there is some irregularity in the rows of anteal varices which succeed the first-formed plain ridged series; they form a few short supplementary rows, several of which are intercalated with the anteal portions of the oblique rows; the nodes nearest to the carina are smaller than those of the carina, with which they are also unconformable in number, as they are slightly more numerous than the carinal nodes. Owing to the enlargement of the nodes nearest to the pallial border and their more lengthened figures, the aspect of that portion of the valve is sometimes remarkable, the nodes being so closely placed that their arrangement in rows is scarcely perceptible unless the shell is viewed from the anterior side and at a little distance, when their order becomes more evident.

In the variety *Orbignyana* the size and figure of the nodes have commonly great variability and inequality, sometimes even in a single row, so that the anteal or pallial extremities of the rows have the nodes large and ovately lengthened, or in other specimens their anteal extremities are attenuated, their nodes becoming small, cord-like, and indistinct; both kind of rows have their anteal extremities conformable with the lines of growth or with the pallial border. For the most part the variety of the Perna bed or typical form has the rows of varices and nodes rather smaller and more regular (Plate XXV, fig. 1). The rows are more separated; the carinal nodes are also more regular and equal, so that the general aspect has little of the confused and crowded nodes so commonly seen in the variety of the crackers.

Young examples, from five to fifteen lines in length, have the general figure depressed, and are shorter or more quadrate; the area is also in proportion larger; the costæ or varices form plain ridges without nodes.

The "Cracker" variety *Orbignyana* has occurred abundantly at Atherfield, in beds higher than those which are characterised by the typical form; the former are also found in a better condition of preservation; their profuse and varied ornamentation command general admiration, and require several specimens to exemplify adequately their several aspects; this variability, although considerable, scarcely affects the general figure, and is nearly limited to the ornamentation of the surface; but with so much irregularity in the nodes, description must to some extent become subordinate to figures. The difference of figure supplies the most constant and clear distinction which characterises the two varieties. In the typical form (Plate XXV) the area constitutes a moiety of the valve, and the pallial or costated portion has somewhat less breadth than in the other; the great length of the siphonal border imparts consequently a more quadrate figure to the shell. The variety *Orbignyana* has the costated or pallial portion larger than the area; the siphonal border passes more obliquely downwards, or forms a more obtuse angle with the upper border; the lower or posteal extremity of the valve is therefore more produced and pointed.

The internal moulds are only known to me from the small specimens of *T. cincta* figured by Agassiz; in these the height is less in proportion than in specimens not deprived of the test.

Dimensions.—Specimen of the typical form from the Perna bed at Red Cliff, Sandown.

Length measured upon the marginal carina	30 lines.
At right angles to the carina across the pallial surface	16 ,,
Across the area	15 ,,
Length of the siphonal border	18 ,,
Length of the escutcheon	22 ,,

Dimensions of a specimen from Hythe:

Length upon the marginal carina	40 lines.
At right angles to the carina across the pallial surface	21 ,,
Across the area	19 ,,
Length of the siphonal border	24 ,,
Length of the escutcheon	$25\frac{1}{2}$,,

Variety *Orbignyana*:

Length upon the marginal carina	30 lines.
At right angles across the pallial surface	21 ,,
Across the area	15 ,,
Length of the siphonal border	18 ,,
Length of the escutcheon	24 ,,
Diameter through the valves	21 ,,
Breadth of escutcheon	$8\frac{1}{2}$,,

Stratigraphical position and Localities.—*T. nodosa* constitutes two varieties in Britain, the differing characters of which may be stated synoptically as follows :—*The typical form.*—Area forming a moiety of the surface of the entire valve; nodes of the marginal carina small and regular in figure, size, and arrangement; rows of pallial varices regular, without bifurcations, and becoming attenuated near to the border. *Position.*—Perna bed at Redcliff, Sandown Bay, Isle of Wight. Hythe, dark grey sandstone bed with *T. ornata*. *Maidstone*, Molluskite bed. Tealby, brown pisolite bed. *Variety* ORBIGNYANA.—Surface of the area never exceeding two fifths of the entire valve; nodes of the marginal carina unequal in size and irregular in figure and arrangement; rows of pallial varices variable in figure and arrangement, at the border either large and simple or bifurcating and attenuated, the nodes either distinct or crowded. *Position.*—The Cracker beds at Atherfield, Isle of Wight; rarely in the Perna bed at Redcliff, near Sandown.

History, affinities, and comparisons with allied testacea.—For specimens illustrative of *T. nodosa* from the Neocomian sandstone quarries near Hythe, which was the locality of the type-specimen figured by Mr. Sowerby, I am indebted to Mr. Mackeson, of that place, who, in compliance with requests from Rev. T. Wiltshire and myself, obtained several examples; these are obscure moulds of external casts in coarse sandstone. Mr. Mackeson had also the good fortune to procure upon a slab of sandstone two external casts of uncompressed specimens; these have afforded good reproductions of the shell by the aid of gutta-percha pressings, and are found to agree with the form procured at Lympne and in the Perna bed at Redcliff, Sandown Bay. I have also been favoured with information respecting the condition of the species in the quarries of Kentish Rag near Maidstone, in a communication from Mr. J. Bensted, jun., the son of the proprietor of the quarries, whose intelligent remarks I have pleasure in quoting. "It is not uncommon in the Iguanodon quarries, and is found principally in one particular bed of sandstone called *Molluskite bed*, but simply as a faint white mark formed by the lime of the original shells, which are quite valueless as specimens. I have a single specimen only, which owes its preservation to the fact that the space occupied by the shell appears to have been filled with flint, which was to a certain extent able to resist the pressure which squeezed the others flat." The same gentleman also favoured me with a drawing of the fossil which demonstrated its identity as a species with the gutta-percha pressings from Hythe, the specimen from Tealby, and those from the Perna bed at Redcliff; from the latter locality the Royal School of Mines has a large specimen, presented by Dr. Fitton, as an example of *T. dædalea*, but as the surface of the area is ill-preserved it is not fitted for the artist.

Upon the same plate with the fragment figured by Parkinson for his *T. dædalea* ('Org. Rem.,' vol. iii, pl. xii) is another more doubtful fragment, to which he attached the name of *T. rudis*; it supplies a warning example of the impropriety of publishing such materials when no satisfactory description can possibly be founded upon them.

This figure, so worthless in itself, has been the source of confusions and errors of identification in the Cretaceous *Trigoniæ* for upwards of sixty years. One of the most singular and notable of these arrangements is seen in the great work of D'Orbigny above referred to. I am only able to correlate with some doubt Parkinson's fragment of *T. rudis*. The confusedly scattered tubercles upon the area, and the apparent absence of distinct carinal nodes, associate it better with the large variety of *T. dædalea* than with any other species. Professor Morris ('Catalogue,' p. 229) placed *T. rudis* doubtfully with *T. spectabilis*. A slight error in Sowerby's delineation of *T. dædalea* ('Min. Con.,' tab. lxxxviii) gives the appearance of rounded nodes upon the marginal angle of the valve, larger than those upon the adjacent rows of pallial varices; this has been a frequent source of error connected with *T. nodosa*, or rather with its variety *Orbignyana;* specimens of this variety have been freely dispersed over Europe, and have been regarded as examples of *T. dædalea;* it was adopted as such by Ibbetson and Forbes ('Proc. Geol. Soc.,' vol. iv, p. 414), by Fitton ('Quart. Journ. Geol. Soc.,' vol. iii, p. 317), and also by the latter author in his elaborate stratigraphical list of Lower Greensand fossils It is also the *T. dædalea* of Mantell ('Geology of the South-East of England,' p. 179) and of De la Beche ('Geol. Manual,' p. 287). It may afford some explanation of these errors to mention that Sowerby's original figure of *T. nodosa* does not accord very closely with any actual known specimen; the effects of vertical pressure will explain the appearance of flattening at the umbo, the partial exposure of the striated hinge processes, the apparent absence of the angulated costæ upon that portion of the shell. Mr. Sowerby's collections of fossils, now in the British Museum, does not contain the original specimen of *T. nodosa* figured in the 'Mineral Conchology,' and no information concerning it can be obtained; two very indifferently preserved portions of the Trigonia in the collection are probably the "inside casts" mentioned in his description.

Agassiz, in common with other palæontologists, appears to have experienced much difficulty in the determination of this species ('Trigonies,' p. 27, tab. vii, figs. 21—23; table viii, figs. 2—4;) the figures upon his plates are named *T. nodosa;* subsequently he was led to regard the species of Sowerby as distinct, and described the supposed new species under the appellation of *T. cincta*. Agassiz grounds the distinctness of *T. cincta* upon its smaller and more regular varices, upon the greater breadth of its area, and upon the ornamentation of the area, which in *T. nodosa* appears to be nearly smooth, excepting that it possesses carinal nodes. The condition of the Hythe specimens is such that we should not expect to have the ornamentation of the area preserved; usually the size of the area is nearly equal to a moiety of the entire surface of the valve, and the rows of varices in Sowerby's figure do not agree very strictly with other Neocomian examples of the same species. The only specimen with the test preserved figured by Agassiz is in an indifferent condition of preservation, the general figure resembles the typical form as exemplified by our specimen from the brown pisolite of Tealby, and also the variety from Ventoux figured by D'Orbigny ('Pal. Fran.,' pl. 289, fig. 5), but the characters of the surface are obscure, and do not afford an adequate idea of the species.

The very characteristic figures given by D'Orbigny ('Paléont. Fran., Terr. Crét.,' Atlas,' tom. iii, plate 289, figs. 1—5) represent three Neocomian Trigoniæ of differing aspects, which are intended as illustrative figures of *T. rudis*, Park.; *T. nodosa*, Sow.; *T. spectabilis*, Sow.; *T. cincta*, Ag.; and *T. palmata*, Desh. The specimen figs. 1, 2 has affinities with certain ill-preserved examples of *T. nodosa*, from the coarse sandstone at Hythe, in the more horizontal direction of the rows of pallial varices, and in the great length of the siphonal border; two imperfect specimens in Mr. Sowerby's collection, mentioned by him in the 'Mineral Conchology,' now in the British Museum, illustrate these features; also our specimen from the lowest or Perna bed at Redcliff, Sandown Bay, excepting that the figure is more lengthened. Fig. 5, from Ventoux, has affinities with our Tealby specimen in the surface of the area and in the rows of small pallial varices with their rounded nodes, but differs in the ornamentation upon the position of the marginal carina; the *T. cincta* of Agassiz also approximates to the Ventoux variety. Figs. 3, 4 are strictly identical with the more common aspect of specimens from the beds of Crackers, Isle of Wight, which I have distinguished as *T. Orbignyana*. None of the figures present even a remote approximation to *T. spectabilis*, Sow., or to *T. palmata*, Desh. The former of these is the species next described; the latter is referred to as a variety of *T. dædalea*; neither of them occurs in the Neocomian formation.

Our variety *Orbignyana* is well exemplified by the shell figured for *T. dædalea* by Pictet and Renevier ('Foss. du Terr. Aptien de la Perte du Rhone,' plate 12, fig. 1), which offers no material difference when compared with the Atherfield specimens. Their example of *T. nodosa*, fig. 2 upon the same plate, differs so materially from all varieties of the species, whether British or foreign, that it cannot be accepted as pertaining to that species. The escutcheon in the present shell is so large, both in length and breadth, that when we find it furnished with a numerous series of regular transverse costellæ, which are visible even when the shell is laid upon its side, it is impossible to associate it with the narrow form, horizontal surface, and crowded, obliquely nodose varices which characterise the escutcheon in *T. nodosa*; the unusually lengthened form, the row of rounded nodes upon the whole length of the median carina, together with the short perpendicular rows of equal nodes upon the pallial portion of the valve, are equally distinctive; they also separate it from a large, imperfect, clavellated *Trigonia* figured by Pictet and Roux ('Grès Vert,' plate xxxv, fig. 5) for *T. nodosa*, in which the rounded nodes in the rows become symmetrically small and inconspicuous as they approach the area; they even pass across the area in an attenuated form. The general aspect of this species would associate it with the *Clavellatæ*, but as the limits of the area are not clearly defined, as it is without bounding carinæ, and as some of the rows of nodes pass across it, it should apparently be arranged with the *Quadratæ*. The escutcheon is not seen.

Foreign Localities.—The small examples of *T. nodosa* or *T. cincta* figured by

Agassiz are from blue Neocomian marls at Neuchatel. D'Orbigny records the occurrence of his *T. rudis* at Ventoux (Vauclose); Mortean (Doubs); Saint Souveure (Yonne).

TRIGONIA SPECTABILIS, *Sow.* Plate XXVI, figs. 1, 2, 3, 4.

TRIGONIA SPECTABILIS, *Sow.*   Min. Conch., vol. vi, tab. 544, p. 83, 1829.
— — *Pusch.* Polens Paläontol., p. 60, 1837.
— — *Agassiz.* Trigonies, p. 8, 1840.
— — *Morris.* Catal., p. 229, 1854.
— NODOSA, *Pictet et Renevier.* Grès Vert, p. 484, pl. xxxv, fig. 5, 1857.

Shell subquadrate, short, depressed in the young state, thick and moderately convex when fully developed; umbones small, pointed, scarcely elevated above the superior border, of which they form the anteal extremity; the anterior border is truncated, it descends almost perpendicularly, but is curved at its junction with the lower border; hinge-border straight, forming nearly a right angle both with the anterior border and with the lengthened siphonal border. The surface of the area is equal to three sevenths of the entire valve; it is slightly convex, but depressed at the well-marked mesial junction of the two portions; the superior or more depressed half has a few unequal and imperfect rows of small nodes, or, in other instances, the few nodes are scattered irregularly; the other or inner portion of the area has at its umbonal extremity several subangular transverse ridges, which are continuations of those upon the other portion of the valve; each of these forms a prominent node at the position of the marginal and also of the median carina to the number of four or five rows, posteally to which the area has only a few irregular and obscure nodosities, which near to the siphonal border are effaced by plications of growth. The escutcheon is lengthened, very narrow, flattened, and inconspicuous; it is slightly overwrapped at its outer border by the nodosities of the inner carina of the area which are extended upon the escutcheon. The other portion of the shell has, mesially, four or five rows of very large, depressed, rounded varices; they become small and curved near to the pallial border, and enlarge considerably towards the angle of the valve; each row has about eight or nine large, depressed, ovate, and scarcely separated nodes, the longer diameter of these is across the varices; or the nodes are sometimes only obscurely defined or partially united in the rows; the first-formed four or five varices are entire, narrow, angulated, and transverse; their anteal extremities are bent upwards suddenly and perpendicularly; there is also an additional short varix adjoining the anteal extremity of the fourth row. The two last-formed rows of pallial varices are comparatively small, depressed, and cord-like; their general direction coincides with the lines of growth. In fully developed specimens there is a space posteal to the varices which is occupied solely by prominent rugose folds of growth.

Internally the margins of the valves are smooth, excepting the pallial pits and eminences, which are small and only two or three in number. The prominences within the border of the escutcheon are also only two or three, and rather obscure. The cardinal teeth are large and diverge widely. All the specimens examined are single valves. The internal mould has not been ascertained.

Dimensions.—The largest example figured upon Plate XXVI is surpassed in size by specimens in both of the National Metropolitan Geological Museums; as, however, its condition of preservation is good, and as it sufficiently represents an advanced stage of growth, it may be accepted as a good illustration of the species. The length transversely is 25 lines; the height 34 lines; the length of the siphonal border and of the escutcheon are nearly equal, or 24 lines; the thickness through the single valve 5 lines. A smaller example has the length and height equal, but usually the length exceeds the height.

Comparisons.—No example of the *Quadratæ*, either British or foreign, appears to possess any near affinity with *T. spectabilis* in the combination of its few, simple, leading features, viz. the short, depressed, subquadrate figure, the few unusually large, nodose, unequal, but sometimes connected or cord-like pallial varices, together with the smaller and for the most part few and irregular nodes upon the expanded area. Our smallest figure, which illustrates the very young condition, is also well characterised by its high-ridged acute costæ and flattened form. Fully developed specimens have sometimes much rude irregularity in their surface ornaments. These occasionally altogether disappear, and the posteal half of the valve is then occupied solely by large rugose plications of growth. Our figures are truthful and characteristic, but do not fully exemplify these conditions, which may be better appreciated by examining the series of specimens upon the tablets in the British Museum. Pictet and Renevier, in their work above cited, have given a good figure of the present species under the designation of *T. nodosa;* no other author after Sowerby appears to have figured it. D'Orbigny united it to his *T. rudis*, together with *T. nodosa*, *T. cincta*, and *T. palmata* ('Pal. Fran., Terr. Crét.,' vol. iii, p. 137); but his illustrative figures represent Neocomian forms only, and are altogether distinct from *T. spectabilis*.

Stratigraphical position and Localities.—It is associated in the Greensand of Blackdown with the other *Trigoniæ* of those beds, but, owing to the fragility of the test, entire valves are rare. I am indebted to the liberality of Mr. Vicary, F.G.S., of Exeter, for the gift of specimens to illustrate *T. spectabilis* in the present Monograph; the species is well exemplified in both of the National Metropolitan Museums; more frequently private collections have it only in fragments.

TRIGONIA TEALBYENSIS, *Lyc.*, sp. nov.

A single very imperfect specimen in the Woodwardian Museum, Cambridge, is the only one known to me of a beautiful species of the *Quadratæ* obtained in the Neocomian formation at Tealby, Lincolnshire. Should no other more suitable specimen occur it is proposed to figure this fossil upon a future plate; unfortunately the shells in the bed of hard limestone at that locality are converted into fragile crystalline lime, which breaks into fragments by the concussion of a blow with a hammer; there is, therefore, little probability that any example devoid of injury will be obtained from that bed. In the present instance the posteal half of the valve has been broken away, and the figure of the shell longitudinally is, therefore, rather doubtful.

Diagnostic characters.—Shell short, inflated; umbones much arched inwards, slightly recurved, not prominent, obtuse; anterior side short, its border curved elliptically with the lower border; superior border convex. Escutcheon broad and flattened, its upper border elevated, its outer or carinal border depressed with closely arranged, narrow, diverging scabrous plications or rugose folds. Area narrow, flattened, forming a considerable angle with the other portion of the valve. The ornamentation of the umbonal portion of its surface is minute and delicate; there is a small median furrow; the marginal and inner carinæ are also very small and slightly knotted. The surface of the area has numerous regular, faintly defined, small, transverse indented lines. The other and by much the larger portion of the shell has the rows of nodose costæ numerous, concentric, and closely arranged; the nodes upon the rows are very numerous, small, perfectly regular, closely arranged, prominent, and only partially rounded, as if compressed laterally in the rows or moulded by the lines of growth, which are densely arranged and conspicuous over the anteal and middle portions of the valve; the rows have a slight horizontal flexure at their posteal extremities and also near to the anterior border, where their nodes become indistinct or cord-like. The fragment referred to has upwards of twenty rows of costæ, which do not represent the entire number. The convexity of the valve is very considerable; the faintly marked features upon the anteal portion of the area indicate that the posteal portion is altogether without ornamentation. The short, subglobose figure suggests the possibility that it may be identical with *T. paradoxa*, Ag. ('Trig.,' p. 46, tab. x, figs. 12, 13), known only from two internal moulds, which exhibit no trace of the external ornaments; in common with our shell they appear to have no near affinities with any one of the Cretaceous Trigoniæ; both are from the same formation. The French specimens of *T. paradoxa* are from the Neocomian of Besançon.

§ VI. SCABRÆ.

This, the predominating section of the genus in the Cretaceous rocks, is also special to them; the entire series has crenulated or scabrous costæ and costellæ; the area and escutcheon, devoid of bounding carinæ, are each limited by an angularity of the surface; the transverse costellæ of the escutcheon for the most part also pass across the area. It is decisively separated from all other sections by the position of the ligamental fissure, which is strictly inter-umbonal, so that in some species when the valves are united the ligament is concealed, or when the umbones are not closely placed the lengthened narrow fissure extends anteally to them.

In Britain the *Scabræ* form three sub-groups, which may be designated the *aliformis*, the *pennata*, and the *spinosa* groups. The first of these is remarkable for the prolonged and attenuated posteal extremity; the siphonal border is, therefore, very short; the incurrent orifice, the excurrent and anal orifices, were arranged near to each other, but separated by a short internal rib, as in other examples of the genus; in Britain its representatives are *T. aliformis*, *T. Vectiana*, *T. caudata*, *T. scabricola*, *T. Etheridgei*, *T. Meÿeri*, and *T. Fittoni*; there are also a numerous series of foreign analogous forms.

The second or *pennata* group has the attenuated posteal extremity of the first group; it is characterised by a feature which somewhat resembles the section of the *Glabræ*; it has an ante-carinal diagonal depressed space, which is either smooth or has a faintly defined series of perpendicular costæ. *T. pennata* and *T. sulcataria* are the only British species; two other species occur in France.

The third or *spinosa* group has the borders of the valves comparatively short and rounded. The pallial costæ and costellæ of the area and escutcheon diverge from the angle of the valve. *T. spinosa*, *T. ornata*, *T. Archiaciana*, *T. Agassizii*, *T. Vicaryana*, and *T. Cunningtoni* are its British representatives. The foreign analogues of this group are even more numerous.

A fourth group, which does not appear to be represented in Britain, is characterised by possessing the usual ornamented escutcheon of the *Scabræ* together with the area of the *Clavellatæ*, with their bounding carinæ and delicate transverse plications; the pallial costæ are extremely variable. *T. Lusitanica*, Sharpe, *T. elegans*, Baily, and *T. Constantii*, D'Orbigny, may be referred to this group.

TRIGONIA ALIFORMIS, *Park.* Plate XXV, figs. 3, 3 *a*, 4, 4 *a*, 5, 6.

> TRIGONIA ALIFORMIS, *Parkinson.* Org. Remains, vol. iii, p. 176, tab. xii, fig. 9, 1811.
> — ALÆFORMIS, *Sowerby.* Mineral Conchology, vol. iii, tab. 215, 1818.
> — ALIFORMIS, *Deshayes.* Coq. car., p. 33, tab. x, figs. 6, 7, 1831.
> LYRIODON ALÆFORME, *Bronn.* Leth. Geog., vol. ii, p. 700, tab. xxxii, fig. 15, 1837-8.
> TRIGONIA ALÆFORMIS, *Pusch.* Polens Paläontologie, p. 60, 1837.
> — ALIFORMIS, *Fitton.* Quart. Journ. Geol. Soc., vol. iii, p. 289 (pars), 1843.
> — — *Agassiz.* Trigonies, p. 31, tab. vii, figs. 14—16, tab. viii, fig. 12, 1840.
> — ALÆFORMIS, *Morris.* Catalogue, 2nd ed., p. 228 (pars), 1854; exclude L. G. S. of Sandgate and Boughton.
> — ALIFORMIS, *Pictet.* Paléont. Suisse, tom. i, pl. xiv, fig. 1, 1857; exclude fig. 2.

> Exclude the following figures of *Trigoniæ* named *T. aliformis*:
> Leopold de Buch, Petrefacta recuillies en Amerique, par A. de Humboldt et par Ch. Degenhardt, fig. 10, 1839.
> Forbes, Geol. Trans., 2 ser., vol. vii, part iii, p. 151, 1846.
> D'Orbigny, Pal. Fran., Terr. Crét., vol. iii, p. 143, pl. 291, figs. 1—3.
> *Lyroden aliforme,* Goldfuss, Petref., vol. ii, tab. 137, fig. 6, 1836.
> Pictet and Roux, Grès Vert, pl. xxxv, figs. 2 *a*, *b*, 1847-53.
> Pictet and Renevier, Terr. Aptien de la Perte du Rhône, pl. xiv, fig. 2, 1857. (Fig. 1 is from a Blackdown specimen.)

Trigonia aliformis is placed at the head of the first group; it is characterised as follows:

Shell sublunate, inflated anteally, produced, attenuated and depressed posteally; umbones much elevated, antero-mesial, pointed, much recurved and incurved; anterior side produced, its border rounded; lower border rounded, but somewhat excavated posteally; hinge-border lengthened, concave, terminating posteally in a rostrated and attenuated extremity; ligamental aperture narrow, inter-umbonal. Escutcheon lengthened, deeply concave, occupying the entire upper surface of the shell; its superior or inner border is plain and much raised; its outer border is formed by a narrow, elevated, and rounded area; it is traversed transversely or obliquely by numerous closely arranged, small, serrated costellæ, which are prominent near to the umbones, but become only faintly traced posteally. The area is very narrow, raised, and convex; it is rendered bipartite throughout its entire length by a deep groove and its superior or umbonal portion has a few small, ridge-like, transverse costellæ; the remainder of its length has small, irregular, transverse plications.

The other portion of the surface has a numerous series of costæ, which originate at the border of the area as narrow, rounded, crenulated ridges, and diverge in every direction; about seven costæ nearest to the apex are concentric or are curved obliquely; the next succeeding seven, or more, enlarge or become inflated at their middle portions, and pass obliquely downwards to the pallial border; their crenulations are faintly traced and irregular, forming obtuse, transverse nodes upon the costæ, which become attenuated as they approach the pallial border. The more numerous and smaller costæ occupy the more flattened or posteal portion of the shell; they are small, narrow, rounded, very closely arranged, minutely crenulated, and nearly perpendicular; their extremities render the lower border dentated. The middle or inflated costæ form a slight undulation approaching to a falciform flexure; the interstitial spaces are plain. The narrow posteal portion of the shell, with its closely placed perpendicular costæ and depressed surface, contrasts strongly with the inflated anteal surface with its more widely separated costæ enlarged mesially and attenuated at their extremities. Usually specimens of adult growth have upon the anteal face of the valve and adjacent to the borders numerous small, rather obscure, horizontal ridges, or supplementary costellæ, which occupy the intercostal spaces of the first-formed six or seven costæ, a feature which is only visible in well-preserved specimens. The change from the inflated anteal surface to the depressed and flattened posteal portion is abrupt and strongly characterises the species.

The inner borders of the valves are dentated by the extremities of the pallial costæ; the narrow flattened surface forming the inner border of the escutcheon has a numerous series of small transverse pits; the narrow produced siphonal border is gaping, and contracted mesially by a projecting longitudinal internal rib in each valve placed beneath the mesial furrow of the area and serving to separate the incurrent and excurrent orifices; the other borders of the valves are close fitting.

Dimensions of an unusually fine specimen in the collection of Mr. Vicary and intended to be figured upon a future plate:—Length of the angle of the valve 27 lines; length from the pedal border to the siphonal border 26 lines; height 21 lines; diameter anteally through the united valves 14 lines; breadth across the area and escutcheon 9 lines; length of the siphonal border 5 lines.

T. aliformis also occurs as a distinct variety and in some abundance in the highest Greensands of Wiltshire at Warminster, and of the Isle of Wight at Ventnor; the fossils are invariably deprived of their tests, and are usually flattened from vertical pressure. Two uncompressed specimens are represented (Plate XXV, figs. 5, 6); the surface ornaments are sufficiently distinct excepting upon the area and escutcheon, where they do not appear to differ materially from the corresponding portions of the typical form. Compared with the latter, the figure is more produced and attenuated posteally and less inflated anteally; the change from the small posteal perpendicular costæ to the larger, curved, middle series is much less abrupt, and, as the umbones are less recurved, they are more erect, and are nearer to the anterior side, which is shorter; the escutcheon is also more lengthened. Possessing these differences, which are well exemplified by our figures, it is, perhaps, doubtful whether the variety from the Chloritic Marls and Sands should be named *aliformis*. The question has engaged my attention fully, aided by the comparison of ample materials. The typical form of the Blackdown and Haldon Greensand is remarkably exempt from any considerable amount of variability, so that, whatever may be the number of specimens or their stages of growth, no question can ever arise respecting their identity as a species. With the presumed variety the conditions are altogether different; deprived of the test, the most reliable means of comparison does not exist. The Warminster specimens are not only ill preserved, but they are invariably flattened and sometimes distorted by vertical pressure. Such appears to be the case with all the specimens collected by Mr. Cunnington, who kindly forwarded to me an unusually numerous series for comparison. A varied series of specimens from a similar stratigraphical position near Ventnor affords a nearly similar result, excepting that the fossils, having been enveloped in a finer sediment, have their surfaces occasionally better preserved, and have sometimes, but rarely, escaped compression. With such selected specimens, however, occur numerous others, flattened, but appearing to possess the usual attributes of the typical form in a degree sufficiently marked to prevent them from being assigned to a distinct variety. Having due regard, therefore, to the much higher position of the presumed variety, I propose to regard it as such, and to distinguish it by the name *attenuata*. The typical form is somewhat abundant in the lower beds of the Blackdown and Haldon Greensand, associated with a still more common species of the same group, *T. scabricola*. It has also been tabulated in lists of Lower Greensand fossils by Ibbetson and Forbes, by Fitton, by Mantell, and by Morris. A few badly preserved examples representing fossils of the same group, obtained in the Lower Greensand at the Isle of Wight, in the Kentish beds, and also near Cambridge, have been brought under my notice; for these and for similar examples from France the reader is referred to the next species, *T. Vectiana*.

History, and Comparisons with allied forms of the Scabræ.—The figures of *T. aliformis* given by Parkinson ('Org. Rem.,' vol. iii, pl. xii, fig. 9) and by Sowerby ('Min. Conch.,' vol. iii, pl. 215) are excellent representations of the typical form from the Greensand (Whetstone pits) of the Blackdown district. During some years subsequently, Continental cultivators of palæontology, in the absence of the means of comparison which are now possessed, assigned to Parkinson's species several allied forms of *Trigonia* from various localities and stratigraphical positions. The general absence of sufficiently definite descriptions of the several features which characterise the *Trigoniæ Scabræ*, in these and other authors, has induced me to omit notice of such descriptions when they are not accompanied by illustrative figures, unless they are in other respects sufficiently verified. As *T. aliformis* has been believed to occur at various localities in each of the four continents, the comparisons, in some instances referring to figures of fossils, the originals of which are unknown to me, require much critical care in estimating them.

The figure and brief description of *Trigonia thoracica*, Morton (' Synopsis Org. Rem. of the Calcareous Group of Alabama, United States,' p. 65, pl. xv, fig. 13, 1834), assigned by Von Buch to the *T. aliformis* of Parkinson, fails in all the characteristic features of that species. It appears to be allied to, and perhaps is not really distinct from, a large Bogota species figured by Von Buch for *T. aliformis* ('Description des Pétrifications recueillies en Amerique, par Alex. de Humboldt et par Ch. Degenhardt,' 1839, fig. 10). The figures of these American allies of *T. aliformis* are indicative of an indifferent condition of preservation, which should induce us to distrust the value of any theoretical conclusions founded upon such materials. It would appear also from the general tenor of Von Buch's remarks in his memoir on the characteristic fossils of the Cretaceous rocks ('Betrachtungen über die Verbreitung und die Grenzen der Kreide-Bildungen,' 1849) that he was even inclined to arrange all the lengthened forms of the *Scabræ* as a single species, or, in other words, to refer them all to *T. aliformis;* he also records his astonishment at its great and perhaps unexampled geographical range; that it occurs in the State of Alabama, in the mountain-ranges of Central America, again in the mountains of Santa Fe de Bogota, South America. He even unites to *T. aliformis* the *T. ventricosa* of Krauss from Algoa Bay, South Africa; and remarks that it is found, as if blown by the winds over the vast peninsula of Hindostan, in the south-west, near Pondicherry. The examination which I have instituted leads me to reject altogether a statement so general and unexampled; the figure of *T. thoracica*, Morton, which Von Buch referred to *T. aliformis*, and also the large Bogota shell, to which he gave the same name, have the general figure short posteally, without attenuation. The costæ are also different in figure; the areas and escutcheons are not delineated; they cannot, therefore, bear a strict comparison with any known European species.

T. ventricosa, Krauss ('Nova Acta Acad. C. L.-C. Nat. Cur.,' vol. xxii, part 2, p. 456, tab. xlix, fig. 2, 1850), is so important a fossil, from its wide distribution in South Africa and its great numbers, and from its having been united by Von Buch to *T. aliformis*, and very inadequately figured by Krauss, that I have been induced to subjoin the following figures taken from a specimen in the British Museum, which possesses a remarkably fine and numerous series numbered 49,990. The locality is Sunday's River, District of Uitenhage, South Africa, at a place named Prince Alfred's Rest.[1]

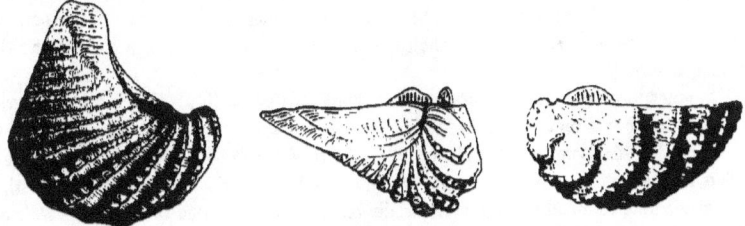

Three views of the valves of *Trigonia ventricosa*, Krauss, from South Africa.

[1] See 'Quart. Journ. Geol. Soc.,' vol. xxvii, p. 500, &c.

Larger than *T. aliformis*, it is so much inflated anteally that the diameter through the united valves is equal to their height and greater than their length in adult specimens; the umbones are remarkably large and arched inwards. The anteal varices (about ten) are large, oblique or almost perpendicular, each fringed with a row of depressed, oblong, or sometimes ovate nodes. The posteal varices (about eight) are small, narrow, closely arranged, and perpendicular; the area is convex, plain, and bipartite; the escutcheon is very wide and boat shaped, with a few transverse costellæ. It occurs very abundantly at several localities on the Sunday's and Zwartkop Rivers, associated with a considerable fauna, including the two gigantic species of *Trigonia*, *T. Herzogii*, Haussman, and *T. conocardiformis*, Krauss, which are also exceedingly abundant. The former of these is well known from the excellent figure of it in the great work of Goldfuss ('Petref.,' tab. 137, fig. 5); of this the British Museum has an unusually fine series of specimens numbered 46,461.

More recently, a considerable number of fossils from South Africa having been added to the rich collection in the Museum of the Geological Society, Mr. R. Tate communicated a memoir descriptive of them and of their distribution ('Quarterly Journal Geological Society,' vol. xxiii, part 3, p. 139, 1867). The comparison of these African forms with their supposed European analogues has led to his assigning them to the Lower Oolitic rocks, but with the reserve which should always qualify conclusions deduced from comparisons of fossils so remote geographically. Admitting the general apparent Jurassic *facies* of certain fossils figured in the plates illustrating Mr. Tate's memoir, I am nevertheless much impressed by the presence in such considerable numbers of *Trigonia Herzogii* and *T. ventricosa*, the former an elongate example of the *Quadratæ*, the latter of the *Scabræ*, sectional forms which in Europe, Asia, and America, are special to and eminently characterise the Cretaceous rocks. *T. ventricosa* more especially is nearly allied to, and possibly is not really distinct as a species from, *T. tuberculifera*, Stol., from the Cretaceous rocks of Southern India ('Mem. Geol. Survey of India,' vol. iii, pl. xv, figs. 10—15, p. 335). A comparison of these figures of Dr. Stoliczka with the very numerous specimens of *T. ventricosa* in the British Museum indicates even a closer affinity than would be looked for, judging from the single specimen of the latter the subject of our wood-engraving; the nodes upon the larger varices occasionally have all the roundness and distinctness which characterise the Indian species.

Another form allied to *T. ventricosa* is a short inflated shell, *T. Delafossei*, Leymerie, from the Cretaceous rocks of Spain ("Mém. sur un nov. type Pyrénéen," 'Mém. Soc. Géol. de France,' 2 sér., tom. iv, pl. viii, fig. 27). This latter form is even shorter and more inflated than the South African fossil; its anteal or larger varices are each fringed with a row of separate rounded tubercles; their direction is more oblique than in *T. ventricosa*. The other gigantic *Trigonia* associated with *T. Herzogii* in Southern Africa is *T. conocardiformis*, Krauss, inadequately represented in the reduced figures given by that author in the 'Nova Acta,' above quoted; it is an abnormal, depressed, and altogether very peculiar example of the *Clavellatæ*, and is only very remotely allied to any other known species of that section. Its general aspect has some resemblance to a lengthened *Pholadomya*; its hinge characters are well exposed, otherwise it might be mistaken for another genus. The *Clavellatæ*, although eminently Jurassic, are not exclusively so, as the section is represented in British Neocomian rocks by two species.

The supposed presence of *T. aliformis* in the Cretaceous rocks of India referred to by Von Buch is founded upon a memoir by Professor E. Forbes on Cretaceous fossils from the hill of Verdachellum, south-west of Pondicherry ('Geol. Trans.,' 2nd series, vol. vii, part 3, p. 151). The liberality of the Geological Society in granting to me the loan of the *Trigoniæ* from Verdachellum upon which the observations were grounded has enabled me to compare them carefully with British species; the results are as follow:—

Six specimens representing species named *T. semiculta*, *T. suborbicularis*, and *T. orientalis*, are examples of the *Glabræ*, and differ from any known European forms; apparently the two latter species should be merged in one; they are sufficiently figured in the plates which accompany the memoir.

The Indian supposed representative of *T. aliformis* is founded upon a single small example of the *Scabræ*, which, when perfect, would be about 12 lines in length, but the posteal portion is broken away, which reduces the length to $9\frac{1}{2}$ lines, in a matrix of reddish-brown, concretionary rock. A delicate valve of a small *Placunopsis* lies across the area and escutcheon, the surfaces of which, however, are not obscured by the parasitic valve. A mere first glance is sufficient to assure us that the comparison could not have been made with any British example of *T. aliformis*, and that the Indian species cannot even be arranged as a variety of that form. The figure not less than the ornamentation is distinct; it has nothing of the posteal flattening and sudden anteal inflation which is so characteristic of Parkinson's species, no separation of the costæ into two kinds, viz. the small, perpendicular posteal, contrasted with the oblique anteal, inflated mesially, with their delicate, almost evanescent crenulations. On the contrary, the contour of the surface is uniform, and the costæ are all of one kind, narrow, elevated, and straight for the most part, but having a few of the first-formed curved towards the anterior border; all are fringed with large, obtuse, prominent, irregular, and unequal nodes, but the few first-formed or umbonal rows have the little nodes or tubercles regular and bead-like; all are united to the angle of the valve, which forms a distinct, narrow-knotted ridge. The area and escutcheon are together large and concave; the area, much wider than in any British species of the *aliformis* group, has transverse costellæ conformable in number with the costæ; they are large and suddenly form a row of prominent narrow varices, which separate the area from the more depressed and concave escutcheon, across which the costellæ also pass in a more attenuated form. The whole of these features differ essentially from the corresponding parts in *T. aliformis*; and the border of elevated narrow varices separating the area and escutcheon form also a prominent distinguishing feature.

Rejecting the comparison with *T. aliformis* as presenting only remote affinities, there remains the possibility that *T. Vectiana* of the Neocomian beds (which has been confounded with *T. aliformis*) may possibly have formed the subject of comparison. As the Indian species of Forbes appears to have been unknown to the Geological Survey of India, being absent in the Cretaceous species described by Dr. Stolickza, I have been induced to give the following engravings of it, magnified one diameter.

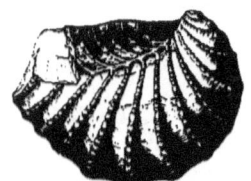 

Trigonia Forbesii, Lycett, from Verdachellum, India.

It will be perceived that the general figure differs from *T. Vectiana* so considerably that their identity as a species is quite precluded.

D'Orbigny assigns an American species named *aliformis* to his *T. limbata* ('Prodr. de Paléont.,' vol. ii, p. 240, No. 592), and also the Indian species of Forbes. To separate the latter from the species of D'Orbigny it is only necessary to direct attention to the narrow, almost linear area, the produced anterior side, and the curved costæ of *T. limbata*, as depicted in the figures given by D'Orbigny.

T. aliformis, D'Orbigny ('Pal. Fran.,' p. 143, pl. 291, figs. 1—3) is distinct from the British species; the apices are less elevated and more anterior; the shell is more produced posteally, having a distinct, plain area, which extends even to the apex; the costæ are crossed by a few widely separated sulcations; they enlarge towards the pallial border throughout all the rows; they also appear to be destitute of crenulations.

The *Lyrodon aliforme* of Goldfuss ('Petref.,' tab. 137, fig. 6, vol. ii) also differs essentially from Parkinson's species in the general figure and the ornamentation of both portions of the shell. The escutcheon has prominent transverse costellæ over its entire length, and the other surface of the valve has the costæ radiating equally and fringed with closely set, regular, small, bead-like tubercles.

The *Trigonia aliformis* of Pictet and Roux ('Grès Vert,' pl. xxxv, figs. 2 *a*, *b*) apparently does not differ materially from the species figured by Goldfuss.

The *Trigonia aliformis* of Pictet and Renevier ('Terr. Aptien de la Perte du Rhone,' pl. xiv, figs. 1 and 2) :—Fig. 1 is a Blackdown example; fig. 2, from the Rhone, is *T. Vectiana*, to which the reader is referred.

T. aliformis, Agassiz, 'Trigonies,' tab. vii, figs. 14—16; also tab. viii, fig. 12, both represent British specimens.

In fine, it results from the foregoing comparisons of analogous forms, that I have been unable to discover any example of *Trigonia aliformis* obtained at any foreign locality; also that the few figures given by foreign authors, which are correctly attributed to that species are delineations of British specimens.

TRIGONIA VECTIANA, *Lyc.* Plate XXIV, figs. 10, 10 *a*, 10 *b*, 11; Plate XXV, fig. 7.

TRIGONIA ALÆFORMIS, *Mantell.* Geology of Sussex, p. 73, No. 11, 1822.
— ? PLICATA, *Agassiz.* Trigonies, p. 33, pl. x, fig. 11, 1840. (Specimen deprived of the test.)
— ALIFORMIS, *Ibbetson* and *Forbes.* Proc. Geol. Soc., vol. iv, part ii Table to face p. 414, 1844. (Fossils of the Perna Band.)
— ALÆFORMIS, *Fitton.* Stratigraphical section from Atherfield to Black Gang Chine; Table opposite p. 289 (fossil No. 64, bed No. 45), Quart. Journ. Geol. Soc., vol. iii, No. 11, 1847.
— ALIFORMIS, *Pictet* et *Renevier.* Foss. du terr. Aptien de la perte du Rhone, et des Env. de St. Croix, 1857, pl. xiv, fig. 2 *a, b c;* exclude fig. 1, which is a Blackdown specimen of *T. aliformis.*
— — *Judd.* On the Punfield Formation, Quart. Journ. Geol. Soc., p. 220, 1871.

One of the *Scabræ* allied to *T. aliformis*, and possessing a considerable general resemblance to that species; it has been mistaken for it by several palæontologists—by Mantell, by Fitton, also by Ibbetson and Forbes, and more recently by Pictet and Renevier. In addition to possessing well-marked palæontological distinctive features, it is also separated from the other species by its stratigraphical position, which is limited to the Neocomian formation. Its diagnostic features are as follows:—Umbones considerably incurved and recurved, or not erect; the middle and anteal portions of the valves are without that sudden inflation which in *T. aliformis* contrasts so strongly with the abrupt flattening of the attenuated portion of the valve posteal to it; the costæ or varices diverge symmetrically from the angle of the valve with great regularity obliquely downwards, with a slight sinuation towards the angle of the valve and regular curvature; the summits and sides of the costæ are strongly, closely, and regularly crenulated; they therefore form a well-marked contrast to the faintly marked and unequal crenulations of *T. aliformis*, which are also limited to the summits of the costæ. The escutcheon possesses features equally distinctive; it has throughout its length a series of regular, transverse, strongly defined costellæ, which also pass across the narrow area to the divisional angle of the valve in specimens of immature growth, but in fully developed examples the posteal portion of the area is destitute of costellæ. The area and escutcheon together form a wider surface than in *T. aliformis;* the escutcheon is less distinctly bounded, its costellæ are smaller and somewhat more numerous than the costæ upon the other portion of the valve; the separation of the two portions of the surface at the angle of the valve is clearly defined, and for the most part forms a narrow divisional ridge, from the sides of which both costæ and costellæ diverge with great regularity. These features, and more especially the transverse ridges or costellæ upon the escutcheon throughout its length, afford a strong contrast to the delicate or evanescent surface-ornaments upon that portion of *T. aliformis.*

It may be mentioned, as affording an explanation of the supposed identity with *T. aliformis,* that well-preserved specimens of *T. Vectiana* in the Perna-bed at Atherfield are rare, and that the moulds of external casts at Black Gang Chine would readily be mistaken for *T. aliformis* without a careful comparison of the escutcheon in each instance.

A small example of the *Scabræ,* figured by Agassiz ('Trigonies,' p. 33, pl. x, fig. 11) under the name of *T. plicata,* appears to be nearly allied to, and may be identical with, *T. Vectiana;* but, as it is founded upon a single specimen only, and deprived of the test, it is not possible to make any satisfactory comparison with it; apparently the crenulated costæ are somewhat more numerous than in the British species. Agassiz believed that his specimen was obtained in the Portland formation (zone of the Calcaire à Ptérocères) in the environs of Besançon. My whole experience is opposed to this conclusion, neither does there appear any reason to doubt that it was obtained from the Cretaceous rocks.

The Indian *T. Forbesii,* described and figured with *T. aliformis,* has the general figure more ventricose, so that the area and escutcheon have greater breadth, the posteal portion of the shell is scarcely attenuated, the umbones are much less elevated, the costæ are without curvature, fringed with a few nodes, and are without the delicate, closely set crenulations, both upon their summits and sides, which characterise *T. Vectiana.*

The *T. aliformis* of Pictet and Renevier from the "Terrain aptien" of the South of France, above cited, represents a finely preserved example of *T. Vectiana,* much exceeding the size of specimens occurring in the Perna-bed at Atherfield, the dimensions of which are fairly represented by our figures; these are of not fully developed growth. The other figure of *T. aliformis* in the work of Renevier is an undoubted Blackdown specimen of that species.

Stratigraphical positions and Localities.—*T. Vectiana* occurs somewhat rarely in the lowest, or Perna, bed stratum of the Neocomian formation at Atherfield, Isle of Wight. At Black Gang Chine, in the same vicinity, its moulds and casts occur abundantly in ferruginous concretionary masses derived from the bed No. 45 of Dr. Fitton's elaborate and valuable stratigraphical Table above referred to; the position of this bed is nearly at the upper boundary of the Neocomian formation. Another locality is Seend, near Devizes, in a similar stratigraphical position.

TRIGONIA MEYERI, *Lyc.*, sp. nov. Plate XXIII, fig. 6.

Shell ovately trigonal, very convex anteally, attenuated and compressed posteally; umbones large, elevated, antero-mesial, pointed, much recurved; anterior side produced, its border rounded and curved with the lower border, which becomes nearly straight posteally near to the attenuated extremity; the superior border is much excavated and lengthened, terminating posteally abruptly, forming nearly a right angle with the siphonal border, which is nearly perpendicular, its height scarcely exceeding one fourth the height of the valves. The area is narrow, much curved, slightly elevated, separated from the other or pallial portion of the valve by a distinct narrow divisional angle or slight ridge; this does not rise higher than the surface of the area, which is delicately transversely lineated over its posteal half, and traversed by a distinct longitudinal mesial furrow; the anteal portion of the area is traversed transversely by a numerous series of small, closely arranged, wrinkled costellæ, which pass also without interruption across the larger escutcheon. The upper surface of the valve is almost entirely occupied by a large, concave escutcheon, which is conspicuously costellated transversely throughout its length; its breadth exceeds that of the area, from which it is separated only by a faintly elevated ridge. The larger or pallial portion of the valve has a series of about twenty-six rows of small, closely placed, rounded, and slightly crenulated costæ, all of which originate at the carinal angle of the valve and pass downwards nearly perpendicularly; the more posteal ten rows occupy the more flattened or depressed portion of the surface, they enlarge slightly near to the lower border; the other or more anteal rows form a sudden flexure about their middle portions, where they also enlarge as suddenly and pass forwards almost horizontally to the anterior border, becoming somewhat curved and attenuated at their anteal extremities. The anteal or pedal border has upon its upper half a closely set series of small, oblique, supplementary costellæ, two of which occupy each of the intercostal spaces to the number of nine spaces.

This is one of the more inflated of the *aliformis* group, the great convexity, as in that species, being limited to the anteal third of the shell; the sudden flexure of the costæ at that part and their attenuation upwards to the angle of the valve are very conspicuous features and distinguish it from all others of the *Scabræ*, including *T. aliformis*, which, with fewer costæ, has also a smaller flexure and much less abrupt; the anteal varices are also much fewer, more oblique, and more inflated.

Affinities, and Stratigraphical position.—The first few specimens examined were very imperfectly preserved; they were obtained by Mr. C. J. A. Meyer in beds of Chloritic Marl and Greensand at Dunscomb Cliffs, to the eastward of Sidmouth, and were at first regarded as dwarfed representatives of *T. abrupta*, Von Buch, figured by Coquand ('Monogr. de l'Étage Aptien de l'Espagne,' pl. xiii, figs. 4, 5), apparently from a large and fine specimen, nearly three inches in length and more than half that measurement through the united valves; it is, therefore, two and a half times larger in linear dimensions than British examples of its analogue; it also presents some differences in figure; its umbones are much more nearly erect, the anterior border is less rounded, the outline of the siphonal border is much more oblique and less short; the contrast between the anteal, horizontal, inflated costæ, and the posteal, smaller, and perpendicular ones, is much less conspicuous; but, perhaps, the most marked distinction consists in the absence of the numerous transverse costellæ upon the anteal half of the area. As the Spanish specimen is from the Aptian beds, we should scarcely expect to find identity of species in fauna so widely separated, both geographically and stratigraphically. The drawing of *T. abrupta*, Von Buch ('Pétref. recueil. en Amer. par Alex. de Humboldt et par Ch. Degenhardt,' fig. 21), to which Coquand referred his specimen, is defective in the several details which are necessary to enable us to characterise Cretaceous *Trigoniæ*; the general figure also differs so decidedly from all known European examples that, in the absence of M. Coquand's monograph I should not have ventured to allude to the South-American fossil as a form nearly allied to that in the higher beds of the Sidmouth Upper Greensand. The subovate outline especially differs; the shell appears to be without the anteal inflation and posteal attenuation common to several species of the *aliformis* group; the umbones without elevation or curvature, the straightness of the divisional angle of the valve, the absence of all character upon the area and escutcheon, and even the drawing of the costæ upon the side of the valve, cannot be accepted as delineations of the European Aptian form, and still less of the smaller British species; the curvature of the anteal costæ has little of the angularity and abruptness upon which the name is founded. The name was probably retained by Coquand in the belief that the American specimen figured by Von Buch is very defective in the several features above mentioned, and that the species is probably identical with the Spanish Aptian Trigonia. Another interesting South-American Trigonia figured by Von Buch is *T. Humboldtii*, figs. 29, 30, which has costæ radiating from the umbo over the greater portion of the surface; it has affinities with *T. divaricata*, D'Orbigny, and tends to connect the Cretaceous *Scabræ* with the living Australian section, the *Pectinidæ*.

At several localities near Sidmouth our species is associated with *T. læviuscula*, *T. sulcataria*, *T. pennata*, and *T. Vicaryana*; it occurs only within a very limited vertical range; several better preserved specimens, including the original of the figure (Plate XXIII, fig. 6), have proved its distinctness from the Aptian form. Other specimens from Chardstock are in the Museum of the Royal School of Mines; only single valves have been obtained. The internal mould is not known. Dimensions of the specimen figured.—Length 16 lines; height 13 lines; thickness through a single valve 5 lines.

Named after Mr. C. J. A. Meyer, F.G.S., whose researches in the Cretaceous rocks of the southern counties of England have contributed materially to a more exact knowledge of the Greensand formations, and of their relations to similar deposits elsewhere.

TRIGONIA ETHERIDGEI, *Lyc.*, sp. nov. Plate XXVII, figs. 1, 1 *a*, 1 *b*, 2, 3, 3 *a*.

TRIGONIA CAUDATA, *Ibbetson* and *Forbes*. Table showing distribution of Lower Greensand Fossils in the Isle of Wight, Bed No. 3 (exclude Upper Greensand), Proc. Geol. Soc., vol. iv, No. 101, p. 414, 1844.

— — *Fitton*. Stratigraphical section from Atherfield Point to Black Gang Chine (Beds 1, 2; Fossil 63). Quart. Journ. Geol. Soc., vol. iii, No. 11, p. 289, 1847.

Shell sublunate, much inflated, compressed upon the anterior face, which is very wide; umbones much elevated, erect, very large, much incurved and recurved, their apices are slightly separated when the valves are close; their recurvature is so considerable that the ligamental fissure is seen anterior to them (Plate XXVII, fig. 1 *a*, 1 *b*); the anterior border is very short and truncated, slightly curved with the lower border, which is nearly straight, and both borders are nearly of equal length; the upper border is much excavated, its posteal extremity terminating abruptly at the produced posteal portion of the valve. The escutcheon is unusually large and deeply excavated, its superior or inner border is slightly raised, it is traversed transversely by ten or eleven narrow, widely separated, slightly serrated costellæ; the outer border is elevated and rounded; its posteal portion is rendered bipartite by a well-marked mesial furrow, constituting a very narrow and distinct area; near to its caudal extremity are some transverse irregular plications. The fully developed shell has upon its sides about seventeen narrow, elevated, ridged costæ; they occur at regular distances and are very widely separated; about nine costæ originate at the anterior border and curve slightly outwards to the flexure of the valve, thence they are suddenly directed obliquely upwards to the area; the four lower costeal ridges of this anteal series are more than usually elevated at their middle portions or at the flexure of the valves, where they become broad and projecting, each forming a kind of lobe and having two or three rounded, irregular nodosities; the succeeding more posteal costæ (eight in number) are smaller, acute, narrow, and high-ridged, passing perpendicularly downwards to the lower border;

all the costæ are more or less slightly crenulated, the posteal series have their edges acute, rough, and irregularly scabrous; their lower extremities alternate at the border with those of the other valve, producing a peculiarly jagged outline. Upon well-preserved examples the anterior face of the valves discloses a considerable number of minutely knotted radiating lines, which cross the wide intercostal spaces and become evanescent near the lower curvature of the valves; they appear to be irregular in their distribution and remind us of the surfaces of the Arcas. The less wide interstitial spaces of the posteal costæ have occasionally each a single mesial perpendicular line, but more frequently these are evanescent. The escutcheon also exhibits some traces of radiating lines similar to those upon the anteal face of the shell. Young specimens do not differ from adult forms in any material degree; they are, therefore, equally distinctive when compared with contemporaneous species.

This large and remarkable species of the *Scabræ* has been mistaken for *T. caudata*, and tabulated as such in lists of Isle of Wight fossils. Compared with *T. caudata*, our species is much larger, the umbones are much more elevated and larger, the anterior side is more short and flattened, the caudal extremity is shorter; the upper border of the escutcheon is much more concave or depressed, its costellæ do not form fringing prominences as in *T. caudata*; the costæ are narrower, more elevated and acute; they are destitute of obtuse, crenulated, fringing outer borders; the presence of the inflated lobes at the anteal flexure of the valves upon four of the costæ, and of radiating knotted lines upon the anterior flattened surface, gives also distinctive features. The peculiarly narrow and scarcely defined area, destitute of distinct costellæ, also affords a contrast to the prominent area and its well-marked costellæ in *T. caudata*. From *T. aliformis* it is distinguished by the few, widely separated, acute, perpendicular posteal costæ, by the absence of the sudden inflation which characterises the anteal half of the valves, by the wide, compressed anteal surface, by the greater breadth of the united valves, and by the few large strictly transverse costellæ upon the escutcheon, contrasted with the very numerous, delicate, oblique, and almost evanescent costellæ of *T. aliformis*; generally, also, by the wide intercostal spaces and by the radiating knotted lines upon the flattened anterior surface in well-preserved specimens. The lower border is also remarkably distinct. *T. aliformis* has the extremities of the perpendicular costæ upon each valve, corresponding each to that of the opposite valve. In *T. Etheridgei* they alternate, thus producing a peculiarly jagged outline at the border.

Dimensions of an adult specimen.—Height $2\frac{1}{2}$ inches; length, horizontally, 3 inches; diameter through the united valves anteally, $2\frac{1}{2}$ inches.

Stratigraphical position and Localities.—In the Isle of Wight it occurs only in the lowest or *Perna Mulleti* beds of the Atherfield clay. The valves are usually united, and, owing to the large size and inflated figure, it is one of the most remarkable species of that peculiarly rich assemblage of Neocomian fossils. The test is thick, and specimens are usually well preserved; it occurs in some abundance. It will be observed that this is a lower geological position than pertains to *T. caudata*. The name is intended as a trifling recognition of valuable and cordial assistance and information rendered to the author by Mr. Etheridge, F.R.S., and of the interest which he has uniformly taken in the progress of this Monograph.

TRIGONIA CAUDATA, *Ag.* Plate XXVI, figs. 5, 6, 6 *a*, 6 *b*, 7.

	TRIGONIA CAUDATA, *Agassiz.*	Trigonies, p. 32, tab. vii, figs. 1—3, 11—13, 1840.
—	—	*D'Orbigny.* Pal. Fran., Terr. Crét., vol. iii, p. 133, pl. 187, 1843.
—	—	*Forbes.* Quart. Journ. Geol. Soc., vol. i, p. 244, 1845.
—	—	*Marcou.* Jura Salinois, p. 142, 1846.
—	—	*Fitton.* Strata at Atherfield, Journ. Geol. Soc., vol iii, No. 11, p. 313, 1847.
—	—	*D'Orbigny.* Prodrome de Paléont., vol. ii, p. 78, 1850.
—	—	*Buvignier.* Statist. géol. minéral. et paléont. de la Meuse, p. 473, 1853.
—	—	*Studer.* Geol. du Schweiz, vol. ii, p. 281, 1853.
—	—	*Morris.* Catal., p. 228, 1854.
—	—	*Cotteau.* Moll. foss. de l'Yonne, p. 76, 1854.
—	—	*Pictet* et *Renevier.* Pal. Suisse, p. 97, pl. xiii, figs. 1, 2, 1857.
—	—	*Raulin* et *Leymerie.* Statist. l'Yonne, p. 424, 1858.
—	—	*Desor* et *Gressly.* Étud. géol. sur le Haut Jura, p. 37, 1859.
—	—	*De Loriol.* Descr. des Anim. invert. du Haut Salins, p. 73, 1861.
—	—	*Coquand.* Monogr. Étage Aptien d'Espagne, p. 133, 1866.

Shell subcrescentic, much inflated anteally, compressed, attenuated and rostrated posteally; umbones large, elevated, sub-involute, and recurved, their apices are attenuated and contiguous when the valves are closed; the anterior side is short, forming a wide and somewhat depressed surface, arcuately curved with the lower border, which is irregularly dentated by the extremities of the costæ; it is somewhat excavated or sinuate posteally. The ligamental space is narrow, lengthened, and inter-umbonal, extending both anteally and posteally to the umbones. The escutcheon is large even when compared with other lengthened examples of the *Scabræ*, forming a deep concavity occupying almost the whole of the upper surface of the shell; its upper border is raised and dentated by the extremities of a series of costellæ, elevated, narrow, and wrinkled or scabrous, passing obliquely across its surface; immediately between the umbones the transverse costellæ pass across a groove, which forms the commencement of the area, and are united to the costæ on the other portion of the valve. The area commences as a furrow, which in its course downwards becomes bourded by gradually widening rounded elevations, traversed transversely by numerous lines or plications, which upon approaching the widening posteal extremity become large, irregular, and rugose. The rows of costæ upon the sides of the valves are numerous (24 to 27), narrow, and much elevated, passing from the area obliquely downwards and forwards, large in their middle portions and attenuated at their two extremities; nearly straight upon the anterior face, they disappear at the border, their other extremities form a slight undulation as they approach the area. The few last-formed or posteal costæ are small, narrow, nearly perpendicular, or slightly waved; their lower extremities project, forming an irregular dentated lower border; their upper extremities are united to the transverse costellæ on the area. All the costæ have narrow, fringing, obtuse, nodose elevations, more or less irregular in their size and prominence, minute and evanescent near to the extremities of the rows. The lines of growth are conspicuous on the anteal face of the shell, and in well-preserved specimens each of the intercostal spaces has a small median elevated line; frequently, however, this feature is only obscurely indicated. The typical form occurs in the Neocomian formation; delicately preserved examples are obtained in the Isle of Wight at Atherfield in the beds called "Crackers," and also in other beds and localities in a less favorable condition of preservation. Few of the specimens exceed two inches in length, measured upon the area. Internal moulds, which probably belong to *T. caudata*, are smooth, and their borders are without indentations; their apices are more erect than in specimens with the tests preserved.

British specimens differ materially from the figures of this species given in the 'Paléontologie Française' by D'Orbigny, where a specimen about two inches in length with eighteen costæ upon the sides of the valves is represented with an area which retains bounding narrow ridges and regular transverse costellæ throughout its entire length, or, in other words, the peculiarities of the immature state are continued in the more advanced stage of growth, an abnormal condition to which we find no approximation.

The specimens figured by Agassiz are all apparently immature, and are only moulds upon which some portions of the surface characters are visible. His description is limited in accordance with such unsatisfactory materials.

Stratigraphical position and Localities.—The Cracker beds of the Neocomian formation at Atherfield have yielded it rather abundantly.

Foreign Localities.—France: Bettancourt, Auxerre, Saint Saveur, Comble, Morteau. Switzerland: Neuchatel.

TRIGONIA SCABRICOLA, *Lyc.* Plate XXVII, figs. 4, 5, 5 *a*, 5 *b*.

TRIGONIA SCABRA, *Morris.* Catal., p. 229, 1854. (Non *T. scabra*, LAM.)

This large and abundant *Trigonia* of the Blackdown and Haldon Greensand has usually been referred to either *T. scabra* or *T. caudata*; to the latter species it is undoubtedly nearly allied; more especially small and immature examples of each are often difficult to distinguish. Adult forms present some differences of figure; *T. scabricola* is more inflated anteally, and its anterior face is more flattened; the umbones are larger, more produced and recurved, so that the caudal or posteal portion of the shell forms a smaller relative proportion of the whole; the costæ have their upper or posteal portions less attenuated, they are also distinctly crenulated, and the crenulations upon the costæ generally are more unequal, larger, and more irregular, thus forming a peculiarly roughened surface. The posteal portion of the narrow area in *T. caudata* forms a series of large, transverse, very irregular and prominent costellæ; in *T. scabricola* the area has the rugæ regular and inconspicuous; the wide interstitial spaces in *T. caudata* have each a plain median small costa, a feature which is altogether absent in *T. scabricola*. There is required, therefore, only a fairly represented series of each form to demonstrate their distinctness as species. The escutcheon is peculiarly large and deeply excavated, exceeding those features in *T. caudata*; its transverse costellæ are similar in character, but are more rugose, their upper extremities forming a peculiarly jagged irregular outline upon the upper border. Measurements of a specimen not of the largest dimensions:—Length upon the angle of the valve 31 lines; across the area and escutcheon 6 lines; across the pallial surface 18 lines; thickness anteally through a single valve 11¼ lines.

Two foreign allied forms of the genus require comparison and separation. The large species figured for *T. aliformis* by Goldfuss ('Petref.,' tab. 137, figs. 6, 6 *a*) has the posteal portion of the shell wider and less attenuated; the upper or attenuated portions of the costæ are without the flexure seen in *T. caudata* and *T. scrabricola*; the anteal portion of the shell is more produced and its costæ are concentric; they are, therefore, without the abrupt obliquity of *T. scabricola*.

T. plicato-costata, Nyst and Galeotti ('Bull. de l'Académie Roy. Bruxelles,' tome vii, 2 partie, p. 221, fig. 1, 1840), from limestone at Tehuacan, in Mexico, was referred by them to Jurassic strata, an error which was corrected by D'Orbigny ('Prodrome de Paléont.,' vol. ii, p. 240, No. 605), who placed the species in his Étage 22 Sénonien, or highest stage characterised by Cretaceous *Trigoniæ*. The figure, which appears to be very well drawn, has the anterior side much more produced and the posteal portion less attenuated than in *T. scabricola*; the costæ, which are similarly crenulated, are nearly concentric anteally, are slightly curved in the same direction as the smaller or posteal rows; their direction, therefore, differs materially from the corresponding parts in the British species. The escutcheon and area, on the other hand, appear to correspond very nearly with *T. scabricola*.

Stratigraphical position and Localities.—*T. scabricola* accompanies *T. aliformis* in the beds of the Blackdown and Haldon Greensand. Certain internal moulds brought to my notice by Mr. Cunnington from the Upper Greensands of Wiltshire probably belong to the same species; they are without the crenulations upon the pallial border which characterise both *T. aliformis* and *T. Vectiana*.

TRIGONIA FITTONI, *Desh.* Plate XXIII, figs. 4, 4 *a*, 4 *b*, 5.

TRIGONIA FITTONI, *Deshayes.* Leymerie, Mém. de la Soc. géol. Fr., tom. v, pl. ix, fig. 6, 1842.

— — *D'Orbigny.* Paléont. Fran., Terr. Crét., vol. iii, p. 140, pl. 290, figs. 1—5, 1843.

— — *Ib.* Prodrome de Paléont., vol. ii, p. 137, No. 243, 1850.

— — *Cotteau.* Moll. foss. de l'Yonne, p. 77, 1854.

— — *Raulin* et *Leymerie.* Statistique de l'Yonne, p. 473, 1858.

Shell ovately oblong, very convex anteally, depressed and somewhat attenuated posteally; umbones large, much elevated, slightly recurved, much incurved, their apices are in contact when the valves are closed, antero-mesial; anterior side short, its border curved elliptically with the lower border; superior border concave; siphonal border short, nearly perpendicular, rounded at its extremities. Escutcheon small and lengthened, moderately wide and concave anteally, narrow and pointed posteally; its superior border is much raised upon the anteal half of its length bounding the ligamental fissure, which is lengthened, extending both anteally and posteally to the umbonal apices. Area moderately wide, destitute of bounding carinæ, traversed by a deeply indented longitudinal mesial furrow; its surface, in common with the escutcheon, has a numerous (about 20) series of delicate, narrow, regularly papillated costellæ, which descend almost perpendicularly from the superior border; they enlarge slightly as they cross the area, and again curving upwards outwardly meet the extremities of the pallial costæ at the divisional line of the valve, forming with them acute angles; about three fifths of the length of the divisional line is occupied by the extremities of these costellæ, which extend upon the superior half of the area even to its posteal extremity; the posteal portion of the lower moiety of the area, therefore, is traversed only by delicate lines of growth. The other portion of the surface is occupied by about twenty rows of elevated, rounded, concentric, regularly papillated costæ, attenuated at both of their extremities; the last-formed seven rows curve upwards almost perpendicularly to the angle of the valve, their lower extremities forming a jagged outline at the lower border, alternating with the projecting extremities of the costæ upon the other valve. The anterior face of the shell has the wide intercostal spaces occupied by numerous small, short, horizontal, rather obscure, supplementary costellæ, two or three of which occur in each space, a feature also seen in *T. aliformis;* each of the papillæ or little nodes fringing the costæ has a little pillar or plication passing downwards, a feature which is distinct only upon the first-formed ten or twelve rows of costæ. The lines of growth are delicate, they are conspicuous upon the anterior face of the shell, and also upon the area and escutcheon. The test is thin and fragile. The internal mould has its general figure and outline so nearly resembling specimens with the test preserved that it is recognised without difficulty; its lower border has indentations produced by the extremities of the costæ; the surface is usually plain, but sometimes it has faint impressions of the costæ; the mesial furrow of the area is well defined.

The specimen figured by D'Orbigny (pl. 290, figs. 1, 2) is of larger dimensions than occurs in Britain; the drawing is not altogether free from objection; fig. 2 has the escutcheon too wide posteally; it appears to form a deep concavity and gives no sufficient indication of the considerable elevation of the superior border at the junction of the valves; the curved costellæ upon the escutcheon and area are only about half as numerous as in British specimens; these features indicate a distinct variety if they are faithfully depicted; the small supplementary costellæ upon the anterior face of the shell are not noticed either in the figures or the description.

Of the highly ornamented section of the *Scabræ* few forms can surpass in beauty and delicacy the surface of the present species; the area and escutcheon more especially present, under a magnifier, the semblance of a series of symmetrical, closely arranged bead-like rows of necklaces, curving upwards at both of their attenuated extremities. It will not readily be mistaken for any other *Trigonia*.

Proportions.—The height is equal to four fifths of the length; the diameter, through the united valves, is equal to half the length.

Stratigraphical position and Localities.—*T. Fittoni* is limited to the lower portion of the Gault, which at Folkestone has produced a considerable number of the internal moulds, to some of which the test is attached, but for the most part in portions only. Well-preserved examples, therefore, rank as the most rare fossils of the Gault.

France: in the sandy beds of Géraudet, Evry, Apothement, Macheroménil, Seignélay.

TRIGONIA PENNATA, *Sow.* Plate XXIV, figs. 4, 5.

TRIGONIA PENNATA, *Sowerby.* Min. Conch., vol. iii, p. 63, tab. 237, fig. 6, 1819.
— — *Pusch.* Polens Palæont., p. 60, 1837.
— — *Agassiz.* Trigonies, pp. 9 and 52, 1840.
— — *Morris.* Catalogue, p. 229, 1854.

Trigonia pennata and *Trigonia sulcataria* are the only known British examples of a small group of the *Scabræ*, which is represented in France by other species and by individuals the dimensions of which far surpass their British analogues; the British Museum possesses a fine series of French specimens pertaining to this group. Like certain of the Upper-Jurassic *Glabræ* they have a space, anterior to the divisional angle of the valve, devoid of the costæ, which occupy the anteal portion of the shell, but possessing a feature peculiar to this group, inasmuch as the space is occupied by a series of faintly defined perpendicular costellæ which are directed upwards, some from the pallial border, others from the posteal extremities of the costæ, to the angle of the valve; the area and escutcheon have also their transverse costellæ; the whole of this ornamentation is slightly crenulated; it may be designated the *pennata* group.

Trigonia pennata, Sow., shell ovately trigonal, convex; umbones antero-mesial; only slightly recurved, not prominent; anterior border rounded, produced; lower border lengthened, curved, its posteal extremity pointed; siphonal border sloping obliquely; escutcheon rather wide, depressed, its border nearly straight. Area wide, flattened, its surface forming a considerable angle with the other portion of the valve; its surface is traversed transversely by small rows of scabrous costellæ, which curve obliquely from the angle of the valve, pass uninterruptedly across a slight elevation which forms the upper boundary of the area and are continued across the escutcheon. The other portion of the valve has the costæ (about thirty in fully developed forms) prominent, regular, nearly horizontal, and closely arranged; posteally their extremities form nearly right angles with the lower extremities of a much smaller, more faintly defined, perpendicular series of costellæ, which terminate upwards at the divisional angle of the valve throughout its entire length; both costæ and costellæ are minutely crenulated. The posteal costellæ and those of the area diverge from the angle of the valve, as stated by Mr. Sowerby, like the rays of a feather.

D'Orbigny ('Pal. Fran.,' Terr. Crét.,' vol. iii, p. 150) made *T. pennata* a synonym of *T. sulcataria*, Lam., an error which probably originated in the very incorrect drawing of Sowerby's species in the 'Mineral Conchology,' where the apex of the valve is represented as considerably produced and recurved; there is also apparently a divisional line or ridge crossing the valve obliquely and intersecting the rows of costæ at the point where they are united to the extremities of the smaller or perpendicular costellæ. Neither of these characters exist; in fact, the artist appears in some measure to have made up the great imperfections of the type-specimen with an ideal representation. The type-specimen now in the British Museum is very defective, and contrasts strongly with the original drawing; its size agrees nearly with the smaller of our figures. Compared with *T. sulcataria*, *T. pennata* is a much smaller species, differing from the former both in the figure and ornamentation; both have the series of scabrous costæ anteally, but *T. pennata* has the general figure less inflated and more lengthened; the umbones are less elevated and less recurved, the area supplies the most striking distinctive feature; it is regularly transversely costellated almost its entire length. In Lamarck's species it is smooth, excepting near to the umbones. The latter species has the costæ shorter, more oblique, and more distantly arranged; the series of small perpendicular or posteal costæ are smaller; they occupy a larger ante-carinal space and are more faintly defined.

Dimensions. A specimen of the largest size has the length 10½ lines; height 9 lines; usually the dimensions are much less, or nearly coincide with the type-specimen of Sowerby.

Stratigraphical positions and Localities. This small and well-characterised species appears to be rare. Mr. C. J. A. Meyer has obtained it at Dunscomb Cliffs, to the eastward of Sidmouth, in Sandy Chloritic Marl; Mr. Cunnington has a fine specimen reputed to have been obtained in the vicinity of Folkestone; Mr. Vicary has collected several specimens in the pebbly bed (Upper Greensand) which overlies the Greensand at Great Haldon. The type-specimen of Mr. Sowerby, now in the British Museum, is from Teignmouth. *T. pennata* is usually accompanied by the following congeneric forms— *T. sulcataria*, *T. Vicaryana*, and *T. Meyeri*. It therefore pertains to a higher position than the Greensand at Blackdown and Haldon, or to the Upper Greensand and Chloritic Marls of South Devon.

TRIGONIA SULCATARIA, *Lam.* Plate XXVI, fig. 8.

TRIGONIA SULCATARIA, *Lamarck.* Anim. sans Vert., tom. vi, p. 92, No. 9, 1819.
— — *Defrance.* Dict. des Sc. Nat., tom. lv, p. 295, 1828.
— — *Deshayes.* Edit. *Lamarck*, tom. vi, p. 517, No. 9, 1835.
LYRODON SULCATUM, *Goldf.* Petref. Germ., vol. ii, tab. 117, fig. 7, 1836.
TRIGONIA SULCATARIA, *Agassiz.* Trigonies, p. 33, pl. xi, fig. 17, 1840.
— — *D'Orbigny.* Paléont. Fran., Terr. Crét., vol. iii, p. 150, pl. 294, figs. 5–9, 1843.
— — *D'Orbigny.* Prodrome de Paléont., vol. ii, p. 161, No. 325, 1850.

This, our second and larger species of the *pennata group*, differs from *T. pennata* in the general figure, which is much shorter posteally, larger and more inflated anteally; the umbones are more nearly mesial, more elevated, and more recurved; the costæ are somewhat fewer (about 24), larger, more distantly arranged, shorter, and directed somewhat obliquely downwards posteally; the ante-carinal space, which is occupied by the posteal perpendicular costellæ, is more depressed; it extends upwards even to the apex of the valve. The area is shorter, so that the angle of the valve has a greater curvature; its costellæ are very delicate and closely arranged; they are limited to the upper half of the area, the lower portion of which is plain; it is traversed by a distinct mesial furrow. The escutcheon is large, slightly concave, and only indistinctly separated from the area, the delicate costellæ of which traverse the escutcheon also, transversely and with increased prominence.

Agassiz placed *T. sulcataria* in the section of the *Undulatæ* probably from the angles formed by the two series of costæ; but on the other hand he placed *T. pennata* with the *Costatæ*. The present arrangement of the group with the *Scabræ* is founded upon the crenulations of the costæ and the presence of crenulated transverse costellæ, which cross both the escutcheon and area, together with the absence of bounding carinæ to the area; whereas in the *Undulatæ* the escutcheon is invariably plain. Agassiz has also made *T. sinuata*, Park., a synonym of *T. sulcataria*, 'Trigonies,' pp. 33, 34. I do not perceive even a remote resemblance in Parkinson's shell, which is well figured in his 'Organic Remains,' to the *T. sulcataria* of Lamark.

Stratigraphical position and Localities. In Britain *T. sulcataria* is rare and usually badly preserved. Mr. Meyer has procured specimens in Chloritic Marls at Dunscomb Cliffs associated with *T. pennata* and *T. Meyeri;* Mr. Vicary has also obtained it associated with *T. pennata* in a pebbly bed with Greensand fossils overlying the Greensand at Great Haldon. Hitherto no British specimen having the test preserved and its outlines perfect has come under my observation; their condition is that of well-preserved moulds of external casts which have no traces of the perpendicular plications upon the costæ nor of the lines of growth upon the ante-carinal space; the costellæ upon the area cover a larger portion of its surface than is shown in the figures of Goldfuss, of D'Orbigny, and of Agassiz; the figure given by the latter author presents differences both in the postcal or perpendicular costæ and in the area, which indicate a distinct species. It is intended to figure a second specimen of *T. sulcataria* from Great Haldon upon a future plate.

French specimens of much larger dimensions and in a fine state of preservation occur in the Greensand of Le Mans. The British Museum has a good series of examples from that locality, numbered "34,888." There are also two fine examples of an allied species named *T. Nereis*, D'Orb., 'Prodr. de Paléont.,' vol. ii, p. 162, No. 322. Of equal size to *T. sulcataria*, it differs in having the surface ornaments far more minute and delicate; it differs also from the brief description given by D'Orbigny in having the ante-carinal space similar in character to that in *T. sulcataria;* in *T. Nereis* the space is stated to be plain.

TRIGONIA SPINOSA, *Park.* Plate XXIII, fig. 10; Plate XXIV, figs. 8, 9; moulds of external casts, Plate XXVIII, figs. 1, 2.

TRIGONIA SPINOSA, *Parkinson.* Org. Rem., vol. iii, pl. xii, fig. 7, 1811.
— — *Sowerby.* Min. Conch., vol. i, tab. lxxxvi, p. 196, 1815.
— — *Pusch.* Polens Palæont., p. 60, 1837.
— PYRRHA, *D'Orbigny.* Prodrome de Paléont., vol. ii, p. 161, No. 326, 1850.
— SPINOSA, *Morris.* Catalogue, p. 229, 1854.
— — *Phillips.* Geology of Oxford, p. 439, 1871.

Shell subovate; moderately convex mesially; rather depressed near the circumference at the borders of the valves; umbones small, obtuse, little produced, and slightly recurved; anterior border produced, curved elliptically with the lower border; hinge-border slightly convex, moderately lengthened, its extremity forming an obtuse angle with the siphonal border, which is somewhat curved, its length being equal to two thirds of the hinge-border; its lower extremity is curved with the pallial border. The oblique divisional angle of the valve is well defined; from it diverge on each side a series of narrow, elevated, closely arranged, spinose costæ, which pass each other with a slightly oblique curvature across the opposite portions of the valve; those which cross the area and escutcheon are the smaller; and, in conformity with the general figure of the valve, are also the more closely arranged; the pallial costæ increase in size downwards towards the lower border, the more anteal ones curving moderately forwards; the spines upon the high-ridged pallial costæ are obtuse and erect; each has its perpendicular plication upon the sides of the narrow ridge; the spines upon the area and escutcheon are much smaller and have little prominence. The surface of the escutcheon is narrow and slightly depressed; a small elevation separates it from the surface of the area, which is somewhat concave, and destitute of any mesial furrow.

Dimensions of a large specimen. Length, measured upon the divisional angle of the valve, 25 lines; across the valve at right angles to the divisional angle, 21 lines; convexity of a single valve 5 lines.

Affinities and differences. I am unable to identify with Parkinson's species the *T. spinosa* of D'Orbigny ('Pal. Fran., Terr. Crét.,' vol. iii, p. 154, pl. 297). The latter has the general convexity much greater; the umbones are more produced, the area is more concave, and is pointed at its lower extremity; it has also a mesial furrow; the escutcheon is wider, shorter, and more concave, so that, when the valve is placed upon its side and viewed from above, the escutcheon is but partially seen. The pallial costæ have much greater curvature; they have delicate crenulations, but are without prominent obtuse spines; it does not appear to agree strictly with any British species.

The same author in his 'Prodrome de Paléontologie,' vol. ii, p. 161, separated from *T. spinosa* a supposed allied form under the name of *T. Pyrrha*, characterised by the few following words: "espèce voisine du *T. spinosa*, mais avec des côtes et des tubercles bien plus gros."

Having due regard to the erroneous figure and description of *T. spinosa* above mentioned, taken in connexion with the few words assigned to *T. Pyrrha*, an impression is conveyed that the latter form is the real representative of the species indicated by Parkinson and Sowerby.

Agassiz figured for *T. spinosa* ('Trigonies,' pl. 7, figs. 4—6) a little mould of an external cast from Upper Greensand of the Undercliff, Isle of Wight; the reader will find it figured and described in the present Monograph under the name of *Trigonia Archiaciana*, p. 140, Plate XXV, fig. 10 (mould).

Another allied form which has sometimes been mistaken for *T. spinosa* is *T. ornata*, D'Orb.; the latter species, which is limited to the Neocomian formation, will also be found figured and described in the present Monograph. Compared with *T. spinosa*, and represented by an uncompressed specimen, it will be seen to have much greater convexity; the umbones are larger and much more recurved; the divisional angle of the valve has greater prominence; the surface of the area is steeper and narrower; the escutcheon is wider and more excavated; the pallial costæ have much greater curvature; they are without obtuse spines, having their rounded surfaces only crenulated; their attenuated extremities also form a sinuation as they pass upwards to the angle of the valve; the perpendicular plications occupying the intercostal spaces are also unusually prominent.

A little *Trigonia* figured by Nilsson under the name of *T. pumila* ('Petref. Sulc.,' tab. 5, fig. 7) has been referred to by Pusch as probably representing a young specimen of *T. spinosa*; the figure does not appear to be a satisfactory representation of Parkinson's species; the small portion of the area visible indicates that its surface forms a considerable angle with the other portion of the valve; the costæ differ in being without distinct spines, and in having a considerable curvature. The general aspect agrees better with a small example of *T. Archiaciana*.

T. Lamarckii, Matheron ('Catal. des Corps org. foss. du Départ. des Bouches du Rhône,' pl. 22, figs. 5—7), possesses some general resemblance to *T. spinosa* in the outline and in the arrangement of the spinose costæ; but the convexity is more considerable; the area is more excavated; the escutcheon has greater breadth; the costæ have greater curvature and much less prominence, their spines are much smaller, more numerous, and more pointed.

The British Museum possesses the unusually fine specimen of *T. spinosa* figured in the 'Mineral Conchology,' numbered " 50,003 ;" but it is equalled in size by other examples in the same collection. Our figures represent much smaller but equally characteristic shells; the two specimens on Plate XXIV are from the Greensand of the Blackdown region; apparently it is absent in the more western outliers of the same formation at Great and Little Haldon.

The specimen figured (Plate XXIII, fig. 10) is from the Upper Greensand of the Isle of Wight; it is somewhat more lengthened and has greater convexity, but differs in no other feature; it is the sole example having the test preserved and obtained in the Upper Greensand which has come under my notice; specimens from that formation without the test, of various dimensions, are not uncommon, both in the Isle of Wight and near to Devizes; from the latter place Mr. Cunnington kindly forwarded numerous specimens. The usual condition of these external moulds is very defective. For the most part they are flattened from vertical pressure; their costæ have little prominence, and their spines are indicated only by slight irregularities upon their upper borders. It is intended to figure two of the larger of the moulds on Plate XXVIII.

Upon the whole, well-preserved specimens of *T. spinosa* rank with the rarer Testacea of the Greensand. No specimen obtained at any foreign locality has come under my observation.

TRIGONIA ORNATA, *D'Orb.* Plate XXIV, figs. 6, 7.

TRIGONIA SPINOSA, *De la Beche.* Geol. Manual, p. 287, 1832.
— — *Mantell.* Geol. South-east of England, p. 179, 1833.
— ORNATA (*spinosa*, var.), *Fitton.* Stratigraphical Section in the Isle of Wight, Quart. Jour. Geol. Soc., vol. iii, No. 11, p. 289, fossil No. 65, 1847.
— — *D'Orbigny.* Prodr. de Paléont., vol. ii, xvii, Étage, p. 106, No. 709, 1850.
— — Pal. Fran., Terr. Crét., 3, p. 136, pl. 288, figs. 5–9, 1843.
— — *Morris.* Catalogue, p. 229, 1854.
— — *Pictet.* Paléont. Suisse, vol. i, pl. xii, fig. 4, 1857.

Shell sublunate or crescentric, convex; umbones antero-mesial, prominent, obtuse, and recurved; anterior and lower borders rounded elliptically; hinge-border short, somewhat concave, sloping downwards, and forming a distinct, obtuse angle with the posteal border of the area, which is of moderate breadth and terminates downwards, forming a right angle with the lower border of the valve. In conformity with the considerable convexity of the shell the area is much curved; it is of moderate breadth; it has a slight median furrow; its inner and outer border are rendered conspicuous by the terminations of its transverse costellæ, which are large, slightly waved, and striated. The escutcheon is of moderate breadth, much excavated outwardly; its superior border is raised; it is traversed by transverse costellæ similar to those of the area; together with the area the posteal or superior surface forms a considerable angle with the other portion of the valve. The costæ are numerous (about twenty-one), large, rounded, and closely arranged; they are somewhat less numerous than the costellæ upon the area, the extremities of which they meet at the prominent dividing ridge or angle of the valve, with which they form an angle greater than a right angle; they pass downwards, nearly in a straight direction, enlarging rapidly about the middle of the valve, and then curving forwards gracefully; they curve somewhat upwards as they meet the anterior border; the last-formed seven costæ diminish in size symmetrically, and pass almost perpendicularly downwards to the lower border; the costæ have their sides regularly plicated, forming rounded elevations upon the summits of the costæ. Its nearest ally is *T. Archiaciana*, compared with which it has greater convexity; the umbones are much more prominent and are more recurved; the area has less breadth and forms a greater angle with the other portion of the valve; the costæ are different in figure; their anteal enlargement and postcal attenuation and slight sinuation upwards towards the angle of the valve are also distinctive features. The costellated, wider, and flattened area serves to separate it from *T. crenulata* and also from the little *T. Vectiana*, which is much more produced and attenuated postcally; the apices are also far more recurved.

Stratigraphical position and Localities. *T. ornata* occurs somewhat rarely in the Perna bed of the Neocomian formation at Atherfield; the valves are disunited; the test is preserved. At Hythe the Neocomian Sandstone has produced it in great profusion, for the most part indifferently preserved and flattened from pressure.

France: St. Dizier, Vassy, Aucerville, Auxerre, Perte-du-Rhône (Ain).

Note.—In the introductory portion of the present Monograph, p. 8, *T. Picteti*, Coq., is mentioned as one of the British species; the subsequent acquisition of uncompressed examples of *T. ornata*, with the test preserved, has convinced me that our specimens should be referred to the latter species. It may also be remarked that the figures of *T. ornata* given in the 'Paléontologie Française' are not good representations of British specimens, and that the figure in Pictet's work is a much nearer approximation to them.

TRIGONIA ARCHIACIANA, *D'Orb.* Plate XXIII, fig. 7; Plate XXV, fig. 10 (mould).

? TRIGONIA	PUMILA, *Nilsson.*	Petref. Suec., tab. v, fig. 7, 1827. (Young example.)
—	SPINOSA, *Sowerby.*	Geol. Trans., 2nd ser., vol. iv, pl. xiii, fig. 3, p. 338, 1836.
—	— *Agassiz.*	Trigonies, p. 30, tab. vii, figs. 4–6, 1840. (Mould.)
—	ARCHIACIANA, *D'Orbigny.*	Paléont. Fran., Terr. Crét., vol. ii, pl. 290, figs. 6, 8, 10, 1843.
—	—	*Pictet et Roux.* Descr. Moll. foss. Grès vert, pl. xxxv, fig. 4, 1847.
—	—	*D'Orbigny.* Prodrome de Paléont., vol. ii, p. 137, No. 241, 1850.
—	—	*Morris.* Catalogue, p. 228, 1854.
—	—	*Pictet et Renevier.* Paléont. Suisse, Terr. aptien de la Perte-du-Rhône et des Env. de St. Croix, p. 95, pl. xli, fig. 3, 1857.

Shell with nearly the general outline and figure of *T. ornata*, but smaller, with the umbones more pointed and less recurved; the general convexity is also less; the area is more expanded, its surface forming a smaller angle with the surface of the other portion of the valve; the costæ are elevated, but narrower and more closely arranged than in *T. ornata*; they curve with great regularity obliquely downwards and forwards, but are without the sinuation which their attenuated carinal portions form in *T. ornata*—a feature which characterises that species. The divisional line of the valve forms a distinctly elevated narrow ridge, from the sides of which the costæ and costellæ diverge, each one forming a considerable angle with its corresponding ridge; the costellæ are therefore large, but more closely arranged than in *T. ornata;* the escutcheon is smaller or narrower, the costellæ of the area pass across it in a similar manner. The intercostal perpendicular plications are small and densely arranged ; they render the upper borders of the costæ prominent and obtuse; they constitute a much less prominent feature than in the allied species *T. ornata* and *T. Vicaryana.* For comparison with *T. Upwarensis* see that species.

The foregoing description is founded upon specimens from the Greensand of Great Haldon, in which rock the test is so fragile that an entire shell is rarely obtained. Our figured example is of the largest dimensions.

The Upper Greensand of Sidmouth is also a locality for *T. Archiaciana*, where the specimens are usually ill-preserved.

A little mould figured by Agassiz ('Trigonies,' tab. 7, figs. 4—6) for *T. spinosa* is probably a small specimen of *T. Archiaciana;* it has impressions of the costæ, but the area and escutcheon are represented only by the scar of the posterior adductor muscle. The deficiencies of the specimen rendered Agassiz' description meagre and insufficient; it was obtained in the Upper Greensand of the Undercliff, Isle of Wight. Our little example (Plate XXV, fig. 10) represents a specimen obtained at the same locality and in a condition precisely similar. Specimens in a like state of preservation also occur in the Upper Greensand of Warminster. The general figure and characters of the costæ resemble *T. Archiaciana*, but a rigid scrutiny is impracticable ; there can be no doubt that it is altogether distinct from *T. spinosa.* Of this latter species additional specimens from the Upper Greensand, and deprived of the test, will be given on Plate XXVIII.

The small and insufficient figure of *T. pumila*, Nilsson, above referred to, only enables me to quote it doubtfully as probably representing a small specimen of *T. Archiaciana.* It was regarded by Pusch (' Pol. Palæont.,' p. 60) as a young example of *T. spinosa;* the costæ have their features very imperfectly represented.

French localities given by D'Orbigny are Varennes, Saulce-au-Bois, Mont Blainville Seignelay.

Trigonia Vicaryana, *Lyc.*, sp. nov. Plate XXV, figs. 8, 9.

Shell ovately elongated, convex, produced and pointed at the umbones, depressed posteally; umbones subanterior, elevated, pointed, recurved ; anterior side short, its border curved elliptically with the lower border ; superior border nearly straight, rounded posteally with the wide siphonal border. Area wide, flattened, its surface together with the escutcheon equal to about two fifths of the entire valve; it has a faintly defined, mesial, oblique depression, and is covered by a very numerous and delicate series of obliquely curved scabrous costellæ which nearly disappear upon its posteal portion; the escutcheon is of moderate breadth, separated from the area only by the border of its concave surface, and by the greater prominence of its costellæ, which are continuations of those upon the area; these form a slight angle at the border of the escutcheon and traverse it in a direction directly transverse to its superior border, which is somewhat raised. Upon the divisional line of the valve the extremities of the costellæ are united to the superior or attenuated extremities of the pallial costæ, forming with them acute angles. The rows of costæ, which are very numerous and small, are curved obliquely downwards; their upper borders form projecting obtuse nodosities, the extremities of a multitude of perpendicular regular plications or little pillars which cross the costæ. Fully developed specimens have the plications slightly waved and somewhat irregular near to the pallial border, where the extremities of the costæ become more distant. No specimen altogether entire has come to my notice.

It will be observed that the specimens figured on Plate XXV constitute two varieties, which pertain to different localities and beds of the Upper Greensand. The left-hand figure (8) is from the Chloritic Sand at Chardstock; it also occurs at Dunscomb cliffs eastward of Sidmouth, and near to Axmouth in the Chloritic marly beds; the costæ are smaller than in the other variety, and more closely arranged; they also form a much smaller angle with the costellæ upon the area; their upper borders have also less prominence. Specimens are in the collection of the Royal School of Mines, in Mr. C. J. A. Meÿer's collection, and in my own cabinet. The right-hand figure (9) is apparently less rare, and has been obtained by Mr. Vicary in a pebbly bed overlying the Greensand at Great Haldon. No specimen altogether perfect has been obtained.

This, which I arrange as the typical form, has the costæ somewhat larger anteally and more widely separated; they become much attenuated at the divisional angle of the valve, and form considerable angles with the costellæ upon the area; the intercostal perpendicular plications are also larger and more conspicuous. Some specimens are delicately silicified, and exhibit the most minute surface ornaments with great clearness. It is intended to give additional figures of the typical form in a future plate.

Affinities and Differences. Defective specimens of the variety with the smaller costæ were at first mistaken for *T. tenuisulcata*, Dujardin ('Mém. Soc. Géol. de France,' vol. ii, pl. 15, fig. 11); the examination of more satisfactory examples has led to their separation from Dujardin's species, which has the costæ more minute, more closely arranged, and peculiarly straight near to the pallial border; the area and escutcheon are also smaller and less expanded than in the British species.

T. Lamarckii, Matheron ('Catal. des Corps org. foss. du Départ. des Bouches du Rhône,' pl. 22, figs. 5—7) has affinities with our species in the general figure and the surface ornaments, excepting that the pallial costæ are much more widely separated and have greater curvature, forming much smaller angles with the costellæ upon the area; their upper borders are also sharply spined; the escutcheon is not seen upon the same figure, which proves that its surface is much depressed; the umbones are less produced and pointed than in our species.

T. crenifera, Stoliczka ('Mem. Geol. Survey of India,' vol. iii, p. 318, pl. 15, figs. 13, 13 a), from Cretaceous Rocks of Southern India, is nearly allied to our species in its surface ornaments, but differs considerably in the general figure, which is much shorter, or subquadrate; the dimensions are also much smaller.

The name is intended as an acknowledgment of important assistance afforded me by Wm. Vicary, Esq., F.G.S., in the loan of *Trigoniæ* from the Greensands of the Blackdown and Haldon districts, of which his extensive and valuable collections of Devonshire fossils have supplied ample and instructive materials.

TRIGONIA UPWARENSIS, *Lyc.*, sp. nov. Plate XXIII, figs. 8, 9.

TRIGONIA SPINOSA, *J. F. Walker.* On some Coprolite Workings in the Fens; Geological Magazine, vol. iv, p. 310, 1867.

Shell suborbicular or subovate, convex; umbones large, obtuse, little elevated, and slightly recurved; angle of the valve well defined, not prominent, much curved; hinge-border somewhat concave, sloping obliquely, its extremity forming an obtuse angle with the siphonal border, the length of which is about equal to the hinge-border; the anterior and lower borders are curved elliptically, forming an extremity somewhat pointed posteally at the junction with the siphonal border. The escutcheon is short and concave, of less breadth than the area; its surface and also that of the area is traversed by numerous, closely arranged, transverse, depressed, but rounded, scabrous costellæ; their outer extremities are in contact at the angle of the valve with the posteal attenuated extremities of the costæ; the latter are somewhat less numerous and larger than the costellæ, with which they form considerable angles. The sides of the valves have the rows of costæ (twenty-four in full-sized forms) narrow, little elevated, their summits rounded, their sides with closely placed perpendicular plications; the extremities of the rows are attenuated and curve upwards; the posteal extremities more especially curve upwards perpendicularly to the angle of the valve. Occasionally the lines of growth are strongly defined near the pallial border; they impress the costæ and obliterate the perpendicular plications; they also become conspicuous on the area.

Young specimens have less convexity and are more pointed at both of their extremities; the concentric curvature of the costæ is less considerable than in adult forms.

The specimens examined are moderately numerous, consisting of valves either united or separated; their tests are converted into carbonate of calcium, and measure from six to twelve lines across; they vary somewhat in the closeness with which the costæ are arranged and in the size of the costellæ upon the area and escutcheon; the costellæ also vary in their number; thus, some examples have the extremities of the costæ and costellæ meeting at the angle of the valve with corresponding regularity over nearly the entire length of the area; in other specimens there is no near corresponding order, the costellæ are then more numerous and smaller.

Affinities and Differences. In general aspect this small species has a considerable resemblance to *T. Archiaciana;* but the outline is more nearly circular; the umbones are less produced or more obtuse; the general convexity is greater; the costæ are more numerous and more closely arranged; they have less prominence and greater curvature anteally; their intercostal spaces are therefore narrower, and their plications have much less prominence near the pallial border. The costellæ on the area and escutcheon are also smaller and more numerous; for the most part they touch the extremities of the costæ at the angle of the valve, which does not form a distinct narrow dividing ridge as in *T. Archiaciana;* the escutcheon is shorter, wider, or more horizontal; the siphonal border is more lengthened.

Compared with *T. spinosa*, the valves are of smaller size and have greater convexity; the umbones are larger and more prominent; the area is smaller, its slope is comparatively steep, its surface forming a more considerable angle with the other portion of the valve; the differing features presented by the costæ are also very conspicuous; the smallness and close arrangement of the rows, their little prominence, their rounded upper borders, and their considerable or concentric curvature, so distinct from the high-ridged nearly straight costæ and obtuse spines of *T. spinosa*.

Dimensions. A large specimen has the length, measured upon the carina, of 12 lines; across the valve at right angles to the carina, 10 lines, of which the area occupies $3\frac{1}{2}$ lines; length of the escutcheon 7 lines; length of the siphonal border $5\frac{1}{2}$ lines; diameter through the united valves 6 lines.

The test appears to have considerable thickness; the hinge-processes are usually large for so small a species.

Stratigraphical position and Locality. All of the specimens known have occurred in a bed of phosphatic nodules in the Fen - district of Cambridgeshire. The position and organic contents of this bed have been investigated by palæontologists connected with the University of Cambridge, and have been referred by them to the Lower Greensand; the results of their observations are embodied in several descriptive notices in the 'Geological Magazine' from 1866 to 1868 inclusive, consisting of communications by Mr. J. F. Walker, Mr. H. Seeley, and Mr. H. Keeping; to the first of these gentlemen I am indebted for the loan of good illustrative specimens, and also for the subjoined note descriptive of the geology of the Upware district.[1]

[1] Upware is situated on the River Cam, about twelve miles below Cambridge; an outlier of the Coralline Oolite occurs here, and the phosphatic bed has been formed on the shore of the Coral Island, from whence a plentiful supply of calcareous matter having been obtained, will account for the vast number of Brachiopoda (about twenty-five species), many of them new, which have been found in this deposit. The phosphate bed is often divided into two or more layers, the "coprolites" in the upper one are lighter in colour, having been acted upon by water, &c.; the shells, &c., proper to the deposit are found more abundantly at the base of the bed; masses of these shells are sometimes found cemented together by calcareous matter. The phosphatic nodules are water-worn, and have been probably derived from the denudation of the Oxford and Kimmeridge clays. Bryozoa, Serpulæ, &c., occur, and some of them have in their growth followed the outline of the nodules, which shows that the nodules have existed in a hardened condition at that period. Besides these nodules and phosphatized shells derived from these clays, several fossils derived from the Coralline Oolite are found; among these are *Diadema pseudodiadema*, *Hemicidaris intermedia*, casts of *Chemnitzia*, &c. Many remains of Fishes occur in this deposit, viz. *Sphærodus gigas*, Ag.; *Gyrodus*, sp.; *Asteracanthus ornatissimus*, Ag.; *Pycnodus gigas*; *Hybodus* (spine and sphenanchis); *Psammodus reticulatus*, Ag.; *Edaphodon*. Of Reptiles—*Pliosaurus*, *Ichthyosaurus*, *Plesiosaurus*, *Dakosaurus*, and *Iguanodon*.

The fossils proper to the bed consist of a friable carbonate of calcium; among the Brachiopoda are *Terebratula Davidsoni*, *T. Fittoni*, *T. sella*, *T. prælonga*, *T. depressa*, *Rhynchonella Woodwardi*, *R. Upwarensis*, &c. Other Bivalves are not very plentiful; among them are *Opis neocomiensis*, *Cardium* sp., *Cyprina* sp., *Pecten Robinaldinus*, *P. Carteronianus*, *P. Cottaldinus*, *P. atava*, *Plicatula Carteroniana*, *Ostrea macroptera*. There are several Gasteropoda. Most of the Sponges which occur at Farringdon are found in this deposit, as *Manon macropora*, *M. porcatum*, *Verticilites anastomosans*, &c. The sections of the deposit vary in different fields. Many *Trigoniæ* were found in the lower part of the first section given.

'*Geological Magazine*,' vol. iv, p. 309, July, 1867.—*First field worked.*

	ft.	in.
Surface, black peaty soil, containing bones of Red-deer, Horse, &c.	about 1	0
Layer of light-coloured "coprolites"	„ 1	0
Sand (called by the workmen "silt")	„ 1	6
Vein of dark-coloured "coprolites"	„ 0	9
Silt	„ 1	6
Vein of dark "coprolites"	„ 1	0
Clay (not pierced).		

Section (Mr. H. Keeping), '*Geological Magazine*,' vol. v, p. 273, June, 1868.—*Another field.*

Non-fossiliferous Gault.	Lower Greensand with few fossils.
Phosphatic bed in Gault.	Lower phosphatic bed of the Lower Greensand, rich in fossils.
Gault, about one foot thick.	
Upper layer of Lower Greensand.	Pure Kimmeridge Clay.
Upper phosphatic bed.	Kimmeridge Clay mixed with Coral Rag.
	Coral Rag.

The age of the bed is the same as that of the deposits at Potton, Faringdon, and Godalming, viz. Upper Neocomian, containing fossils proper to that deposit, and fossils derived from the denudation of the Kimmeridge and Oxford Clays and of the Coral Rag.

TRIGONIA CUNNINGTONI, *Lycett*, sp. nov. Plate XXIII, fig. 11.

Shell ovately trigonal or subtrigonal, moderately convex; its outline forms nearly an equilateral triangle, with the angles rounded; the umbones are submesial, erect, and obtuse; the anterior side is short, its border is truncated or nearly straight; the lower border is gently curved; the hinge-border is nearly straight, sloping obliquely downwards, and forming an obtuse angle with the perpendicular posteal extremity of the area. The area, which is of moderate breadth, forms a considerable angle with the other portion of the valve; it is flattened or slightly convex, having numerous, closely arranged, small, depressed, curved, oblique, scabrous costellæ; there are no traces of carinæ; a simple divisional angle separates the area from the other portion of the surface. The costæ are numerous, closely arranged, depressed, and rounded; they curve obliquely downwards towards the pallial border, and are much attenuated as they approach the oblique divisional or carinal angle; they are traversed by unequal and irregularly arranged horizontal plications of growth, which also pass over the area;! each plication as it crosses the costæ forms a line of small, rounded, depressed nodes, the direction of which is horizontal or only slightly curved; fourteen rows are visible upon one imperfect specimen without including others near the umbo where the test has disappeared; the rows are more closely arranged near the lower border, but their relative distances and the prominence of the rows are extremely variable; about thirteen nodes occupy each row of longitudinal plications.

The internal mould is smooth; it is less trigonal than the test; the apices are elevated, erect, and widely separated; the dental impressions are large, and the line that bounds the area is distinct; the pallial border is deeply crenulated.

The height and the length are equal; the diameter through the united valves slightly exceeds half the height.

Our species is readily distinguished from all other known examples of the *Scabræ* by the peculiar aspect of the closely arranged, depressed, rounded costæ, with their rows of small horizontal nodes and transverse plications.

It appears to have been mistaken for *T. Constantii*, D'Orb. ('Pal. Fr.,' pl. 291, figs. 4, 5), and has been thus named in collections; the latter species is much more lengthened transversely, and more ovate; its narrow ridge-like costæ, and area destitute of costellæ, render the ornamentation altogether distinct.

Stratigraphical position and Locality. The specimen with the test partially preserved was obtained by Mr. Cunnington in the Upper Greensand of Devizes, Wilts; apparently no other example is known to have occurred at that locality. The British Museum has several fine specimens from Normandy; these are nearly destitute of the horizontal plications and nodes.

COSTATÆ. 147

The name is intended as a slight record of my extensive obligations to Mr. Cunnington in the loan *Trigoniæ* of from various formations, but more especially from the Cretaceous Rocks of Wiltshire; the number of specimens figured affords no criterion of the advantages derived from the comparisons of the very numerous forms placed at my disposal; these included every example of the genus contained in a collection very extensive and peculiarly local.

§ VII. COSTATÆ.

TRIGONIA COSTATA, *Sow.* Plate XXIX, figs. 5, 6, 7, 8, 9, 10.

TRIGONIA COSTATA, *Sowerby.* Mineral Conchology, vol. i, tab. 85, p. 195, 1815.
LYRODON COSTATUM, *Goldfuss.* Petrefacta Germaniæ, tab. 137, fig. 3 *a, b*, 1836.
TRIGONIA COSTATA, *Agassiz.* Trigonies, pl. 3, fig. 11, 1840.
— LINEOLATA, *Ibid.* Ibid., pl. 4, fig. 1-5, 1840 (young example).
— COSTATA, *Deshayes.* Traité élémentaire de Conch., pl. 32, figs. 12-14, 1849.
— — *Quenstedt.* Der Jura, tab. 60, figs. 10-12; also wood-engraving, p. 502, 1857.

Exclude the following figures of *Trigoniæ* attributed to *T. costata* :

Knorr, Versteinerungen, Supplement, tab. 5 *c*, fig. 3, 4, 1772.
Parkinson, Organic Remains, vol. 3, pl. 12, fig. 4, 1811.
Smith, Strata Identified, Cornbrash, fig. 4, 1819.
Young and Bird, Geol. Survey, pl. 8, fig. 19, 1828.
Sowerby, in Grant's Memoir on the Geology of Cutch, Geol. Trans., 2nd ser., vol. 5, pl. 21, fig. 17, 1836.
Ziethen, Petrefacta Wurtemb., tab. 137, fig. 3 *a, b*, 1838.
Bronn, Lethæa Geognostica, tab. 20, fig. 4, 1837-8.
Deshayes, Encyclop. Méthod., Supplement, pl. 238, fig. 1 *a, b*, 1836-8.
Dewalque et Chapuis, Paléont. Luxemb., pl. 25, fig. 8, 1855.
Pusch, Polens Palæont., tab. 7, figs. 1, 2, 1837.
Goldfuss, var. triangularis, Petrefacta Germaniæ, tab. 137, fig. 3 *d*, 1836-9.
Ib., ib., tab. 137, fig. 3 *c*, 1836-9.
Quenstedt, Der Jura, tab. 67, fig. 13, 1857.
Ib., tab. 45, fig. 15, 1857.

The foregoing list of authorities excludes descriptions which are not accompanied by figures.

In Britain *T. costata* comprehends two varieties, both of which occur together in the Inferior Oolite in its course through the Southern and Midland Counties of England.

The typical form, which is exemplified by a very good figure in the "Mineral Conchology" of Sowerby, pervades various beds of that formation, and also occurs rarely in the southern Cornbrash; nevertheless it is not generally an abundant fossil. Compared with the other variety, which I designate *lata*, it has great convexity; the form is more lengthened or more pointed, and produced at both of its extremities; the area has greater breadth; its surface forms a more considerable angle with the other portion of the shell, so that when a valve is laid horizontally and viewed from above, the area is only partially visible. The variety *lata*, therefore, has the area somewhat smaller, but more expanded, and the siphonal border is somewhat shorter.

The following description is intended to apply to examples of the species generally, and not as a minute delineation of any individual specimen. This is rendered necessary by the fact, that examples of the same variety from a single bed and locality are not precisely alike in their general proportions or in the lesser details of their surface ornaments; thus, although considerable differences will be found to exist between certain selected specimens, others intermediate render it difficult to arrange such forms into distinct varieties. The separation here adopted will not, therefore, in every instance appear to be well founded. Our figures will, it is trusted, enable the reader to appreciate this subdivision with greater certainty than would be effected solely by a description.

Diagnostic characters. Shell subtrigonal, very convex near the divisional angle of the valve and near the apex, rather depressed posteally; umbo prominent, pointed, incurved, and somewhat recurved; anterior side little produced, its border truncated, near the base it curves with the somewhat shorter lower border; the superior border of the escutcheon is slightly convex, its posteal extremity forms an obtuse angle with the siphonal border, which has a sinuated outline.

The escutcheon is flattened and slightly depressed; its breadth with the valves united exceeds its length; it is well circumscribed and is traversed by large obliquely diverging varices or costellæ. The area is large and flattened; it has some convexity near the marginal carina in the right valve; its superior half is somewhat concave; its breadth at the siphonal border differs considerably, occasionally it is equal to half the height of the shell; it is bounded by two plicated carinæ, but the marginal carina only has any considerable prominence; it is large; its indentations are closely placed and do not deeply impress it, they pass across the costellæ of the area uninterruptedly, giving a reticulated surface to that portion of the shell. The inner carina is broad and depressed, or in another variety nodose, its plications crossing the escutcheon as small, waved striations.

Each portion of the area has from three to five costellæ, uncertain in size and number; the costella which divides the inner from the outer portion of the area is somewhat the larger, forming a median carina; in the right valve it exists only as one of the four or five outer costellæ which are larger than in the other valve.

In common with other examples of the *Costatæ* the marginal carina of the right valve is much larger than that of the other; it overwraps and partly conceals the post-carinal groove.

The other portion of the shell has about twenty-four large, plain costæ, all of which originate at the anterior border; they are small and delicate, their borders are indented or rendered nodulous by oblique, decussating lines of growth, which are conspicuous upon the anteal portion of the shell. At the curvature of the valve in passing to the side the costæ form a considerable downward curvature; they become horizontal about the middle of the valve, and form a second slight downward curvature as they approach the marginal carina, to which their extremities are united in the right valve, but the costæ of the other valve are separated from the carina, their extremities terminating abruptly at the well-defined ante-carinal groove.

Examples of the very young shell, when only four or five lines in length, have the anteal truncation less well defined; the three carinæ upon the area are prominent, acute, and without indentations; the intercarinal costellæ are scarcely formed, or there is a single small costella in each of the intercarinal spaces.

Comparative measurements of the two varieties:

The typical form
$\begin{cases} \text{Diameter through the united valves } 2\frac{7}{8} \text{ inches.} \\ \text{Length upon the marginal carina } 3\frac{6}{10} \text{ inches.} \\ \text{Across the valve at right angles to the carina } 2\frac{3}{10} \text{ inches.} \end{cases}$

Variety *lata*
$\begin{cases} \text{Diameter through the united valves } 1\frac{5}{10} \text{ inches.} \\ \text{Length upon the marginal carina } 2\frac{8}{10} \text{ inches.} \\ \text{Across the valve at right angles to the carina } 2\frac{3}{10} \text{ inches.} \end{cases}$

A good figure of the *left* valve representing the typical form is given by Agassiz ('Trigonies,' tab. iii, fig. 12), but figure 14, which is intended as a delineation of the area of the *right* valve, has apparently been drawn by the aid of a looking-glass from a specimen of the *left* valve, and is consequently altogether incorrect.

An excellent figure of the right valve is given by Quenstedt ('Der Jura,' p. 502).

Positions and Localities. Both varieties of *T. costata* occur together in beds of Inferior Oolite at various localities in the south-western counties, as at Bradford Abbas, from whence good illustrative specimens have been kindly forwarded to me by Professor Buckman; other well-known localities are Burton Bradstock, Chideock, Half-way House Quarry, Yeovil, Dundry, &c. Throughout the range of the Cotteswold Hills one or both of its varieties occur at many places, but apparently only over small areas; the external casts are sometimes clustered in great profusion in the bed called Upper Trigonia-grit, but good specimens with the tests preserved are comparatively rare.

In the extension of the same formation through Oxfordshire, Northamptonshire, Rutlandshire, and Southern Lincolnshire, the species is comparatively rare, and in Northern Lincolnshire it is absent.

In the North Riding of Yorkshire the Inferior Oolite occurs under other and peculiar conditions, and hitherto has not revealed *T. costata*. At Blue Wyke the Dogger yields numerous small valves, which have been attributed to this species, but their condition of preservation is such as to preclude any rigid scrutiny; upon the whole I am inclined to refer them to *T. denticulata*, which occurs in some abundance in a bed of limestone higher in the series upon the same coast. *T. costata* is absent in the Cornbrash of Yorkshire, in which the *Costatæ* are represented by two other large species.

A specimen of the typical form from the Cornbrash of Closworth has been kindly forwarded to me by Colonel Mansel Pleydell; it differs in no respect from Inferior Oolite examples.

The specimens quoted by Agassiz are from the Cantons of Bâle and of Soleure; the large specimens in Quenstedt's work ('Der Jura,' p. 502) is from Ehningen.

In Southern Germany the species also occurs in the highest zone of the Lower Oolites at Ehningen, associated with various Testacea well known in the Cornbrash of Britain.

The localities, both British and foreign, assigned to *T. costata* are very numerous, but as some of them do not appear to have been founded upon trustworthy specimens or upon sufficient critical knowledge of the species, but little confidence can be reposed in such determinations. The following remarks refer to specimens which have been figured:

The figures attributed to *T. costata* in the work of Knorr, Verst., Suppl., tab. 5 *c*, figs. 3, 4, are coarsely engraved, and are scarcely trustworthy illustrations of any fossil species; they are certainly distinct from *T. costata*, but appear to agree with our *T. sculpta*, to which they are referred.

To the same species should be united the *T. costata* of the 'Encyclopédie Méthodique,' Supplement, table ccxxxviii, fig. 1 *a*, *b*.

Also the figure of *T. costata* in the 'Lethæa Geognostica' of Bronn, table xx, fig. 4.

The *T. costata* of Zieten, 'Petref. Würtemb.,' tab. cxxxvii, fig. 3 *a*, *b*, appears to agree with *T. denticulata*.

The *T. costata* of Parkinson, 'Org. Rem.,' vol. 3, tab. xii, fig. 4, is altogether untrustworthy; the costated portion of the valve may represent the typical form, but the ornamentation of the area is a mere work of invention; the same remark will also apply to the surface-ornament of the escutcheon, the outline of which is also erroneous.

The *T. costata* of Smith, 'Strata Identified,' fig. 4, is a good representation of our *T. sculpta*, var. *Rolandi*, from the Cornbrash of the southern counties, and also of Lincolnshire.

The *T. costata* of Young and Bird, 'Geol. Survey York. Coast,' tab. viii, fig. 19, is *T. Meriani*, Ag., from the Coralline Oolite of Yorkshire and of the southern counties.

The *T. costata* of Sowerby, in Grant's memoir "On the Geology of Cutch," 'Geol. Trans.,' 2 ser., vol. 5, pl. 21, fig. 17, appears to agree with our *T. elongata*, var. *lata*.

Pusch, 'Polens Paläontologie,' p. 58, tab. 7, figs. 1, 2, described and figured, as a variety of *T. costata*, a remarkable example of the section in which, as also in the *glabræ*, a diagonal space exists, anterior to the marginal carina, entirely devoid of ornamentation; this species was separated by Agassiz under the name of *T. zonata* ('Trigon.,' p. 36), and by Quenstedt as *T. interlævigata* ('Der Jura,' p. 503, tab. lxvii, fig. 8). Oppel also described it under the latter name ('Juraformation, p. 486, No. 49). Apparently it pertains to the horizon of the Cornbrash at Ehningen, Oeschingen, also near Freiburg.

A shell figured by Goldfuss under the name of *T. costata*, var. *triangularis* ('Petrefacten.,' t. cxxxvii, fig. 3 *d*), is evidently nearly allied to *T. zonata*, but is apparently distinct. The outline presents some differences in the greater height and shortness, and in the greater elevation of the escutcheon; other distinctions consist in the delicate costellæ upon the area, the small carinæ, and the more numerous and delicate costæ; it is from the Black Limestone of Lübke, the geological position of which I am unable to correlate; both of these forms are unknown in British strata.

The *T. costata* of Chapuis and Dewalque ('Foss. Ter. Second. de Luxembourg,' p. 170, pl. 25, fig. 8) represents an elegant species, which differs not only from *T. costata*, but also from every other example of the section known to me; it is remarkable for the great extent to which the anterior side is produced, so that the recurved apices of the valves are placed a little posterior to a line drawn perpendicularly through the middle of the shell; the escutcheon is remarkably large and transversely minutely costulated; the ligamental fossa is unusually lengthened; the area is very narrow, with a minutely reticulated surface, which is represented as alike upon both the valves; the bounding carinæ are small, and accord with the other delicate features of the area; the siphonal border is unusually short: altogether, the drawing differs so materially from the description given in the text as to lead to the conclusion that the latter was founded upon true examples of *T. costata*, and that by some error another costated form was substituted in the plate for the species intended to be represented.

Trigonia Moorei, Lyc. Western Australia.

Another interesting allied species, derived almost from our antipodes, is *T. Moorei*, Lyc. (Moore's memoir on "Australian Mesozoic Geology," 'Quart. Journ. Geol. Soc.,' vol. xxvi, p. 254, pl. 14, figs. 9, 10). Allied in its general aspect to *T. costata*, it differs in having the general figure more depressed; the escutcheon is unusually narrow and lengthened; the area is larger, more convex, and more expanded; it is distinctly bipartite, but has no median carina; the inner carina is slightly nodular and inconspicuous. The costæ are short and curved concentrically; anteally they approach the border almost perpendicularly; there is no distinct anteal truncation; the marginal carina of the right valve is much larger than that of the other. Numerous examples have been brought from Western Australia, but the locality and geological position have not been exactly ascertained.

TRIGONIA DENTICULATA, Ag. Plate XXIX, figs. 1, 2, 3, 4.

> TRIGONIA COSTATA, *Zieten.* Petref. Würtemb., tab. cxxxvii, fig. 3 *a*, *b*, 1838.
> — DENTICULATA, *Agassiz.* Trigonies, p. 38, tab. xi, figs. 1—3, 1840.
> — SCUTICULATA, *d'Orbigny.* Prodrome de Paléont., vol. i, p. 278, No. 314, 1850.
> — DENTICULATA, *Sharp.* Oolites of Northamptonshire, Quart. Journ. Geol. Soc., vol. xxvi, p. 388, 1870.
> — — *Phillips.* Geology of Yorkshire, vol. i, p. 250, 3rd ed., 1875.
> — — *Judd.* Mem. Geol. Surv., Rutland, &c., pp. 153, 281, 1875.

Shell smaller than *T. costata*, more ovately trigonal, and less convex; umbones prominent, pointed, much incurved and more or less recurved; anterior border produced, curved elliptically with the lower-border; hinge-border straight, sloping obliquely; its posteal extremity forms an obtuse angle with the siphonal border of the area. The area is wide and flattened, the plane of its surface forming a considerable angle with the surface of the other portion of the valve; it is bounded by two well-marked denticulated carinæ, having also in the left valve occasionally a small median carina or a costella somewhat larger than the others and separating the area into two portions, the superior one of which is depressed and concave; the intercarinal spaces have numerous small denticulated costellæ which vary much in their prominence in different specimens; in the right valve the costellæ are fewer and more irregular and unequal; there is a median groove but no distinct median carina; the marginal carina is prominent, rounded, and denticulated even to the apex; the escutcheon is lengthened, moderately wide, and slightly depressed; its superior border is somewhat elevated, its ornamentation consists of very small diverging delicately serrated plications. The other portion of the valve has the costæ differing much in numbers, narrow, usually numerous, horizontal, curved upwards to form a slight undulation anteally, so that all the costæ terminate at the anterior border. Specimens differ in the length, measured in the direction of the costæ, and also in the general convexity.

Examples from Cloughton have not uncommonly the epidermal tegument preserved over a considerable portion of the surface; the lines of granules are large and closely arranged; the matrix of soft shale appears to be the cause of this favorable condition of preservation. The lines of growth are peculiarly delicate and densely arranged.

This is an elegant and moderate-sized species of the *Costatæ*; specimens differing considerably in their general outline, and less so in the prominence of their carinæ and intercarinal costellæ; the latter are never large, usually delicate or minutely denticulated. Commonly there is no median carina excepting in the very young shell which usually has the carinæ and costellæ strongly defined; the area is also more concave.

The acquisition of numerous well-preserved valves from the grey limestone (Inferior Oolite) of Cloughton, near Scarborough, has enabled me to compare and separate from them without difficulty, a small, more narrow, costated form which occurs rather abundantly in the Great Oolite of South Lincolnshire; the valves have usually suffered compression and are rarely well preserved; a specimen in unusually good condition is figured, Pl. XXIX, fig. 4. The costæ are usually smaller and more numerous, the escutcheon more narrow, and the hinge-border shorter than in the Inferior Oolite specimens of *T. denticulata*; they are equally distinct from other recognised species: upon the whole it seems proper to arrange them as a variety of *T. denticulata*.

Affinities and Differences. Agassiz described *T. denticulata* from a single specimen, and expressed his indecision whether to regard it as a distinct species or only as a variety of *T. monilifera*; his figures of each of these species represent a single example of immature form in which the characteristic features are but slightly developed, and *T. denticulata*, although figured from a specimen in a fine condition of preservation, possesses but little of the aspect exhibited in specimens of more advanced growth, which have less general convexity, less prominence in their carinæ; and their areas are less concave.

T. monilifera, a much larger species, has its surface-ornaments altogether more prominent; its costæ are larger and more distantly arranged; the escutcheon more especially has its surface-ornaments very distinct.

T. pullus, a small species abundant in the Lower Oolites of Gloucestershire and Wiltshire, has larger costæ; the surface-ornaments of the area are coarse and conspicuous; the escutcheon is also especially distinct.

T. sculpta, including its varieties, has greater convexity, the area much more coarsely and prominently sculptured; the simple flexure upwards of their costæ, anteally, contrasts with the undulation in *T. denticulata*.

Trigonia costata differs from *T. denticulata* in the general form, which is more trigonal, truncated, and erect; in the conspicuous truncation of the anterior border; in the peculiar undulation of the costæ; in the more prominent area with its larger reticulations; it has also larger carinæ, with more conspicuous indentations.

Positions and Localities. *T. denticulata* appears to have a considerable and unusual extent of stratigraphical range, if I am correct in placing with this species costated forms nearly allied to each other, which occur in several widely separated horizons of the Lower Oolites. Possessing little prominence in their characters as species, they have nevertheless much general resemblance, and are incapable of being clearly separated; so that, as compared with other forms, they may be distinguished from them chiefly by negative characters only. The partial indecision which attaches to certain supposed examples of

T. denticulata has not resulted from any insufficient examination or lack of materials, unless indeed it may be of specimens which are required to be exceptionally well preserved to enable us to estimate fairly the amount of variability which they possess. It would have been easy to have increased the number of figures of such specimens upon our plates; the practical utility, however, of this would have been doubtful, and I content myself with offering the present explanation, together with the following brief statement of geological positions which have come under my observation.

A fine specimen was obtained by Mr. Witchell in the highest bed of Supra-liassic Sandstone, at Haresfield Hill, near Gloucester. I obtained the species in the same position, and accompanied by *T. formosa*, in the celebrated Ammonite-bed at Frocester Hill. In the Inferior Oolite of the Cotteswold Hills it is comparatively rare; the only specimens known to me were from the hard limestone of the Upper Trigonia-grit at Rodborough Hill. In the same formation through the midland and northern counties it appears to be a more common species. Upon the coast of Yorkshire at Blue Wyke the Dogger has numerous ill-preserved costated forms, and also the Millepore-bed upon the same coast, in a higher position, which should probably be referred to it, but hitherto only doubtfully. The grey limestone and shale near to Cloughton, higher in position, has produced numerous examples of different stages of growth, delicately preserved in a thinly laminated soft shaly bed. Apparently also the species may be tabulated with the Kelloway Rock, to the southward of Scarborough at Cayton Bay, but valves are rare and ill preserved; a specimen in my cabinet with the valves in position and free from compression offers no distinction when compared with Inferior Oolite specimens. The small, supposed variety from the Great Oolite of South Lincolnshire, having the general figure somewhat shorter, and the habit gregarious, has been already noticed (p. 153).

TRIGONIA ELONGATA, *Sow*.   Plate XXX.   The typical form, figs. 3, 3 *a*, 3 *b*, 6.
— — *ib.*, var. *angustata*, Lyc.   Plate XXX, figs. 1, 1 *a*, 2.
— — *ib.*, var. *lata*, Lyc.   Plate XXX, figs. 4, 5.

Trigonia elongata of various authors; for figures refer to the following works:

TRIGONIA ELONGATA, *Sowerby*.   Min. Conch., tab. cccexxxi, figs. 1, 2 (exclude fig. 3, a distinct variety from France), 1825.
— COSTATA, *var.*   Sowerby in memoir by Grant on the Geology of Cutch, Trans. Geol. Soc., 2nd series, vol. v, pl. xxi, fig. 16, 1837.
— ELONGATA, *Damon*.   Geo. of Weymouth, Sup., pl. ii, figs. 1, 2, 1860.
— — *Lycett*.   Suppl. Monograph Great Oolite Mollusca, Pal. Soc. for 1861, p. 46, tab. xxxix, fig. 6, 1863.

The typical form ovately trigonal, short, very convex at the position of the marginal carina; umbones elevated, pointed, much arched inwards, and somewhat recurved; anterior side short, its border truncated, lengthened, depressed at the junction of the valves, its lower portion curved elliptically with the lower border, which is short and nearly straight; hinge-border very convex and short, forming a considerable angle with the siphonal border, which is equal to it in length and is excavated at its upper or anal portion. Escutcheon raised, convex, cordate; its breadth in the united valves is equal to three fourths of its length; it is well circumscribed by a prominent indented inner carina, and has a series of large, closely arranged, obliquely diverging, dentated, but depressed costellæ. Area very large; together with the escutcheon it is equal in size to the other portion of the valve, with the surface of which it forms nearly a right angle; its greatest breadth measured upon the siphonal border exceeds half the height of the entire valve; it is bounded outwardly by a large, deeply indented, marginal carina; a small but well-defined median carina divides it into two nearly equal portions; the superior portion is depressed and concave; it has a numerous series of minute, delicate, oblique, reticulated costellæ; the other or outer portion of the area has in the right valve only one or two large indented costellæ; the median carina in its lower portion usually divides into two similar costellæ; the lower portion of the other valve has four or five costellæ. The marginal carina of the left valve somewhat overwraps the ante-carinal groove; in the right valve the post-carinal groove is conspicuous, and the marginal carina is much larger than that of the other valve. The transverse striations upon the costellæ of the area are minute and delicate—a feature which affords a contrast to the more deeply sculptured indentations upon some other species of the *Costatæ*. The other portion of the surface has the costæ large, elevated, short, and only slightly oblique in their general direction; anteally they have a small, sudden undulation at the curvature of the valve, and become attenuated near the border; their number in adult forms varies from eighteen to twenty-seven. In the left valve their posteal extremities end suddenly at the border of the ante-carinal groove, where each forms a slight enlargement; in the right valve they pass onwards and are united to the marginal carina.

Dimensions of an adult specimen.—Height 30 lines; diameter of a valve at right angles to the marginal carina 21 lines; across the area of the united valves 21 lines; length of the escutcheon 15 lines; its breadth 9 lines.

There is great uniformity in the surface-ornaments in specimens of different states of development, belonging to the typical form; the area and escutcheon more especially are almost without variation, and differ only in the convexity of the escutcheon, thus rendering the hinge-border either horizontal or oblique. Specimens with the valves united often have them perfectly closed by the oblique opposition of the extremities of the marginal carina, thus indicating the exertion of muscular power when they were overwhelmed by a muddy current unfitted to be introduced into the gills by the incurrent orifice.

Altogether the characters of the *Costatæ* are more prominently developed in the typical form of *T. elongata* than in any other British example of the section; the elevated cushion-like escutcheon and the considerable concavity formed by the upper division of the area separate it both from its varieties and from other allied forms.

Position and Localities. It is not an uncommon fossil in the Oxford Clay of the southern counties of England; numerous and fine examples have long been obtained in the Backwater to the rearward of the town of Weymouth. The figures 1 and 2 of pl. ccccxxxi in Sowerby's 'Mineral Conchology' are good representations of the right and left valves from that locality; apparently figure 3, which is a French specimen, should be united to *T. cardissa*, Ag.

Variety Angustata. A very narrow form, lengthened perpendicularly, and having considerable convexity near the umbones, is depicted in Plate XXX, figs. 1, 1 *a*, 2. The costæ are numerous, more closely arranged than in the typical form, short, and nearly horizontal, excepting upon the anteal face of the valve, where they have a slight horizontal undulation. The marginal carina is comparatively inconspicuous, with small, numerous, transverse indentations. The surface of the area is similar to that of the typical form, excepting that it is not concave. The escutcheon is lengthened, sloping obliquely downwards; it has less breadth than in the typical form; its upper border is convex. Our figures represent specimens of the largest dimensions.

Position and Localities. The variety *Angustata* appears to be limited to the Cornbrash of the north of England. My few specimens are from the vicinity of Scarborough, where it has occurred only rarely; the narrow form, abrupt truncation of the lengthened anterior border, and short, horizontal costæ will usually separate it from another larger and more common variety in the same bed, or *Macrocephalus-zone* of Quenstedt and Oppel.

Variety Lata. This, the largest of the *elongata* group, is almost limited to the Cornbrash, an occasional badly preserved specimen having been obtained in the lowest bed of Kelloway Rock at the same Yorkshire locality. It is moderately abundant, occurring very rarely with the valves united; for the most part it is ill preserved, especially the surfaces of the area and escutcheon; our figures, Plate XXX, figs. 4, 5, appear to illustrate it sufficiently. The general convexity is considerable, but scarcely equals that of the typical form.

There is much variability in the proportions of the general figure; usually the area has less breadth, and is more elevated, than in the typical form; its surface forms a smaller angle with that of the other portion of the valve; its upper or inner division is more flattened, but has some depression; and the median carina is distinct. The costated portion of the valves varies in breadth and in the distinction of its anteal truncation; the costæ are large, their general direction is oblique, and they have an horizontal undulation upon the anteal surface. The marginal carina in each valve is large, but less prominent than in the typical form. The escutcheon is large and usually flattened; it slopes obliquely downwards; its length is much greater than that of the siphonal border; its surface has rugose, irregular, oblique, depressed, large-knotted costellæ; they appear to be variable in character. The interior of the valves have the dental hinge-processes unusually large and prominent.

Positions and Localities. This large variety has been obtained only in the Cornbrash of the north of England; the large valve, fig. 4, is from Southern Lincolnshire, the others are from the vicinity of Scarborough; occasionally specimens in the shortness of their costated surfaces approach to the variety *angustata*, but usually the two forms are sufficiently distinct.

The lines of growth are conspicuous upon well-preserved examples of all the varieties; when they are of fully developed growth the lines replace all the surface-ornaments.

To the Weymouth or typical form apparently should be assigned a *Trigonia*, which occurs in the Elsworth Rock of Cambridgeshire, examples of which have been forwarded to me by Mr. J. F. Walker; their condition of preservation is indifferent.

Trigonia cardissa, Agassiz, so well delineated in the work of that author ('Trigonies,' tab. xi, figs. 4—7), should be arranged as distinct from *T. elongata*. There is much general neatness in the surface-ornaments; the escutcheon is depressed; the marginal carina is comparatively small; the costæ are narrow, somewhat oblique, and curved almost perpendicularly upwards upon the anterior face of the shell, which forms a considerable excavation; this last feature in the costæ separates it decisively from the British group allied to it. Agassiz did not ascertain the stratigraphical position of *T. cardissa*; both Quenstedt and Oppel refer it to the Kelloway Rock of France and Switzerland. D'Orbigny (' Prodrome de Paléont.,' vol. i, p. 338, No. 161) makes *T. cardissa* a synonym of *T. elongata*, but excludes figures 1 and 2 of the 'Mineral Conchology,' which are Weymouth specimens, and unites them to the Neocomian *T. carinata* of Agassiz. These arrangements were made in the absence of sufficient knowledge of British species. *T. cardissa* is not known as a British species.

TRIGONIA SCULPTA, *Lyc.* Plate XXXIV, figs. 1, 2, 2 *a*.
— — *ib.*, var. *Cheltensis*, fig. 3.
— — *ib.*, var. *Rolandi*, fig. 4.

TRIGONIA COSTATA, *Knorr.* Versteinerungen, Supplement, tab. v *c*, figs. 3, 4, 1772.
— — Smith. Strata Identified, Cornbrash, fig. 4, 1816. (Var. *Rolandi*, Cross.)
— — Deshayes. Encycl. Méthod., Suppl., tab. ccxxxviii, fig. 1, *a*, *b*, 1836, 1838.

Lyriodon costatum, *Bronn.* Lethæa Geognostica, tab. xx, fig. 4, 1837, 1838.
Trigonia sculpta, *Lycett.* Handbook Cotteswold Hills, p. 65, 1857.
— — *Sharp.* Oolites of Northamptonshire, Quart. Journ. Geol. Soc., vol. xxix, p. 293, 1873.
— Rolandi (*Lyc.*), *Cross.* Geol. of N. W. Lincolnshire, Quart. Journ. Geol. Soc., vol. xxxi, p. 125, 1875. (Var. of *T. sculpta*).
— sculpta, *Judd.* Mem. Geol. Survey, Rutland, &c., p. 281, 1875.

Shell subovate or ovately oblong, moderately convex; umbones prominent, pointed, subanterior, and slightly recurved; anterior side short, its border curved elliptically with the lower border; superior border straight, lengthened, forming an obtuse angle with the siphonal, the length of which it exceeds by one fourth. The escutcheon is lengthened, flattened, and depressed; it has some oblique irregular plications which take the direction of the lines of growth. The area has some convexity, more especially in the right valve; its greatest breadth is somewhat less than one third the breadth of the entire valve; it is rendered conspicuously bipartite by the considerable depression of the superior half; it is bounded by two deeply dentated carinæ; the intercarinal costellæ are few, large, and somewhat irregular; all are coarsely denticulated and in some specimens the first costella of the lower or outer half is slightly larger than the others, forming a median carina, a feature which is not distinct in the right valve, which has the lower half of the area more elevated and its costellæ larger. The marginal carina is large in both the valves and its denticulations are very prominent. The costæ, about twenty-seven in fully developed forms, are curved obliquely or subconcentric, are somewhat narrow and flattened, with little elevation; anteriorly their extremities are simply curved upwards; their posteal extremities approach the marginal carina nearly at right angles. In the right valve the few last-formed costæ have frequently some irregularity and less prominence, or become imperfect.

The foregoing description applies to the larger or typical form, a species as large as *T. costata*, from which it differs in some important features. The general figure is less trigonal; it has less convexity at the angle of the valve; the umbones are more pointed and terminal; the anterior border, although little produced, has nothing of the truncation of the other species; the area is somewhat less wide; its surface-ornaments, together with those of the bounding carinæ, are much larger or more coarsely sculptured; the costæ are curved obliquely, having a simple curvature upwards towards the anterior border; they are therefore destitute of the anteal undulation and slight double flexure which characterise those features of *T. costata*. The test is thick and the hinge-processes are so large that they occupy nearly one third of the interior of the shell.

Positions and Localities. *T. sculpta* has occurred rarely in the highest or Ammonite-bed of the Supra-liassic Sands at Haresfield Hill, near Gloucester; its more common position is the Gryphite-grit or Lower Trigonia-grit of the Cotteswold Hills, near Stroud and Cheltenham, where it has occurred abundantly; other localities are Dundry

Hill, and the Inferior Oolite of Oxfordshire and Northamptonshire; Mr. Sharp has also collected it in the Lincolnshire Limestone of Tinkler's Quarry near Shamford.

A distinct and smaller variety (*Cheltensis*) occurs in the Cotteswold Hills to the eastward of Cheltenham; the general outline agrees with the typical form, but the valves have somewhat less convexity and are less massive; the costæ are much smaller and more closely arranged; the area and escutcheon possess all the strongly marked characters which distinguish the species. Plate XXXIV, figure 3, exemplifies this variety.

A variety designated *Rolandi* in Mr. Cross's memoir, Plate XXXIV, fig. 4, must also be arranged with *T. sculpta;* it appears to be limited stratigraphically to the upper division of the Great Oolite formation, including the Forest-marble and Cornbrash. It was figured by the venerated author of "Strata Identified," at p. 65 of that work, as a characteristic fossil of the Cornbrash. Its surface-ornaments agree closely with those of the typical form, from which it differs in the lesser breadth of the costated portion of the valve, so that the general figure is shorter, and the area, which is very wide, occupies a much larger proportion of the surface; the carinæ and intermediate costellæ, with their denticulations, possess all the prominence which characterises the two other varieties, and these features are conspicuous even in the smaller specimens. Our figure, Plate XXXIV, fig. 4, represents a specimen of medium size. This variety has occurred at several localities in Wiltshire, Oxfordshire, and Northamptonshire, and also at Appleby, North-Western Lincolnshire; it appears to be somewhat rare.

Affinities.—*The Lyriodon simile* of Bronn ('Lethæa,' tab. xx, fig. 3), afterwards figured by Agassiz under the name of *Trigonia similis* ('Trigon.,' tab. ii, figs. 18—21), also by Quenstedt ('Der Jura,' tab. xlv, fig. 15) under the name of *T. costata*, has affinities with *T. sculpta* in the general figure of the shell and more especially in the costæ; the coarsely sculptured area also possesses some resemblance. It differs in the general uniformity of the area, which is almost destitute of a median carina, and in the much greater angle which the surface of the area forms with the other portion of the surface, from the lesser convexity of the shell; the dental processes are also smaller and less massive, occupying a smaller portion of the interior of the shell.

The Australian *T. Moorei*, Lyc., figured with the description of *T. costata* (p. 151), resembles *T. sculpta* in the general figure and in the costæ; the surface-ornaments of the area, including the carinæ, are, however, much less prominent, the escutcheon is much narrower, imparting a greater depression to that portion of the shell.

The figures named *T. costata* in the 'Versteinerungen' of Knorr and in the 'Encyclopédie Méthodique,' appear to have been drawn from specimens of *T. sculpta;* they are very coarsely engraved, and the surface-ornaments present features apparently much exaggerated, even when compared with the deeply indented sculpture of that species.

TRIGONIA TENUICOSTA, *Lyc.* Plate XXXIII, figs. 7, 8, 9, 9 *a*.

> TRIGONIA TENUICOSTA, *Lycett.* Trigonias from Inf. Ool. of the Cotteswolds, Proc. Cott. Nat. Club, vol. i, p. 252, pl. ix, fig. 4, 1853.
> — — *Morris.* Catalogue, p. 229, 1854.
> — — *Lycett.* Cotteswold Hills Handbook, p. 64, 1857.
> — — *Judd.* Mem. Geol. Surv., Rutland, &c., pp. 153, 170, 281, 1875.

Shell ovately trigonal, very convex; umbones elevated, acute, arched inwards, and recurved; anterior side very short, its border truncated almost perpendicularly, and slightly excavated beneath the umbones; inferior border short, curved elliptically; hinge-border sloping obliquely, and forming an obtuse angle with the siphonal border, which is nearly perpendicular and equal in length to the hinge-border. Area large, concave; its surface forming nearly a right angle with the costated portion of the valve; it is rendered unequally bipartite by a minute but distinct median carina in each valve; the superior or inner portion is much depressed and concave; the entire area has numerous delicate oblique intercarinal costellæ, and is bounded by small, minutely indented, distinctly elevated carinæ. The marginal carina has its transverse plications very narrow, numerous, and nearly regular; their number is equal to thrice those of the costæ. The escutcheon is wide, heart-shaped, with the valves in contact, and slightly depressed; its superior border is convex; its surface is occupied by densely arranged oblique lines of minute granulated lineations. The other portion of the surface has the costæ, about twenty-eight in number, narrow, elevated, nearly horizontal, curving upwards anteally, and there forming a sudden undulation, their attenuated extremities meeting the anterior border horizontally. The lines of growth are minute.

The hinge-processes are large, and project considerably, in common with others of the *costatæ* which have much umbonal convexity.

Dimensions of the larger of the specimens figured.
Length measured upon the marginal carina . 36 lines.
Across the valve at right angles to the carina . 27 ,,
Breadth across the area . . 9 ,,
Across the escutcheon of a single valve . 3¼ ,,
Thickness through a single valve . 9 ,,

The diagnostic characters may be summarised as follows:
Considerable convexity of the valves.
Narrow elevated figure, and prominent umbones.
Anteal truncation.

Wide concave area and escutcheon.
Delicately sculptured small carinæ and intercarinal costellæ.
Small horizontal costæ with their anteal undulation.

During many years only two examples of this form, from the Cotteswold Hills, have come under my notice; and, in the absence of all other information, frequent comparisons with Inferior Oolite examples of the *Costatæ* were made in the expectation that connecting forms might be found tending to unite it with them, but without result. At length five examples of *T. tenuicosta* were placed in the British Museum from the Inferior Oolite of Bradford Abbas; subsequently various specimens in differing conditions of preservation were kindly forwarded to me by Professor Buckman, from the same locality; Colonel Mansel Pleydell has also contributed a small specimen obtained by him in the Inferior Oolite at Walditch near Bridport. Comparisons of these materials have removed all doubts of their distinctness from others of the same section, and justified the separation which had been claimed by me for it in the year 1853.

Position and Localities. At Walditch, two miles from Bridport, the Inferior Oolite is seen to rest upon the Midford or Supra-liassic Sands. At Bradford Abbas, *T. tenuicosta* occurs in a single bed from three to five feet thick, termed by Professor Buckman the *Cephalopod-bed*, from the very numerous and finely preserved species of Inferior Oolite Ammonites which it has produced; it has also yielded a profusion of other Molluscan forms; the associated *Trigoniæ* consist of two varieties of *T. costata*, two varieties of *T. striata*, a variety of *T. formosa*, also *T. bella*, which is the next species described. The Cotteswold examples of *T. tenuicosta* were obtained in the Gryphite-grit of Inferior Oolite at Rodborough Hill, associated with a multitude of valves of Conchifera, including *Trigonia sculpta*, *T. formosa*, *T. Phillipsii*, and *T. hemisphærica*; the two latter species very rarely. At the same locality, by passing upwards some twenty feet, a hard shelly bed called Upper Trigonia-grit is attained, abounding in fossils which are for the most part altogether distinct from those of the lower shelly bed; the *Trigoniæ*, which are also distinct, consist of the following species: *T. costata* (two varieties), *T. signata*, *T. producta*, *T. duplicata*, *T. angulata*, *T. V-costata*, *T. gemmata*, and *T. denticulata*. Both beds are, as a rule, destitute of Ammonites, excepting that the upper bed has rarely been found to contain a specimen of *A. Parkinsoni*. The associations of *Trigoniæ* here enumerated apply to beds of Inferior Oolite in the Cotteswold Hills; their dissimilarity to the *Trigoniæ* of the same formation in the Somersetshire and Dorsetshire district is remarkable, more especially considering the small space by which they are separated.

A nearly allied and remarkable form of the *Costatæ* occurs in the rich fossiliferous bed of Inferior Oolite in the vicinity of Bayeaux; the general figure differs only slightly; it is apparently even shorter and more inflated; the anteal truncation is somewhat less decided. The most striking peculiarity consists in the presence of a minute row of regular beadlike papillæ upon the edges of the costæ, more especially of their posteal portions; these close-set papillæ have each also a slight depression upon its middle portion; the small transverse plications upon the marginal carina have also each a row of similar, more minute, papillary prominences. Apparently this ornamented surface is rarely preserved; my specimen has it only upon the right valve, and it is not distinguishable upon examples in the British Museum. I propose to designate this species *T. fimbriata*. *T. granigera*, Cont., from Upper Jurassic strata near Berne (Calcaire à Corbis), has fringing papillæ upon its costæ, but less regular and distinct; its costæ are smaller and more numerous; the general figure is also very different, with much less convexity.[1]

TRIGONIA BELLA, *Lyc.*, sp. nov. Plate XXXII, figs. 6, 7, 8, 8 *a*.

Founded upon fine examples of shells in different stages of growth, this species is found to possess little variability in its figure and none in the ornaments of its surface; young examples have somewhat less convexity and the figure is more lengthened, as exemplified, Plate XXXII, figs. 8, 8 *a*. Upon the whole the size is smaller than in several of the larger species of the section. Its more salient features consist in the unusually great breadth and prominence of the area, contrasted with the comparatively narrow costated portion of the shell; hence it follows that the posteal or siphonal border of the area has unusually great length, even exceeding that of the escutcheon—a feature which is not observed in any other British example of the *Costatæ*.

Diagnostic characters. Shell convex mesially, much produced and pointed at its umbonal extremity, which is only slightly, or sometimes not at all recurved. Escutcheon narrow, depressed, and excavated, so that no portion of it is seen when a valve is laid horizontally upon its borders and viewed from above; its length exceeds twice its breadth in the united valves; its borders are well circumscribed by the inner carinæ, which form an elevated ridge on each side, fringed with large obtuse nodes. The surface of the escutcheon has a numerous series of very delicate, diverging, slightly indented costellæ, which are remarkable for their distinctness and minuteness.

The area is divided into two nearly equal spaces by an unusually large, elevated, and nodose median carina; the inner or anal space has a considerable and unusual amount of concavity in both the valves; its costellæ, eight or nine in number, are very irregularly knotted or indented; the lower or outer space is more flattened, but also more elevated, having about eight intercarinal costellæ in the left valve; the right valve has only three or four larger costellæ, and its surface is more elevated. The marginal carina is

[1] For a detailed and instructive paper on the Inferior Oolite as exhibited at Bradford Abbas and the vicinity, see the 'Somersetshire Archæological and Natural History Society's Proceedings,' 1874, vol. xx, by J. Buckman, F.G.S., &c.

large, prominent, nearly straight, with deeply indented plications throughout its length. The other portion of the shell has the costæ (28 or 29) moderately elevated, narrow at their upper borders, separated by wider spaces, very oblique in their direction, and have little curvature; when the upper border of a valve is placed in a horizontal position, the costæ have their general direction nearly parallel with it, excepting near to the anterior border, where they are attenuated and curve upwards; they are therefore without the anteal undulation seen in *T. costata, T. tenuicosta,* and others of the same section; posteriorly they terminate abruptly at the strongly defined antecarinal groove of the left valve; they are united to the carina of the other valve. The lines of growth are conspicuous; they decussate and indent the anteal portions of the costæ.

Affinities and Differences. To separate it from *T. costata* it is only necessary to compare the general figure and proportions of the several parts of the valves above described, which will be found to be altogether dissimilar; the escutcheon, small, excavated, with very delicate costellæ, would alone be sufficient to exemplify its distinctness.

T. tenuicosta, another allied species, has the umbones much more narrow and produced; the general convexity is greater, the escutcheon is much wider, the area is more excavated, and its carinæ are small in conformity with the very delicate intercarinal costellæ. From others of the *costatæ*, the large area, the unusual prominence of the median carina, and the great length of the siphonal border afford differences sufficiently conspicuous. *T. carinata* alone has the costæ more oblique, but in other respects is only remotely allied to it.

Dimensions of two specimens in my cabinet:	lines.	lines.
Length upon the marginal carina	42	29
Across the valves at right angles to the carina	33	22
Breadth of the area	13	11
Length of the escutcheon	17	14
Breadth of the escutcheon in the united valves	7	6½
Length of the siphonal border	20	17
Convexity of a single valve	10	9

Position and Locality. This well-characterised and remarkable species of the *Costatæ* has been hitherto obtained only in the ferruginous pisolite or Cephalopod-bed at Bradford Abbas. Apparently *T. bella* is somewhat rare; and, like its congeneric associate, *T. striata,* it has not been collected to the northward of the Carboniferous rocks of the Bristol coal-field. I have been favoured with specimens by Colonel Mansel Pleydell and by Professor Buckman; the examples figured are not of the largest dimensions.

TRIGONIA PULLUS, *Sow.*   Plate XXXIV, figs. 7, 7 *a*, 8, 9.

TRIGONIA PULLUS, *Sow.*   Min. Conch., tab. DVIII, figs. 2, 3, 1826.
— —   *Agassiz.*   Trigonies, p. 9, 1840.
— —   *d'Orbigny.*   Prodrome de Paléont., vol. i, p. 308, 1850.
— COSTATA, var. PULLUS, *Morris* and *Lycett.*   Monogr. Great Oolite, Part II, p. 58, tab. 1, fig. 22 (Palæont. vol. for 1853), 1853.
— PULLUS, *Morris.*   Catalogue, p. 229, 1854.
— COSTATA, *Quenstedt.*   Der Jura, p. 502, tab. lxvii, fig. 13, 1858.
— — *Park.*, var. PULLUS, *Sharp.*   Oolites of Northamptonshire, Quart. Journ. Geol. Soc., vol. xxvi, p. 388, 1870.
— n. sp., near to PULLUS, *J. E. Cross.*   Geology of N. W. Lincolnshire, Quart. Journ. Geol. Soc., vol. xxxi, p. 125, 1875.
— COSTATA, var. PULLUS, *Judd.*   Mem. Geol. Survey, Rutland, &c., pp. 151, 155, 161, 220, 281, 1875.

Shell ovately trigonal, convex; umbones prominent, acute, and recurved; anterior side moderately produced, its border curved elliptically with the lower border; hinge-border nearly straight, sloping obliquely from the posteal extremity of the escutcheon to that of the marginal carina. The escutcheon is wide and concave, delicately impressed by a twofold kind of ornamentation; its anteal portion has a series of small depressed costellæ, which pass across the surface transversely; the posteal portion has more obscure, oblique costellæ, which take the direction of the lines of growth; both series of costellæ are wrinkled. The area is wide, the plane of its surface forms a considerable angle with the costated portion of the valve; it is bounded by two well-marked carinæ, of which the inner carina is small, but distinctly dentated; the marginal carina is large; plain and smooth near the apex; its middle and posteal portions are more or less plicated. The surface of the area is somewhat concave, divided into two portions by a mesial depression; and in the young state it has also a distinct median carina, which posteally can usually only be considered as one of the costellæ which ornament the surface; posteally these costellæ become merged in the folds of growth; the right valve has the area divided into two portions, but has no distinct median carina; the costellæ are fewer and larger than in the other valve. The other portion of the valve has the costæ large, closely arranged, and rounded; their attenuated extremities are simply curved upwards to the anterior border: the largest specimens, about eighteen lines in length, have twenty costæ.

The peculiarities of the escutcheon in *T. pullus* supplies the most clear distinctive feature separating it from other small examples of the *Costatæ*.

Positions and Localities.—In the Inferior Oolite the shelly freestones of Leckhampton contain specimens of *T. pullus*, which are scarcely larger than the head of a pin. The Wiltshire Cornbrash, more especially at Hilperton, near Trowbridge, has produced a multitude of examples in a fine condition of preservation; the species is also present in the northern extension of the Great Oolite at Appleby, Lincolnshire; near to which locality it has been collected and recorded by the Rev. J. E. Cross. It is unknown in the Cornbrash of Yorkshire.

Sowerby figured a group of small costated forms from Cutch as examples of *T. pullus* ('Geol. Trans.,' 2nd series, vol. 5, plate 21, figure 7), but as the escutcheon and area are not clearly exposed the identity of the species remains doubtful.

A small area and escutcheon figured by Quenstedt ('Der Jura,' tab. 67, fig. 13), as *T. costata* from the highest member of the Great Oolite at Einingen, undoubtedly represents *T. pullus*.

TRIGONIA MONILIFERA, *Ag.* Plate XXXI, figs. 1, 1 *a*, 1 *b*, 2, 2 *a*, 10.

> TRIGONIA COSTATA, *Goldfuss.* Petrefacta Germaniæ, tab. cxxxvii, fig. 3 c, 1836.
> — MONILIFERA, *Agassiz.* Mem. sur les Trigonies, p. 40, tab. iii, figs. 4—6, 1840.
> —? PARVULA, *Agassiz.* Idem, tab. xi, fig. 8, young example.
> — PAPILLATA, *Agassiz.* Idem, p. 39, tab. v, figs. 10—14.
> — RETICULATA, *Agassiz.* Idem, pl. xi, fig. 10.
> — MONILIFERA, *d'Orbigny.* Prodrome de Paléontologie, vol. i, p. 365, No. 293, 1850.
> — — *Quenstedt.* Der Jura, p. 759, 1858.
> — — *Damon.* Geology of Weymouth, Suppl., pl. iv, fig. 1, 1860.
> — MARGINATA. Idem, pl. vi, fig. 8 (mould of the interior).
> — COSTATA, *Grewingk.* Gest. u. Geol. Livonia und Courland, Dorpat, 1864.

Shell ovately trigonal, very convex, both mesially and anteally, umbones prominent, much incurved, and more or less recurved; anterior side moderately produced, rounded, its border curved elliptically with the lower border, its superior or umbonal portion is slightly excavated; hinge-border concave, its length is one fourth greater than that of the siphonal border, with which it forms a considerable angle. The escutcheon is very wide and concave, its surface is for the most part delicately reticulated, having two series of numerous small fine ridges; the series occupying only the anteal portion of the surface passes transversely across the escutcheon, the posteal series takes the direction of the lines of growth; these are also delicate and decussate the anteal series. The area is of moderate size, distinctly bipartite, somewhat concave, and is nearly alike in both the valves; the plane of its surface forms nearly a right angle with the costated portion of the valve, it has a prominent median carina, and larger bounding carinæ; the marginal carina is prominent in both the valves, its indentations are small but are conspicuous even to the apex; that of the right valve is much larger than the other, and in specimens approaching to the adult condition its indentations over the posteal half degenerate into irregular large transverse plications which are sometimes united to the posteal extremities of the costæ; the inner carina, although small, is prominent and nodose; the intercarinal costellæ are irregular, unequal, and variable in number; usually those of the right valve are the larger, more especially adjacent to the post-carinal furrow; occasionally the median carina divides into two costellæ and then loses its prominence. The costæ (about 25 in adult forms) are large but somewhat flattened, and (excepting in the young condition) are widely separated; they have an undulation near to the anterior border, and also turn slightly downwards near to their posteal extremities, where also in adult forms the few last-formed costæ become somewhat irregular and broken. The lines of growth are prominent over the whole of the shell, and assume the form of irregular plications when specimens are of advanced growth; they also sometimes render the anteal portions of the costæ granulated.

Fully developed specimens of *T. monilifera* are probably the largest examples of the section; occasionally Dorsetshire shells have the granulated epidermal tegument preserved over the greater portion of the surface, the lines of granules are distinctly separated in the rows, and are sometimes perceptible even to the unaided vision. The convexity of the valves varies considerably in specimens of the same size from the same formation and locality even when there is no appearance of distortion or compression. The following measurements refer to two of our specimens; these are only of medium size compared with some others which have scarcely less than twice their linear dimensions; the latter, however, are usually more or less compressed, and are therefore unsuitable for comparison.

No. 1. { Length measured upon the marginal carina, $3\frac{1}{2}$ inches.
„ across the valve at right angles to the carina, 2 inches.
„ across the united valves, $1\frac{6}{10}$ inch.

No. 2. { Length upon the marginal carina, 4 inches.
„ across at right angles to the carina, $2\frac{1}{10}$ inches.
„ across the united valves to the carina, $1\frac{9}{10}$ inch.

The internal mould is smooth, but has a slight rib, indicative of the position of the marginal carina; the hinge-processes are very large, they project considerably, corresponding with the great breadth of the escutcheon, and considerable incurvation of the umbones.

The specimen figured by Agassiz ('Trigonies,' tab. 3, figs. 4—6), from the terrain à chailles or Oxford Oolite, has the general aspect of a dwarfed variety of the fine species which attains such large dimensions in the lower portion of the Kimmeridge Clay of Wiltshire and Dorsetshire.

Affinitive and Distinctive Characters.—The very considerable convexity of the valves mesially, the ornamentation of the escutcheon, the prominence and general narrow figures of the bounding carinæ, the considerable angle which the surface of the area forms with the other portion of the valve, together with the more lengthened costæ, and the absence of truncation anteally, serve in the aggregate clearly to separate it from *T. costata*, and also from others of the same section.

For comparison with *T. Meriani* the reader is referred to that species.

The *T. reticulata* of Agassiz ('Trig.,' table 2, fig. 10), which is exemplified by a single fragment, may represent a large and compressed example of *T. monilifera*; the apparent absence of a median carina in this instance is similar to a like variation in occasional Dorsetshire examples, in which the median carina divides in fully developed forms into two or three costellæ, and the carina thus disappears.

Without hesitation the *T. papillata* of the same author, 'Trig.,' tab. 5, figs. 10—14, may be referred to *T. monilifera*; here again the median carina divides into costellæ; the prominence of these and of the bounding carinæ may be taken to represent a small Dorsetshire example of our species.

I would exclude *T. monilifera*, Quenst. 'Handbuche der Petrefacten-kunde,' tab. 43, fig. 15, which represents a small Trigonia having very numerous small rows of costæ and also a species, Quenst., Jura, tab. 93, fig. 4, bearing the name *T. costata-silicea*; the latter may possibly agree with a delicately ribbed species of the Upper Oolites, *T. suprajurensis*, Ag., Trigonies, tab. 5, figs. 1—6, page 42. The latter is unknown in the Kimmeridgian strata of Britain.

Positions and Localities. The lower beds of Kimmeridge Clay in the vicinity of Weymouth, of Wotton Basset, and of Swindon, have produced examples, some of which in their general dimensions much exceed any other species of the costatæ; these, however, are usually more or less compressed or distorted; it also occurs in the Coralline Oolite of Wilts and of Weymouth; at the latter locality specimens deprived of the test, and ill-preserved, have recently been obtained in a red pisolitic iron rock at Abbotsbury, and forwarded to me by Mr. J. T. Walker, of York.

Foreign examples recorded by Agassiz and by D'Orbigny have been obtained in the Terrain à Chailles or Lower Calcareous Grit at Argan (Haut Rhin), Birze Environs of Bale. Besançon (Doubs), Neuvizi, Trouville, Nantua, Marans. Also by Grewingk, at Poplinacny, in the Province of Kowno; Lithuania, there known as *T. costata*.

TRIGONIA MERIANI, *Ag.* Plate XXXIII, figs. 1, 2, 3.

    TRIGONIA COSTATA, *Young* and *Bird*. Survey of the Yorkshire Coast, pl. viii, fig. 19, p. 225, 2nd edit., 1828.
    —   —   *Phillips*. Geol. of Yorkshire, p. 228, 1st edit., 1829.
    —   MERIANI, *Agassiz*. Trigonies, p. 41, tab. xi, fig. 9, 1840.

TRIGONIA MERIANI, *d'Orbigny*. Prodrome de Paléont., vol. ii, étage 14, p. 17, No. 262.
— — *Buoignier*. Paléont. du Dep. de la Meuse, p. 265, 1852.
— — *Waagen*. Der Jura in Frank. Schw. u. der Schweig., p. 218, 1864.
— CLAVELLATA (misprint?), *Phillips*. Geol. of Oxford, pl. xiii, fig. 2, 1871 (reduced figure).
— MERIANI, *Phillips*. Geology of Yorkshire, 3rd ed., 1875, vol. i, p. 250.

Shell ovately trigonal, very convex; umbones produced, pointed, arched inwards and recurved: anterior side produced, its border rounded elliptically with the lower border, which is slightly excavated posteally; escutcheon comparatively small, depressed, flattened, with its superior border somewhat raised; its surface has small, closely arranged, delicate oblique plications; it is well circumscribed by the small nodose varices of the inner carina; as the hinge-border slopes obliquely downwards and the siphonal border of the area is also oblique, their junction does not form any prominent angle, but the length of the hinge-border exceeds somewhat that of the other. The area is slightly excavated and flattened, rendered distinctly bipartite by the superior or inner half being more depressed by the other portion; it is bounded by two well-defined small carinæ; the marginal carina is elevated, peculiarly narrow in the left valve, and somewhat larger in the other valve; it has small inconspicuous plications; the inner carina forms a border fringed with closely arranged, small nodosities; there is also a small, flattened, narrow band which represents the median carina; the intercarinal costellæ are small, numerous, closely arranged and unequal; they are nearly alike in both the valves; the entire area has conspicuous, densely placed, transverse lines of growth, which strongly indent the whole of the surface. The sides of the valves have a very numerous series of costæ (forty or more in advanced growth); they are small, somewhat unequal in size, and irregular in their direction; they form a flexure near to both of their extremities; the few last formed costæ are more imperfect and irregular; their anteal portions take the direction of the lines of growth and curve upwards anteally, embracing the extremities of the costæ previously formed; in the right valve their posteal extremities pass across the marginal carina as so many plications; in the other valve they terminate at the small ante-carinal groove. The lines of growth are minute and irregularly crowded over the valves generally.

The defective, irregular figures and partial effacement of the few last-formed costæ, indicate the ultimate stage of growth in the life of the Mollusk when the mantle continues to add to the growth of the valves, but ceases to produce surface ornaments.

The hinge processes are prominent and lengthened; the posteal cardinal process of the right valve extends horizontally nearly the length of the escutcheon. The internal mould has not been ascertained.

A large example in my collection has the length of the marginal carina $4\frac{3}{4}$ inches;

at right angles to the carina across the valve 4¼ inches; convexity of a single valve 1⅛th of an inch.

A smaller Trigonia occurs in the Coral Rag of Wiltshire, which appears to be a variety of *T. Meriani*; the general figure has less convexity and is somewhat more pointed at both the extremities; the other general features are alike in both the forms; our smallest figure represents this variety.

T. Meriani has scarcely hitherto been recognised as a British species; it has occurred only rarely, and in its usually defective condition has been assigned to *T. costata*; its aspect when well preserved is sufficiently remarkable both on account of its large dimensions and also for a certain elegance of figure, together with a minuteness and delicacy in the surface ornaments, which might be expected to attract attention, and offers a considerable contrast to the *T. costata* of the Inferior Oolite. Considering the large dimensions of *T. Meriani*, its more remarkable characters consist in their general diminutive size and small prominence, such as the smallness and partial irregularity of the costæ, the small and nearly plain marginal carina, the inconspicuous median and inner carina, together with the minute and almost evanescent intercarinal costellæ. The smallness and irregularity of the costæ in so large a species is a feature altogether alone in the Jurassic *costatæ*, and is suggestive of a Spanish Neocomian species, *T. peninsularis*, Coquand, ' Monogr. de l'Étage Aptien de la Espagne,' pl. xxiii, fig. 3, in which the same feature is more remarkably conspicuous.

T. monilifera, Ag., from the Coral Rag and Kimmeridge Clay of the southern counties, has in the general figure some resemblance to *T. Meriani* and sometimes occurs even of larger dimensions; it will readily be distinguished by the large, widely separated costæ, by the remarkably prominent and strongly indented carinæ, and by the great concavity both of the area and escutcheon; the costellæ of the escutcheon also offer a minute but not less distinctive feature.

Hitherto *T. Meriani* has been very insufficiently figured; the drawing intended for this shell, named *T. costata* in the work of Young and Bird above cited, is in the usual coarsely executed style of the other figures, and the surface ornaments of the area are altogether erroneous; the general contour of the valve, the characters of the costæ, and of the marginal carina are distinctive and readily recognised.

A single imperfect and insufficiently characterised specimen was figured by Agassiz; his description was founded upon better preserved examples subsequently brought to his notice.

To the present species must also be referred a reduced figure of a costated form in the work of Professor Phillips on the 'Geology of Oxford,' from the Coral Rag of Heddington (Pl. xiii, fig. 32); printed erroneously *T clavellata*.

Position and Localities. Only a small number of specimens of this large Trigonia have come under my observation; they have all been obtained in the Coralline Oolite formation at several widely separated localities, as at Weymouth, in Wiltshire, in Oxfordshire, and in Yorkshire at Malton and Pickering; at the two latter localities it may be understood as the species intended whenever *T. costata* is mentioned in lists of their fossils. Specimens are in the Museum of the Royal School of Mines, in the Woodwardian Museum, Cambridge, in the Geological Museum of the University of Oxford, in the museum of the Philosophical Society, York, in that of the Philosophical Society, Scarborough; also in private collections.

Specimens from the Coralline Oolite of Weymouth are usually of smaller size, and are more lengthened, or measure less across the valve at right angles to the carina, than Yorkshire examples.

In Switzerland, Agassiz records its occurrence in the same geological position at Zwingen (Soleure), at which place he states that it has been obtained in considerable numbers and in a fine condition of preservation. D'Orbigny also records it in the Coral Rag of France at several localities.

TRIGONIA CASSIOPE, *D'Orb.* Plate XXXII, figs. 1, 2, 3, 4, 5.

TRIGONIA CASSIOPE, *d'Orbigny.* Prodrome de Paléont., vol. i, p. 308, 1850.
— — Lycett. Sup. Monog. Gr. Ool. Moll. Pal. Soc., p. 49, tab. xxxvii, fig. 10, 1863.
— — Rigaux and *Sauvage.* Descr. de quelques espèces nouv. de l'Étage Bathonien du Bas-Boulonnais, p. 19, 1868.
— — Phillips. Geology of Yorkshire, 3rd ed., vol. i, p. 250, 1875.
— — Judd. Mem. Geol. Survey Rutland, &c., p. 289, 1875.

Shell variable in its general figure and outline; ovately trigonal; somewhat depressed and comparatively short anteally; usually considerably lengthened and attenuated posteally; anteally and also near to the umbones the valves have much convexity, the general outline becoming subcrescentic. The umbones are small, moderately elevated, and usually much recurved; placed upon a line one third the length of the shell from the anterior border. The anterior side is rounded and inflated, its border curves elliptically with the lower border, which is lengthened and sinuated posteally; the hinge-border is lengthened and concave, forming posteally only a slight angle with the siphonal border, which is oblique and shorter than in any other British example of the section; its posteal extremity forms with the marginal carina a figure much produced and pointed. The escutchon is large, depressed, and somewhat excavated, so that when a valve is placed horizontally the escutcheon is scarcely seen when viewed from above; its surface is very delicately, obliquely plicated, and obscurely reticulated. The area is unusually narrow for so large a species, its surface forms a considerable angle with the costated portion of the shell; its surface is very delicately reticulated; it is bipartite, each portion having very small longitudinal costellæ, one of which is sometimes prominent, forming a small median carina; in other examples, and more especially in the right valve, there is no distinct median carina. The bounding carinæ are very small upon both the valves, but the marginal carina is always distinct; it is minutely, transversely plicated and the antecarinal groove of the left valve is only slightly defined; the umbonal portion of the carina has a considerable and graceful curvature. The inner carina is commonly only indistinct; it is one of the reticulated costellæ of the area slightly separated from the more depressed surface of the escutcheon.

The other, and by much the larger portion of the shell, has the rows of costæ small and numerous; their edges are acute, little elevated, uniform in character; their number in fully developed specimens varies from 24 to upwards of 36, depending upon the more close or distant arrangement of the rows; the first formed or umbonal rows have their general direction obliquely downwards posteally; the rows subsequently formed are directed more horizontally, but curve upwards anteally, where they form a well-marked undulation; near to the lower border they take the direction of the lines of growth or conform to the figure of the border; the left valves sometimes have the posteal extremities of the costæ, each forming a short, downward prolongation; the right valve has several of the last-formed costæ passing across the marginal carina as so many plications. Well-preserved specimens have the costæ crowded with minute perpendicular lines of epidermal granules.

Young examples when less than fifteen lines in length are much depressed, the costæ have no distinct anteal undulation, and the several features which characterise the fully developed shell are scarcely perceptible.

A large, gracefully curved, and transversely lengthened form, remarkable for the large curvature of the small marginal carina, the produced and attenuated posteal extremity, the narrow, excavated, and minutely reticulated area, the considerable angle which it forms with the costal portion of the surface, the largeness of that surface and the rounded and produced anteal side; the combination of these characters will usually readily distinguish it from others of the *costatæ*.

Judging of *T. Cassiope* from our figures only, it might be imagined that the species is divisible into varieties; a more extended knowledge of this Cornbrash form will lead to the more correct inference that these figures represent nearly the extremes of variation in opposite directions; figure 1 exemplifies the more wide and depressed, and figure 2 the more narrow and inflated forms. The greater number of specimens will be found to approximate to figure 4; there is, however, so much variability both in the outline of the valves, in their convexity, and in the size and number of the costæ, that taken in combination with the frequent and, indeed, usual compression or distortion of some portion of the shell, it is scarcely possible to find any two specimens having any close agreement with each other. Occasionally a valve occurs even more inflated than the narrow specimen figure 2, so that when placed horizontally and viewed from above, no portion of the area is seen. The internal mould is not known. The hinge teeth are widely divergent, they have little prominence, and are smaller than is usual with the larger examples of the *Costatæ*.

Position and Localities. Upon the Coast of Yorkshire to the southward of Scarborough it is common in the bed of Cornbrash and also in the small exposure of the same bed to the northward of the Castle Hill, where the specimens are badly preserved or have suffered from compression—a condition which invariably occurs when the valves are in contact. The species has also been obtained rarely in the brown, sandy, lowest bed of Kelloway Rock in the same vicinity. The compression and distortion which is so common in this large Trigonia is a frequently recurring feature in the larger testacea of the Cornbrash; the other large Trigoniæ from the same bed and locality present similar defects in their state of fossilisation.

As the Trigoniæ generally are good stratigraphical guides it will appear remarkable that its representatives which characterise the Cornbrash or Forest Marble throughout the southern and midland countries of England have altogether disappeared to the northward of the Humber, where they are replaced by other species. Thus, upon the Coast of Yorkshire we look in vain for *T. pullus*, *T. costata*, *T. flecta*, *T. undulata*, var. *arata*, *T. Moretoni*, and the short variety of *T. sculpta* designated *Rolandi*; these are replaced in the northern Cornbrash by *T. Scarburgensis*, *T. Cassiope*, and by two varieties of *T. elongata*, all of which occur in some abundance.

The changes thus exhibited by the genus Trigonia will be found scarcely less remarkable in other associations of Molluscan forms, more especially in the numerous and varied series of *Conchifera*, some of which are identical with Kelloway Rock species and are unknown in the southern Cornbrash. These data tend to the conclusion that the Cornbrash of Yorkshire represents a deposit of marine testacea more recent than that of the southern countries or more transitional and tending to connect more nearly the mollusca of the Lower and Middle Oolites.[1]

The foreign localities assigned to *T. Cassiope* are Luc (Calvados); Vezelay (Yonne); Grange Henry rear Nantua; it also occurs rarely in the Great Oolite of the Bas-Boulonnais (Rigaux et Sauvage).

Mr. R. (now Professor) Tate, 'Quart. Jour. Geol. Soc.,' 1867, p. 171, records its occurrence in rocks (supposed to be Jurassic) in Southern Africa. The single specimen upon which this identification is founded has scarcely half the linear dimensions of Yorkshire examples; the distinctive characters are not prominent, and in the absence of more satisfactory materials for comparison, the specific identity may be regarded as doubtful.

[1] See a note by the present author on the association of generic forms of Mollusca in the Yorkshire Cornbrash compared with those of the Cornbrash of the southern counties, 'Supplementary Monograph on the Mollusca of the Stonesfield Slate, Great Oolite, Forest Marble, and Cornbrash, Palæontographical Society,' volume for the year 1861, p. 117. 1863.

TRIGONIA CULLENI, *Lyc.*, sp. nov.   Plate XXXI, figs. 9, 9 *a*.

The Marine intercalated Millepore bed, a subordinate member of the Lower Sandstones and Shales at Cloughton, a few miles to the northward of Scarborough, has associated with *Trigonia recticosta* and other testacea of the Inferior Oolite, a small costated species of Trigonia, and as the matrix is usually very hard and the tests of the Trigonia are either delicate or indifferently preserved, it rarely happens that a specimen can be separated sufficiently entire to allow of the clear development of its characteristic features; thus it happens that our sole illustrative specimen is imperfect; it appears, however, possible by its aid to describe the species and separate it from other allied forms of the *Costatæ*.

The most prominent features consist in the form unusually lengthened transversely and posteally; in the considerable recurvature of the umbones together with the produced and rounded anterior side; also in the minuteness of the lengthened horizontal rows of costæ, the specimen figured having upwards of twenty-seven; they are elevated, and so narrow as to appear almost linear, viewed from the anterior side, and with the valves united. The horizontal rows are, however, distinctly separated, and are almost without any anteal undulation.

The upper border of the shell is concave; it forms an obtuse angle with the siphonal border, which is oblique, and has considerable length; the lower border is lengthened and is only slightly curved.

The escutcheon and area are large and have no clear separation, the former is deeply excavated; the inner half of the area is also concave; the costellæ upon their surfaces are minute and very irregular. The marginal carina is small, but is distinctly transversely plicated.

The specimen figured has the length 15 lines; the height 10 lines; the diameter through the united valves 9 lines.

Named after its discoverer, Mr. Peter Cullen, a veteran and intelligent collector of fossils at Scarborough, the greater portion of whose life has been occupied in the development of the palæontological treasures of the district in which he resides.

Affinities and differences. *T. hemisphærica* with similar minute costæ has the general figure much shorter and less produced anteally, the umbones are much more nearly terminal and have little recurvature; the surface ornaments of the area and of its carinæ are much larger and more deeply sculptured. Another small gregarious example of the *Costatæ* also occurs at Cloughton; the test has usually decomposed, but the ribbing is distinct; its fewer and larger costæ separate it from the present species; it may represent the very young condition, or a variety of *T. denticulata*.

TRIGONIA HEMISPHÆRICA, *Lyc.* Plate XXXI, figs. 4, 5, 6, 7, 8. Var. *Gregaria,* Plate XXXIII, figs. 4, 5, 6.

TRIGONIA HEMISPHÆRICA,	*Lycett.*		Ann. and Mag. Hist., vol. xii, tab. xi, fig. 2, young specimen, 1850.
—	—	*Morris.*	Catalogue, p. 228, 1854.
—	—	*Lycett.*	Cotteswold Hills, Handbook, p. 137, pl. iv, fig. 4, 1857.
—	—	*Phillips.*	Geology of Oxfordshire, p. 162, 1871.
—	—	*Sharp.*	Oolites of Northamptonshire, Quart. Journ. Geol. Soc., pp. 267, 293, 1873.
—	—	*Cross.*	Geology of North-west Lincolnshire, Quart. Journ. Geol. Soc., pp. 121, 124, 1875.
—	—		(drawf variety). Ibid., pp. 121, 125.
—	—	*Judd.*	Mem. Geol. Sur. Rutland, &c., p. 153, 157, 281, 1875.

Shell ovately trigonal, convex, umbones prominent, pointed, slightly recurved; anterior side short, its border curved elliptically with the lower border, posterior side moderately lengthened. Escutcheon depressed, its superior border somewhat raised, moderately wide, obliquely closely plicated, its length greater than that of the siphonal border of the area with which it forms an obtuse angle. The area is of considerable breadth, or equal to one third of the entire valve, it is for the most part flattened, is depressed and slightly concave; the right valve has the lower half or that adjoining the marginal carina raised or convex. The marginal carina is large, with elevated, deeply indented irregular plications, these are in some instances scarcely more numerous than the costæ; the inner carina is also conspicuous from its prominent close set row of obtuse nodes. There is no median carina in the right valve; the left valve has occasionally some approximation to a median carina in the greater elevation of one of its costellæ. The costellæ of both valves are prominent and closely arranged with large, deeply indented reticulations; the lower half of the area has eight or nine, the upper half six or seven costellæ. In the very young state, the right valve has the costellæ of the lower half of the area fewer and larger, which is a feature not unusual with other species of the *Costatæ.*

The other portion of the surface has a very numerous regular series of narrow, elevated, closely arranged, rounded, or sometimes subacute costæ, varying from 40 to 55 or even exceeding the latter number, occasionally in specimens measuring two inches or upwards upon the marginal carina; those of the left valve are slightly enlarged at their posteal extremities and terminate abruptly at the narrow deep antecarinal groove; anteally their extremities become attenuated, and have a single curvature upwards to the border; the costæ of the right valve curve slightly downwards, where they are united posteally to the plications upon the marginal carina which is larger than that of the other valve; the last formed one or two costæ in adult specimens are more depressed or become squamous.

It is doubtful whether the internal mould has been recognised.

Dimensions of a full sized Cotteswold specimen in the Geological Museum of the University of Oxford.

Length of the marginal carina	27 lines.
Across the valve at right angles to the carina	23 „
Convexity of a single valve	9½ „
Greatest breadth of the area	12 „
Dimensions of a full sized Santon specimen in my cabinet.	
Length of the marginal carina	25 lines.
Across the valve at right angles to the carina	21 „
Convexity of a single valve	7 „
Greatest breadth of the area	9 „

History, and local variability.—The specimen first described by me, and figured in the year 1850, was a very young Cotteswold example, and although measuring only three lines across, the valve had about twenty costæ, and was supposed to be an adult specimen; its convexity was so considerable that the figure was almost hemispherical; larger examples have much less convexity. The hinge-processes and their sulcations are large, they equal in prominence the corresponding parts of any other species of the section having equal general dimensions. The surface ornaments present no well-defined differences between the very young and advanced conditions of growth. A more than common amount of variability occurs in the size and number of the costæ, even in specimens from the same locality; one of our Santon examples, Plate XXXI, fig. 6, having the costæ more distantly arranged illustrates this feature, it does not constitute a distinct variety, as it is connected with the typical form by other examples of intermediate ribbing; neither does this variation extend to the ornaments of the area and escutcheon, which are conspicuous for their strongly defined sculpture.

The largest specimens of *T. hemisphærica* have been obtained in the Cotteswold Hills; it has occurred only in the middle portion of the range in the district to the eastward of Cheltenham extending southwards to Stroud and including Rodborough Hill; from the latter locality all the earlier known specimens were obtained; its position is the bed termed Lower Trigonia Grit, associated with the more abundant *T. sculpta* and *T. formosa*, but our species ranks as one of the most rare testacea of the district.

The museum of the Royal School of Mines has several unusually small examples obtained by the officers of the Survey in the South Lincolnshire district at Stamford,

Collyweston and Wakerley, in Lincolnshire Limestone; these have the costæ unusually minute, numerous, and faintly defined; they appear to be examples of dwarfed adult forms. Other localities noted by Mr. Sharp in his memoir on the Oolites of Northamptonshire (' Quart. Jour. Geol. Soc.,' May, 1873), are Ravens Wood, Glendon, near Kettering, Ponton and Denton, near Grantham.

In North-West Lincolnshire at Santon Bridge, near to Appleby, a deep road cutting exposes a good section of the lowest beds of Inferior Oolite, reposing upon dark-coloured clays of Upper Lias; the Midford Sands are altogether absent. The lowest bed of Inferior Oolite is a dark-coloured marly rock only sparingly fossiliferous; resting upon it is a thick bedded pale brown, hard shelly limestone, containing a numerous and finely preserved series of conchifera; *Trigonia hemisphærica* is not uncommon, always as single valves which never attain the dimensions of fully developed specimens in the Cotteswold Hills; it is accompanied rarely by valves of *T. Phillipsii*.

In the same vicinity, higher in position, a small variety of our species (*gregaria*) occurs abundantly in a large quarry of Lincolnshire Limestone adjacent to the railway station at Appleby; in size the specimens agree with others from a similar position at the opposite extremity of the county near to Stamford; compared with the typical form the Lincolnshire Limestone specimens present some differences, the size measured upon the marginal carina varies from six to fourteen lines, but the greater number are from ten to twelve lines.

The costæ anteally near to the border form a slight undulation, more especially the more umbonal rows; the area in its surface and carinæ has its ornaments less conspicuously sculptured. It is difficult to separate specimens of this small variety in a condition suitable for the cabinet, the valves are often in position, but distortion is also common. The first few ill-preserved specimens are mentioned at page 11 in the introductory portion of this monograph as *T. gregaria*, a name which should be retained only as indicating the present small variety; figures of which are given upon Plate XXXIII. The matrix, a pale buff-coloured, tough limestone, is a coralline mud rock, identical in lithological characters with beds which in the Cotteswolds immediately underlie the Oolite Marl, and into which it passes insensibly; locally they abound in the Cotteswolds with clusters of *Nerinæa*, *Chemnitzia* and other univalves, more rarely also with Trigoniæ, of which the following seven species have been separated, *T. angulata*, Sow., *T. subglobosa*, Lyc., *T. costatula*, Lyc., *T. pullus*, Sow., *T. gemmata*, Lyc., *T. Phillipsii*, Mor. and Lyc., *T. tuberculosa*, Lyc.

At Brough, near Hull, the shelly Inferior Oolite also contains this variety, obtained by Mr. J. F. Walker, of York.

The testacea of the rock at Appleby consists chiefly of numerous genera and species of Conchifera, almost the whole of which are identical with Cotteswold forms, it is therefore the more remarkable that the Trigonæ at the northern locality are represented by a single species of the *Costatæ*, and one of the forms which in Gloucestershire occurs most rarely; the two quarries, the one Santon, the other at Appleby, together scarcely exceeding fifty feet in thickness have produced a greater number of specimens of *T. hemisphærica* than the whole of the Cotteswolds. As the geology of the vicinity of Appleby was but little known previously to the memoir by the Rev. J. E. Cross ('Quart. Jour. Geol. Soc.,' May, 1875), my thanks are especially due to that gentleman under whose guidance I visited the quarries of that vicinity in the year 1870; the specimens figured in this monograph were the gift of Mr. Cross. At Cloughton upon the coast to the northward of Scarborough, the Inferior Oolite Millepore bed contains some dwarfed examples of the variety *gregaria* associated with *T. recticosta*; owing to the hard matrix and delicate condition of the testacea; specimens are rare and usually ill-preserved, they measure from six to ten lines across the valves.

Affinities.—The distinctive features of *T. hemisphærica* are so conspicuous, that it will not readily be mistaken for any other species. The only form in the lower oolites which appears to approach to it is *T. tenuicosta*; when comparing specimens of equal dimensions it will be seen that *T. tenuicosta* has greater convexity, that it is shorter and truncated anteally; its umbones are more elevated, narrow and pointed, its area is much more excavated; its carinæ and the surface of the area generally has a much smaller and more delicate ornamentation; the costæ also have a considerable anteal undulation different from the simple anteal curvature of *T. hemisphærica*.

In the Upper Jurassic rocks of France one of the *Costatæ* resembles *T. hemisphærica* in having very numerous small costæ; *T. Etalloni*, De Lor. ('Paléont de la Haute Marne,' P. de Loriol, et E. Royer et H. Tombeck ('Mém. Soc. Lin. de Normandie,' tom. 13, Pl. XVII, figs. 13, 14, 15), has the numerous costæ curved concentrically throughout their course; the marginal carina is minute, and therefore wholly unlike the large and deeply sculptured carina of the British species.

Our species is not limited to a single stratigraphical position of the inferior oolite. It occurs in the middle portion of the range of the Cotteswold Hills, between Cheltenham and Stroud; its bed (Lower Trigonia Grit) overlies nearly two hundred feet of the Inferior Oolite. In North West Lincolnshire, at the Santon cutting, only four or five feet separate the shelly bed with *T. hemisphærica* from the dark clays of the Upper Lias; both this and the position of the Lincolnshire Limestone are widely separated from the Trigonia bearing Ragstones of the Cotteswolds; the latter occupy a much higher position, and have their ornaments more prominently sculptured than the Lincolnshire specimens, in every stage of their growth.

The steep escarpments upon the eastern side of the Nailsworth Valley, with its numerous quarries and other smaller rock exposures, has in its beds of hard white limestones and their enclosed Conchifera exact counterparts of the fauna and deposits of North Lincolnshire, reproducing upon a smaller scale the limestones which, in their more northern prolongation, acquire so much greater thickness and importance. The bed of oolite marl has disappeared in the Nailsworth valley, merged in the more indurated limestones which are altogether without *T. hemisphærica*.

The exact position of the shelly *hemisphærica* bed of the Santon section is perhaps rather doubtful, probably it should be referred to the Northampton sands, nearly at the base of the formation.

(See series of comparative vertical sections, of members of the Lower Oolites, by J. W. Judd, 'Mem. of the Geological Survey of England and Wales, Rutland, &c.,' Plate I.) A most comprehensive and instructive series of sections drawn to a scale of one inch to fifty feet.

The Museum of the Royal School of Mines has a varied series of specimens of *T. hemisphærica*, from the Cotteswolds, from the district of Rutland, from South Lincoln, and from North West Lincoln. A large Cotteswold specimen is in the geological museum of the University of Oxford. The collections of Dr. Wright, Cheltenham ; Mr. Witchell, Stroud ; Mr. S. Sharp, Northampton ; Rev. J. E. Cross, Appleby, and my own cabinet also, fully illustrate its several aspects.

It appears to be absent in the Inferior Oolite of Northampton, Oxford, Somerset and Dorset ; neither has it been recorded at any foreign locality.

[1] In the foregoing description I have expressed the conviction long since entertained by me, founded upon a comparison of fossils of Lincolnshire limestone with those of the Oolite Marl and subjacent beds in the Cotteswolds, that the fossil fauna of the two deposits are identical, and differ only in species which inhabited the variations of sea bottom, such as may be expected to occur in a deposit which extends from the estuary of the Severn to that of the Humber broken only by the upraised portions of Oxfordshire and Northamptonshire.

The general scope of this monograph will not admit of any detailed comparisons between the fauna of widely separated deposits ; it may, however, be useful to indicate a few typical forms which by their considerable extension laterally connect the more distant deposits. At both localities the Ammonites are almost absent.

Amongst the Gasteropoda will be found the gigantic *Natica Leckhamptonensis*, which is rare near to Stroud and to Cheltenham, but reappears abundantly in South Lincolnshire, and is well exemplified in the Grantham Museum ; it is also present in the limestone at Appleby, exemplified in the collection of the Rev. J. E. Cross.

Of the Conchifera, *Lima bellula*, Lyc., formerly obtained in the Nailsworth Valley at Culverhill in considerable numbers in a quarry long disused, is specially deserving of notice ; it retains its colour partially preserved ; examples are now in the Museum of the Royal School of Mines. As the rock of this quarry is extremely hard the stone is not employed commercially and the shells are rare. A small specimen of the species obtained by Professor Morris in the limestone of the Ponton cutting, South Lincolnshire, was figured in the Great Oolite monograph of this Society, vol. for 1853, pl. iii, fig. 9. More recently it has been found in South Lincolnshire, with the colours partially preserved. At the north-western extremity of the County at Appleby it is met with abundantly in a beautiful condition, in tint a dull purple with narrow concentric zones of cream colour ; specimens marking all stages of growth import a peculiar aspect to the Limæ in the collection of the Rev. J. E. Cross.

Of Brachiopoda the characteristic *Terebratula fimbria* of the Cotteswolds is found to have disappeared in the limestone of Lincolnshire ; it retains the accompanying *T. submaxillata*, Dav., which is also locally abundant in the Cotteswolds ; at Appleby the specimens are peculiarly large, surpassing the Cotteswold forms.

In contrast to the general accordance of species of Conchifera which exists between the Lincolnshire and

§ VIII.—BYSSIFERÆ.

TRIGONIA CARINATA, *Ag.*   Plate XXXV, figs. 3, 4, 4 *a*, 5, 5*a*, 6, 6*a*.

 TRIGONIA CARINATA, *Agassiz.* Trigonies, p. 45, tab. vii, fig. 7, 10, 1840.
 — SULCATA, *Agassiz.* Trigonies, p. 44, tab. xi, fig. 16; tab. viii, figs. 5, 11 (Moulds).
 — HARPA, *Deshayes.* Leymerie, Mem. de la Soc. Geol. Fran., tom. v, pl. ix, fig. 7, 1842.
 — CARINATA, *Forbes.* Quart. Journ. Geol. Soc., tom. i, p. 244, 1845.
 — — *d'Orbigny.* Pal. Fran. Terr. Cret., vol. iii, p. 132, pl. 286, 1843.
 — HARPA, *Matheron.* Catal. de Corps, org. foss. du Depart des bouches du Rhone, p. 166, 1842.
 — CARINATA, *Marcou.* Rech. Geol. sur l'Jura Salin., p. 142, 1846.
 — SULCATA, *Marcou.* Idem.
 — CARINATA, *Fitton.* Quart. Journ. Geol. Soc., vol. iii, p. 289, 1847.
 — — *d'Orbigny.* Prodrome de Paléont., vol. ii, p. 78 (pars); exclude *T. elongata*, 17 etage, 1850.
 — HARPA, *Archiac.* Hist. Progr., tom. iv, p. 322, 1851.
 — CARINATA, *Corneul.* Bull. Soc. Geol. de Fr., tom. viii, p. 435, 1851.
 — — *Bavignier.* Statist. Geol. de la Meuse, p. 473, 1852.
 — — *Studer.* Geol. Schweiz., tom. ii, p. 281, 1853.
 — — *Morris.* Catalogue, p. 228, 1854.
 — — *Cotteau.* Moll. foss. de l'Yonne, p. 76, 1854.
 — — et SULCATA, *Trilobet.* Bull. Soc. Natr. Neuchatel, tom. iv, p. 73, 1856.
 — — *Pictet et Renevier.* Paléont Suisse, Terr. Aptien, p. 101, 1857.
 — — *Archiac.* Bull. Soc. Geol. de Fr., tom. vi, p. 496, 1857.
 — — *Lycett.* On a Byssiferous Fossil Trigonia, Ann. and Mag. Nat. Hist., p. 17, 1870.

Shell ovately elongated, or somewhat spindle shaped in young forms; very convex in the adult state, umbones prominent, sub-terminal, acute, much arched inwards, and slightly recurved. Anterior side very short, truncated, inflated, its border forming with the other valve a large depressed space; the posterior border is somewhat convex, curved gradually towards the lower extremity of the prominent marginal carina. Escutcheon of great breadth and length, slightly concave, its borders raised in every direction, traversed by strong diverging rugose scabrous ridges. Area very wide, conspicuously

Cotteswold districts it is remarkable that of the seven species of Trigonia obtained in the Cotteswolds above enumerated, two only, *T. pullus* and *T. Phillipsii*, occur in South Lincolnshire, and even these have disappeared in the limestone at Appleby.

bipartite, with two elevated, strongly indented, bounding carinæ; the upper half of the area or portion adjacent to the inner carina has much concavity, with six or seven small knotted costellæ which take the same direction as the carina, and disappear upon the posteal half of the area, obliterated by the large irregular plications of growth, which there cross the area and efface the surface ornaments; the other portion adjacent to the marginal carina is slightly convex, with three or four knotted costellæ which are larger than those of the upper or concave space; there is no distinct median carina, but the upper or inner costellæ acquire the semblance of a median carina from its prominent position adjacent to the more depressed concave space. The marginal carina is narrow, elevated and ridge-like upon the umbo, becoming more rounded and depressed posteally, it is crossed by large irregular sub-nodose plications, which are effaced posteally by the plications of growth; the inner carina is also prominent near to the umbo, it is occupied by a line of unequal, irregular, nodose varices formed by the lines of growth. The ante-carinal groove of the left valve is narrow, and similar in character to the post-carinal groove of the right valve. The other portion of the surface has about eighteen rows of elevated narrow plain costæ, all of which originate at the anterior border, their direction is at first horizontal upon the depressed anterior sides, but with the curvature of the valve, each bends obliquely downwards towards the marginal carina until it arrives at the groove of the left valve, when its extremity becomes suddenly attenuated and passes downwards to the extremity of the costa next in succession; in the right valve the attenuated extremities of the costæ touch the marginal carina. In fully developed shells the two or three last formed costæ have less obliquity, and follow more nearly the direction of the lower border.

The lines of growth have great prominence; they produce reticulations where they cross the costellæ of the area, and upon the sides of the valves they sometimes even break the continuity of the costæ, giving to them a wrinkled or broken aspect.

One other feature, altogether unusual in this genus, and which imparts a distinctive character to the present section, consists in the presence of a small, distinct byssal aperture, with a slightly thickened margin upon the anterior border of the valve, a feature which pertains only to examples of fully developed growth.

In fully developed specimens the length of the marginal carina is equal to twice the diameter of the united valves, and is somewhat greater than twice the diameter of a valve measured at right angles to the marginal carina.

It is distinguished from all others of the costatæ by the unusually large escutcheon, by the great prominence of the bounding carinæ to the area, by the great obliquity and straightness of the costæ; lastly, by the presence of the byssal aperture.

English Neocomian specimens, when of full size and retaining the test, are usually defective at the posteal portion of the area and escutcheon, the siphonal border and adjacent thin portions of the shell having been broken away retain the sharp fractured edges, a condition which indicates that the mollusk was destroyed suddenly, and probably perished from the attack of some predaceous species. As this does not coincide with the usual condition of fossil tests of this genus, a special cause is required to account for it. The byssal aperture indicating a fixed or sedentary condition offers an explanation; the adult mollusks fixed by their byssal appendages had not the free saltatory motions of other *Trigoniæ*, and became victims to the marine flesh-eaters which attacked the siphonal or thinner and more exposed upper portions of the valves. The appearance of the lines of growth indicate that the byssal orifice was formed at the completion of adult growth; the shell generally at that period, and more especially the area, exhibits commonly an abraded aspect similar to the appearance of adult Bysso-Arcas, and probably produced by similar conditions of existence. Usually, therefore, it is only upon immature specimens that the surface ornaments of the area have been fully preserved, and even in such instances the posteal portion had no longer distinct carinæ or costellæ, their positions are occupied by transverse rugose folds.

Some alteration in the figure of the area and escutcheon took place during the progress of growth; the young specimen, figure 4 *a*, has the escutcheon comparatively small and concave; the surface of the area equals the other portion of the valve, so that the length of the siphonal border is equal to the hinge-border; in the adult state the escutcheon extends fully three fifths of the posteal slope, and the siphonal border is comparatively short.

The figures of *T. carinata*, given in the work of Agassiz ('Trigonies,' tab. vii, figs. 7—10) represent two small specimens deprived of their tests; the surfaces of the area and escutcheon are not preserved, and the costæ are only faintly indicated. *T. sulcata* of the same work (tab. viii, figs. 5—7) represents an internal mould of *T. carinata*. Fig. 8 is a large example deprived of the test; it has traces of the oblique costæ, but has no portion of the surfaces of the area and escutcheon; figure 10 represents a very young specimen destitute of any portion of the surface; the species, therefore, is very inadequately illustrated by the figures of these two supposed species.

The figures of *T. carinata* given by d'Orbigny (' Paléont. Fran. Terr. Crét.,' plate 286), although affording beautiful examples of lithographic art, are not altogether satisfactory representations of the species. The larger of the specimens is not of adult growth, but should represent the changes which had taken place over the posteal portion of the area in the effacement of the carinæ and intercarinal costellæ by rugose irregular plications; on the contrary, the carinæ, including a delicate median carina and the intercarinal costellæ, continue prominent even to the siphonal border; there are no differences in the intercarinal spaces; the escutcheon assumes the aspect of a deep concavity, and the costæ although prominent are without the posteal terminal attenuations. These several features differ essentially from the more perfectly preserved examples of the species from Atherfield, so celebrated for the fine condition of its Neocomian Testacea. The description in the text of the same work is so concise that it does nothing to rectify the incorrect details of these figures.

Affinities and Differences. The *Lyrodon sulcatum* of Goldfuss is one of the Cretaceous *Scabræ*, and altogether a different species; it is better known as *Trigonia sulcataria*, to which the reader is referred. *T. peninsularis*, Coquand (' Monogr. de l'étage Aptien de l'Espagne,' pl. xxiii, fig. 3), is the only other costated *Trigonia* known in the Cretaceous rocks; its general aspect is that of sectional degeneracy, and has no near affinity with *T. carinata;* it is chiefly remarkable for the very irregular and unequal costæ, some of which are angulated mesially; the few last formed become irregular plications.

By a singular error, which could only have arisen from an imperfect acquaintance with British *Trigoniæ*, D'Orbigny (' Prodr. de Paléont.,' vol. i, p. 338, No. 161; also vol, ii, p. 78, No. 288) made Sowerby's figures 1, 2, of *T. elongata*, from the Oxford Clay of Weymouth, to be identical with the Neocomian *T. carinata*. Our figures of each of these species render any detailed comparison between them unnecessary.

Positions and Localities. *T. carinata* occurs in the Upper Neocomian formation, in the lowest or *Perna Mulleti* bed at the Atherfield section in the Isle of Wight; the test is usually only partially preserved; specimens are of every stage of growth, but it is somewhat rare, and its condition of preservation compares unfavorably with the numerous valves of Conchifera at that celebrated locality; other localities and Neocomian positions are Hythe, Lympne, and Maidstone, at which places the test is usually absent, the fossil consisting of glauconitic sandstone. It is unknown in the Middle Neocomian Stage at Speeton, Tealby, Norfolk, and Cambridgeshire.

Our figures represent Atherfield specimens excepting Plate XXXV, fig. 3, from Upper Greensand, near to Ventnor. The latter, although deprived of the test, has the surface ornaments well preserved. Apparently also the range of *T. carinata* is even more considerable than is indicated by the Isle of Wight specimens. Certain internal moulds, very imperfect and unfit to be submitted to the artist, were forwarded to me by Mr. Cunnington from the Upper Greensand of Devizes and from the Chloritic Marls of Warminster; their occurrence at the latter locality and position in the highest beds of the glauconitic series extends the range of *T. carinata*, and exceeds that of any other known example of the genus.

It is also noteworthy that, although the range both stratigraphically and geographically is so considerable, no separation into varieties occur; the more close or more distant arrangement of the costæ represent the limits of variation. It will be observed that the small specimen from Upper Greensand of Ventnor, Plate XXXV, fig. 3, offers no distinction from the specimen, fig. 4, which is of similar size, and obtained in the lowest bed of the Neocomian formation at the Atherfield section, separated from the newer position by upwards of 1000 feet of strata.

T. carinata is absent in the Gault both of Kent and of the Isle of Wight, and equally so in the Greensand of the Blackdown and Haldon regions and in the Chloritic Marls of the South Devon Coast.

Foreign localities cited are all Neocomian; these are Saint Saveur, Vaux-sur-Blaize, Brillon, Gréaux, Vorey, near Besançon. Switzerland—Hautervive, near Neuchatel.

ADDENDA.

Species and Varieties of Trigoniæ figured in this Monograph which were not received sufficiently early to allow of the descriptions being placed in their proper sectional order.

TRIGONIA SCAPHA, *Ag.* Pl. XXXVIII, fig. 6 (SCAPHOIDEÆ).

<blockquote>
TRIGONIA SCAPHA, *Agassiz.* Trigonies, p. 15, pl. vii, figs. 17—20, 1840 (internal moulds).

— — *d'Orbigny.* Prodrome de Paléont., vol. ii, p. 79, No. 293, 1850.

— — *Pictet.* Paléont. Suisse, pl. cxxviii, figs. 6—8, 1866.

— HUNSTANTONENSIS, *Seeley.* On the Fossils of the Hunstanton Red Rock, Ann. and Mag. Nat. Hist., 3 series, No. 82, p. 276, October, 1864 (mould).
</blockquote>

Shell scaphoidal, anterior side short, somewhat truncated, posterior side produced, rather depressed; umbones elevated, somewhat recurved, superior border lengthened, nearly straight, curved posteally with the siphonal border, which is of moderate length, its lower extremity curving elliptically with the lower border. The escutcheon is moderately lengthened and depressed; the area is wide and slightly convex. The only specimen at my disposal is deprived of the test; the area therefore exhibits only the muscle scar; the other portion of the shell or pallial surface retains traces of the surface ornaments; there are a few nearly perpendicular subnodose costæ which pass downwards from the angle of the valve to the middle, where they meet the extremities of another oblique anteal series of similar but smaller costæ. The dental hinge processes are large.

Dimensions of the mould above described:

Length, 26 lines; height, 14 lines; thickness, 9 lines. The convexity is less than in the mould figured by Agassiz, which is probably the effect of vertical pressure.

The only British specimen known is in the Woodwardian Museum: Cambridge, the surface ornaments are obscure. It was deposited there by Mr. H. Seeley, and catalogued by him in his list of Hunstanton fossils under the name of *Trigonia Hunstantonensis.* 'Ann. and Mag. Nat. Hist.,' October, 1864. Subsequently in a communication to the same periodical Mr. Seeley expressed his belief that the fossil was not derived from the red rock of Hunstanton, but that it came from the drift of Norfolk. The general aspect of the mould agrees with the latter conclusion.

Position and Localities. The moulds figured by Agassiz are from the Neocomian formation at Voray, near Besançon. The specimen figured by Pictet ('Paléont. Suisse,' plate 128, figures 6—8) has the test preserved; the broad area with its median groove, large transverse plications, and bounding tuberculated carinæ is conspicuous; the first-formed six rows of pallial varices are nodose and transverse; they pass across the valve without interruption, the succeeding rows are broken and angulated about the middle of the valve; all are nodose.

Within the period during which these sheets were passing through the press I have been favoured by the Rev. T. Wiltshire with moulds of two Trigoniæ from his fine collection of Hunstanton fossils obtained by him in the Red Chalk (Gault) of that locality. The moulds represent a short suborbicular form distinct from *T. scapha*, but their condition of preservation will not allow of comparison with any other recognised species.

TRIGONIA EXALTATA, *Lyc.*, sp. nov. Pl. XXXVIII, fig. 2 (SCAPHOIDEÆ).

Shell scaphoidal, somewhat depressed, truncated anteally, much produced posteally; umbones subanterior, much elevated, incurved, and recurved, anterior side very short, forming a flattened surface anteally, curved elliptically with the lower border; superior border lengthened, concave. Escutcheon lengthened, plain, concave, its upper border raised. Area narrow, flattened, traversed longitudinally by an oblique mesial furrow and crossed by very numerous, irregular, wrinkled, and rugose plications which become prominent at the position of the inner carina and at the median furrow; at the outer angle of the valve near to the apex is a distinct small and narrow carinal elevation, which with advance of growth changes to a line of small papillated oblique elevations; these become evanescent over the posteal moiety of the valve which has no carinal elevation. The other portion of the shell is characterised by a series of nearly straight, large, nodose varices, about fourteen in number; the nodes are large, and rounded near to the angle of the valve; the more posteal six rows have the nodes more cordlike or unequal and irregular near to the pallial border; the first-formed few rows of varices are oblique and near to the anteal curvature of the valve are replaced by or are united irregularly with a smaller, more numerous, shorter series of varices, the few first formed of which become attenuated and pass across the flattened anterior surface of the valve to the pedal border; the junctions of the extremities of the two series are throughout very irregular; and the last-formed three or four posteal perpendicular rows are also very irregular. The lines of growth are large and widely separated.

Dimensions of the large specimen figured upon our plate :

Length 5¼ inches ; height 3½ inches ; thickness through the single valve 10 lines.

ADDENDA.

Affinities and Differences. The nearest analogue of this grand species is *T. navis,* Lam.; the latter has the general figure shorter and more convex; the anteal flattening of the valve is more pronounced; the escutcheon is wider, more excavated, and shorter; the smooth area and pallial rows of varices, attenuated upwards, are also very distinctive features.

T. scapha, Ag., a much smaller and shorter species, is characterised by the angularity formed by the two portions of the varices about the middle of the valve, by the smaller elevation of the umbo, by the absence of any distinct anteal flattening or truncation, and by the greater breadth of the valve at the siphonal border.

T. Robinaldina, D'Orb., from the Neocomian formation of Saint Sauveur (Yonne), one of the *Scaphoideæ,* has the general convexity much more considerable, so that the diameter through the united valves is equal to three fifths of the length; the area is without tubercles at the angle of the valve; it has no median furrow and has no clear separation from the pallial surface; the escutcheon is short and has great breadth.

Position and Locality. Our figure represents the largest known example of the *Scaphoideæ;* the original is a specimen in the British Museum which bears the inscription Drift of Norfolk; its history is unknown. The general aspect indicates that its original seat was the Middle Neocomian formation of West Norfolk, a locality from whence the specimens of *T. ingens* in the Lynn Museum were derived and probably also the mould of *T. scapha* in the Woodwardian Museum last described.

TRIGONIA PULCHELLA, *Ag.* Pl. XXXVIII, figs. 10, 11, 12, 12 a (SCAPHOIDEÆ).

TRIGONIA PULCHELLA,	*Agassiz.*		Trigonies, p. 14, tab. ii, figs. 1—7, 1840.
—	—	*Marcou.*	Recherches Géologiques sur le Jura Salinois, Soc. Géol. de France, § 47, 1846.
—	—	*Terquem.*	Paléont. du Dép. de la Moselle (Extr. de la Stat. de la Moselle), p. 23, 1855.
—	—	*Quenstedt.*	Der Jura, p. 311, tab. xliii, fig. 1, 1857.
—	—	*Oppel.*	Jura-Formation, Würtemb. Natur. Jahreshafte, 12—14 Jahrg., p. 406, No. 426, 1857.
—	—	*Quenstedt.*	Handbuch der Petrefactenkunde, tab. xliii, fig. 14, 1867.
—	—	*Lepsius.*	Beiträge zur Kentniss der Jura-Formation im Unterelsass, p. 48, 1875.

Shell small, subquadrate, moderately convex mesially, depressed anteally and posteally; umbones small, anterior, little elevated; hinge-border lengthened, straight; anterior side perpendicular, abruptly truncated; lower border slightly curved, having near to its posteal extremity two or three short perpendicular folds corresponding with as many depressions and prominences in the interior of the shell; its extremity forms nearly a right angle with the siphonal border, the superior extremity of which forms a similar angle with the posteal extremity of the lengthened, narrow, elliptical escutcheon. The area is somewhat convex, its size is nearly equal to the other portion of the valve; it has a series of large acute transverse costellæ, which are somewhat broken at the position of the usual median furrow, which is not distinct. The marginal carina is represented by a row of regular, minute, rounded papillæ. The other or pallial portion of the valve has a small series of subnodose varices which are variable in figure; occasionally the anteal varices are curved and entire, more frequently they are broken mesially, when their anteal portions become horizontal; posteally there are two or three short perpendicular rows. The interiors of the valves have the nacreous layer of the test preserved; the hinge processes are unusually large; the borders of the valves are plain excepting the depressions and prominences near the posteal extremity of the pallial border.

The anteal truncation and the characters of the varices ally it to the *Scaphoideæ*; the internal pits near to the posteal extremity of the pallial border resemble a similar feature in the Cretaceous *Quadratæ*, from which section, however, it is separated decisively by its plain escutcheon.

The *T. pulchella* of Reuss ('Die Versteinerungen der Böhmischen Kreideformation,' tab. 41, fig. 3) is a minute form, having no affinities with the species to which Agassiz had previously given the same name; only two lines in length, it is allied to and may possibly represent the very young condition of *T. disparilis*, D'Orb. ('Pal. Fran. Terr. Crét.,' vol. 3, plate 299), which it resembles in its radiating, knotted, unequal costæ, and in the transverse costellæ which cross both the area and escutcheon; it has not occurred in Britain.

The stratigraphical range of *Trigonia pulchella*, Ag., appears not to be limited to a single geological position; it was first obtained by M. Gressly at Urweiler and at Mühlhausen (Department of the Haut-Rhin), in beds which were assigned by himself and by Agassiz to Upper Lias, but regarded by them as representing a peculiar and local development of that formation.

M. Terquem obtained this *Trigonia* in the bed of Marly Sandstone or Grés Supraliassique of the Department of the Moselle, associated with *Trigonia navis*, *T. litterata*, and a series of Testacea, several of which have their equivalents in some portion of the Supra-liassic Sands of England, in the Cotteswold Hills, and in the southern counties. The more northern or Yorkshire development of these sands presents differences both lithological and palæontological.

Professor Quenstedt from an extensive knowledge of the strata of Urweiler and its numerous Testacea exemplified in plates 42, 43 of his 'Der Jura,' determined the position of *T. pulchella* to be the lowest zone of Inferior Oolite or that of *Ammonites torulosus*, associated with *Trigonia navis*, *Ammonites opalinus*, *A. Hircinus*, *Venulites trigonellaris*, *Trigonia similis*, &c., a remarkable association of Testacea which is almost wholly distinct from any British assemblage of Inferior Oolite fossils. Several of these forms have their analogues or are even identical with Cotteswold species of the Supraliassic Sands.

Professor Oppel also ('Juraformation,' p. 406) referred *T. pulchella* to the lower portion of the Inferior Oolite, of which he regarded it as a characteristic species; he recorded its occurrence at Gundershofen (Bas Rhin), Milhau (Aveyron), and Metz (Moselle).

It is only recently that *T. pulchella* has become known as a British species; its discovery in a lower stratigraphical position has resulted from the persevering researches of Mr. Keeping, of the Woodwardian Museum, Cambridge, who obtained specimens in the Upper Lias at Bracefield brick-pits, near Lincoln, associated with *Ammonites serpentinus*, *A. bifrons*, *A. communis*, and other Ammonites special to that zone. The Lincoln specimens are smaller than those figured by Agassiz, and by Quenstedt in his 'Der Jura;' they agree better with the figure given by the latter author in his 'Handbuch der Petrefactenkunde,' tab. 43, fig. 14. It has not been obtained at any other British locality.

TRIGONIA AFFINIS, *Miller*. Pl. XXI, fig. 7, also Pl. XL, fig. 2 (GLABRÆ).

TRIGONIA AFFINIS, *Sow*. Min. Conch., tab. ccviii, fig. 3, vol. 3, 1818.

The description of this species and also of *T. excentrica* given at pages 94 and 95 require to be supplemented by the revised present description.

Shell short, convex mesially, umbones not prominent, erect, anterior side produced and rounded, hinge-border short, its outline slightly convex, curved elliptically, with the posteal extremity, which is obtusely rounded. The sides of the valves have numerous small regular subconcentric costæ which have prominence anteally even to the border; they disappear rather suddenly about the middle of the valve, and the posteal half of the shell is plain; the posteal slope is slightly flattened and is distinct only near to the apex. The lines of growth are only slightly defined; there are three arrests of growth, which are also only obscure. The hinge characters have not been exposed.

The specimen, Plate XL, fig. 2, contributed by Mr. Vicary, has the length 31 lines, the height 24 lines, the convexity of a single valve $7\frac{1}{2}$ lines.

Compared with *T. excentrica* the general figure is larger, much shorter, or more equilateral; the umbones have less prominence, the costæ are more numerous, imperfectly developed, and nearly approach the concentric figure. Specimens of advanced growth acquire much thickness and their costæ become evanescent. For the most part the specimens upon the tablets of our Greensand Collections, both public and private, are very indifferently preserved and are not separated from specimens of *T. excentrica*.

Positions and Localities. The specimen figured, Pl. XL, fig. 2, was obtained by Mr.

Vicary at Great Haldon in the pebble-bed which overlies the Blackdown and Haldon Greensand, associated with other characteristic *Trigoniæ* of the Upper Greensand. Another imperfect specimen was obtained by Mr. Meÿer in the Chloritic Marls in the vicinity of Axmouth. It appears to be rare.

TRIGONIA DUNSCOMBENSIS, *Lyc.*, sp. nov. Pl. XL, figs. 5, 6 (GLABRÆ).

Shell subovate or ovately oblong, very convex, umbones antero-mesial, prominent, not recurved; superior border slightly concave, its posterior extremity curved elliptically with the siphonal border, which is short and curved with the inferior border; the anterior border is produced and rounded. In the young condition the costæ are very closely arranged or almost linear; they pass horizontally across the valve and are slightly undulated at each of the extremities in a manner resembling *T. excentrica*, but less pronounced. With advance of growth the costæ become narrow, prominent, and much more distantly arranged, their prominence continues in well-preserved specimens even to the lower border; they disappear posteally, so that about one third of the valve is plain. The valves have no decided arrest of growth, but the lines of growth are irregular and strongly defined.

The nearest ally is *T. excentrica*, compared with which the general figure is shorter or less produced posteally; the convexity is much more considerable; the costæ are more prominent, less excentric anteally, the rows are also more widely separated, more especially upon the middle portion of the valve and near to the lower border, at which part of the valve in *T. excentrica* the costæ are usually evanescent.

Dimensions of the specimen, Pl. XL, fig. 5 :

Height 21 lines; length 26 lines; diameter through the single valve 9 lines.

The test is of considerable thickness and is often found separated into two layers, of which the inner layer is of much the greater thickness; its surface consists of a series of minute, closely set perpendicular lines, often of unequal size; they are more or less impressed by distantly arranged concentric lines of growth.

Numerous specimens have been placed at my disposal by Mr. Meÿer, collected by him in the Chloritic Marls of the South Devon Coast, near Sidmouth; more especially in the cliffs at Dunscombe, Branscombe, and Beer Head. All the specimens are more or less imperfect or fragmentary and their outlines are rarely preserved entire; they afford little information concerning the hinge processes or of the interior of the shell; as, however, large portions of the surface are often very well preserved, the costæ can be compared with and separated from the *T. excentrica* of the Blackdown Greensand. The general very defective condition of the *Trigoniæ Glabræ* in the Upper Greensands will account for my having mistaken some of them for *T. excentrica* as at p. 96. I therefore

ADDENDA. 189

take the present opportunity of stating that in no instance does it appear certain that an example of *T. excentrica* has occurred in a higher position than the Blackdown Greensand.

TRIGONIA DEBILIS, *Lyc.*, sp. nov. Pl. XL, figs. 8, 8 *a* (GLABRÆ?).

Shell small, subovate, moderately convex, umbones pointed, produced, antero-mesial; costæ upon the sides of the valves, depressed, rounded, very closely arranged, horizontal, or transverse, each having an undulation and becoming smaller posteally where their extremities are suddenly turned upwards, meeting the divisional line of the valve at a considerable angle. The small posteal slope has the costellæ crowded, transverse, and scabrous.

The single specimen figured is broken at the siphonal and lower borders; the height and length are nearly equal, or 5 lines. The costæ are thirteen in number, but apparently five others would be required to render the shell perfect to the apex.

Probably this is only the young condition of a much larger species; it has some resemblance to young shells of the *Glabræ*, but is quite distinct from either of the known Greensand species in their young states.

Collected by Mr. Meÿer in the bed No. 10 of the Chloritic Marls of Dunscombe Cliffs.

TRIGONIA CRENULIFERA, sp. nov. Pl. XL, figs. 1, 1 *a*, 1 *b*, 7, 9, 9 *a* (SCABRÆ).

Shell near to the general figure of *T. scabricola*, but shorter and wider, or more expanded posteally; the umbones are remarkable for their great elevation and their recurvature; the anterior side is very short, but is curved elliptically with the lower border; the anterior face of the shell has little convexity with considerable breadth. The hinge-border is concave and short; the escutcheon is very large and concave, its borders are raised; it is traversed transversely by a series of scabrous costellæ which are not altogether regular and have their direction somewhat oblique; they curve irregularly where they pass across to the area, which has greater breadth than in species generally of the *aliformis* group; it has much convexity, is conspicuously bipartite, its boundaries, separating it from the escutcheon and from the sides of the valves, are well defined; it has a deep mesial furrow which interrupts a series of prominent crenulated costellæ which are disposed somewhat irregularly or nearly in zigzag order; in the adult condition the posteal portion of the area widens and becomes more flattened; the costellæ are ultimately replaced by irregular transverse plications.

The sides of the valves are covered by a very numerous and closely arranged series of crenulated costæ sometimes twenty-five or more in number, small at the divisional angle of the valve; they enlarge downward, the summits of the rows are everywhere narrow, and delicately crenulated; their direction is obliquely downwards over the middle portion of the valves; the first formed or more umbonal rows are curved forwards anteally, becoming attenuated and almost horizontal upon the anteal face of the shell.

The internal mould is usually seen in the South Devon specimens; the dividing ridge of the area is strongly defined; the lower borders of the valves are slightly dentated and the impressions of the costæ are more or less visible, resulting from the considerable attenuation of the test, a feature in which it differs from allied species of the *aliformis* group.

Dimensions. Length of the largest specimen figured 32 lines. Height 30 lines; convexity of a single valve 10 lines.

Affinities. Compared with *T. crenulata* from Le Mans, our species has the costæ usually more prominent, rugose, and less numerous; the area and escutcheon are more steep, or form a more considerable angle with the other portion of the shell; the posteal slope in *T. crenulata* is therefore more expanded and more fully exposed when a valve is placed horizontally and viewed from above; the median furrow is distinct, but is without the deeply impressed groove of *T. crenulifera*; but the chief distinction consists in the prominent zigzag costellæ upon the area and escutcheon which imparts a characteristic aspect to that portion of our shell.

T. crenulata is exemplified in specimens from Le Mans in the 'Paléontologie Française' of D'Orbigny, vol. 3, pl. 295, which has the area destitute of costellæ. The splendid specimen figured by Agassiz, 'Trigonies,' pl. vi, figs. 4, 6, obtained at the same locality, is depicted with very small irregular crenulated transverse costellæ upon the umbonal or anteal portion of the area—a varietal character forming some approximation to, but distinct from, the prominent and peculiar costellæ of our species. Compared with *T. scabra*, Lam., the latter form has the costæ upon the sides of the valves much larger; they pass across the area and escutcheon continuously, much reduced in size, but are not interrupted by the elevated boundaries which mark the limits of that portion of the shell; they are therefore without the large irregular or broken costellæ of our species.

The peculiarities of the costellæ possess some resemblance to a similar feature exhibited upon the area and escutcheon in one of the *Scabræ* from Bogota, figured and described by D'Orbigny under the name of *T. subcrenulata*, ' Coquilles foss. de Colombie,' pl. iv, figs. 7, 8; the latter shell has the general figure much more inflated, with a much smaller and more concave area, which, however, is altogether without the deeply impressed median groove of *T. crenulifera*, the junction with the escutcheon is ill defined, the transverse costellæ have an undulation not broken by a median groove as in *T. crenulifera*; the crenulated costæ upon the sides of the valves are very small and widely separated.

ADDENDA.

Stratigraphical positions and Localities. Numerous specimens more or less defective in their general condition have been obtained by Mr. Meÿer, in hard rock of the Chloritic Marls in cliffs between Beer Head and Sidmouth, also from Pinhay Cliff near Lyme Regis; they present some variability in the prominence of the lines of growth, and also in the costellæ upon the area and escutcheon. Mr. Meÿer states that specimens occur of much larger dimensions in the cliffs near Dunscombe; but from the hardness of the rock he has found it impossible to get them out.

Associated with them are the following species of *Trigonia* — *T. sulcataria, T. pennata, T. Meyeri, T. Vicaryana, T. Archiaciana, T. scabricola;* and in the cliff a large specimen has been seen of what Mr. Meÿer believes to be *T. quadrata,* Ag., and also indications of another large species. Ample and valuable imformation upon the position of our species in connection with its localities and the associated Testacea will be found in a paper by Mr. Meÿer, 'Quart. Jour. Geol. Soc.,' August, 1874. The beds 10 to 12 of the classified section in the memoir contain one species, there assigned doubtfully to *T. crenulata.*

The specimens from Pinhay Cliff are usually more rugose, their costæ are more highly ridged, their crenulations indistinct, the costæ are of larger size and fewer in number, probably it should be regarded as a variety, but their usual condition of preservation forbids any rigid comparison.

With the numerous Cretaceous *Trigoniæ* kindly forwarded to me by Mr. Cunnington, were several examples from the Upper Greensand of Potterne, near Devizes, which although partially deprived of the test retained some portions of their surface ornaments upon the escutcheon and area, and should apparently be referred to the present species; they are now deposited in the Museum of the Royal School of Mines.

T. crenulata, Lam., has also been recorded by Professor R. Tate, in the Hibernian Greensand of the North East of Ireland, 'Jour. Geol. Soc.,' 1864, p. 30. As the position accords with that of our species, it is not improbable that both of them represent the same *Trigonia.*

TRIGONIA CYMBA, *Cont.* Pl. XXXVIII, fig. 1 (CLAVELLATÆ).

> TRIGONIA CYMBA, *Contejean*. Étude de l'Etage Kimmeridien dans les env. de Mont-
> beliard, et dans le Jura de France et Angleterre,
> Extr. Mém. de la Soc. d'emulation du Doubs, pl. xiv,
> figs. 1, 2, 1859.

One of the *Clavellatæ*, remarkable for the considerable elongation of the valves posteally, for the small curvature of the rows of costæ which are nearly horizontal, for their inconspicuous tubercles, and for the small development of the ornaments upon the valves generally.

The general figure has some resemblance to the more lengthened forms of the *Pholadomyæ*; the umbones are large, elevated, and nearly erect, placed within the anterior third of the valves; the anteal portion of the shell has considerable convexity, the posteal and more lengthened portion is comparatively much depressed. The escutcheon is lengthened, narrow, flattened, and slightly depressed. The area is narrow, bounded upon each side by a row of minute tubercles over the anteal or umbonal half of its length; over the same portion there is also a distinct median furrow and delicate transverse plications; these ornaments disappear about the middle of the valve; the posteal half of the area has only transverse rugæ which are not very strongly defined; it is also much depressed. The other portion of the shell has rows of clavellated costæ about fifteen or sixteen in number, small, nearly horizontal, or coinciding in their direction with the lines of growth; anteally they do not extend to the border of the valve. The tubercles in the rows are small, nearly equal, compressed, little elevated, and near to the lower border they become small scabrous elevations.

Length of the specimen figured $4\frac{1}{4}$ inches; height 2 inches; convexity of the single valve 7 lines.

The imperfect example herewith figured was obtained by J. C. Mansel Pleydell, Esq., in Portland Sand of the cliffs in Kimmeridge Bay, Dorsetshire. The minuteness and delicacy with which the characters of the surface have been preserved leave little cause to regret the absence of the test, but the surface ornaments have much less prominence than in the fine specimen figured by M. Contejean, from Mont. Beliard, at which locality he states that it is abundant.

The unusually lengthened and much curved figure, together with the numerous, small, nearly horizontal rows of costæ, and inconspicuous closely set tubercles, serve to separate it from other clavellated *Trigoniæ* of the Upper Oolites. The convexity anteally is considerable and contrasts with the lengthened and depressed posteal portion of the valves; these comparisons refer more especially to *T. Pellati*, Mun., *T. Cottaldi*, Mun., *T. Alina*, Cont., and *T. muricata*, Goldf.

ADDENDA.

TRIGONIA ALINA, *Cont.* Pl. XXXVIII, fig. 3 (CLAVELLATÆ); also variety, Pl. IX, fig. 2 (corrected).

TRIGONIA ALINA, *Contejean.* Étude de l'Etage Kimmeridien dans les environs de Montbeliard et dans le Jura de France et Angleterre, Extr. Mém. de la Soc. d'Emulation du Doubs., pl. xiv, figs. 3—5, 1859.

— — *P. de Loriol, E. Royer, et H. Tombeck.* Descr. Géol. et Paléont. des Étages Jurassiques supérieurs de la Haut Marne, Mém. Soc. Linn. de Normandie, tom. xiii, fig. 5, pl. xvii, 1872.

Shell ovately trigonal, convex, short antcally, lengthened posteally; umbones anteal, prominent, obtuse, much arched inwards. Hinge-border lengthened, slightly raised, its posteal extremity forming an obtuse angle with the short siphonal border; anterior and lower borders curved elliptically. Escutcheon large, lengthened, and concave, its length being equal to the height of the valve. Area narrow and flattened, its width not exceeding that of the escutcheon; it has the usual median furrow, and has transverse irregular plications which enlarge slightly at their extremities, forming small bounding elevations or carinæ.

The other portion of the shell has numerous, closely arranged rows of small nodose varices (our specimen, a small one, has seventeen rows); their general direction have but little curvature, but are not altogether regular and symmetrical in their course; their posteal extremities form right angles with the angle of the valve which they touch; the nodes in the rows are small, rounded, closely arranged, often touching each other, and vary little in size, excepting that the three or four rows last formed have the nodes larger and less closely arranged; but as our sole specimen is not of adult growth this feature has but little significance; the few first formed or umbonal rows are nearly plain or slightly knotted.

Compared with other allied examples of the *Clavellatæ*, the distinctive characters consist of the narrow and nearly smooth area, the large escutcheon, the general figure of the shell curved and lengthened posteally, the considerable number and close arrangement of the small nodose costæ, together with the small curvature which the rows make upwards near to the angle of the valve.

Upon comparing the original figure in the work of Contejean with that of de Loriol, Royer, and Tombeck, some differences exist; the latter has the rows of costæ less numerous, the tubercles are fewer or more widely separated in the rows, the diminution in the size of the few last-formed costæ is also a distinctive feature; the general figure of the shell has greater height, and is more attenuated at the posteal extremity; it would appear also that the rows of costæ have a greater curvature upwards towards the carina than obtains in the specimen figured by Contejean.

Stratigraphical position and Locality. The specimen upon which our description is founded is deposited in the Museum of the University of Oxford, and was obtained in the Portland Limestone of Shotover Hill; it is of less advanced growth than the specimen figured by Contejean, and still less so than the one figured by De Loriol; the area and its bounding carinæ are only indifferently preserved, so that their little plications are not shown.

For this addition to the *Trigoniæ* of the Portland formation I am indebted to the liberality of the late Professor Phillips, who obtained and forwarded to me well-executed plaster casts of all the *Trigoniæ* in the Geological Museum of the University of Oxford.

Increased knowledge of the *Clavellatæ* of the Kimmeridge Clay has led me to regard the specimen figured, Plate IX, fig. 2, and there given as a supposed variety of *T. incurva*, as referable rather to a variety of *T. Alina*, Cont., having fewer costæ than the shell of the Portland Oolite, but possessing no other distinctive feature.

TRIGONIA HUDLESTONI, *Lyc.*, sp. nov. Pl. XXXIV, figs. 5, 6; Pl. XXXIX, figs. 1 *a* and 2 (CLAVELLATÆ).

Shell ovately trigonal, depressed, excepting the posterior slope which is steep and convex. Umbones prominent, pointed, nearly erect; anterior side very short, its border curved elliptically with the lower border; hinge-border straight, lengthened, sloping downwards, forming less than a right angle with the anterior border. Area narrow, flattened, transversely delicately plicated; marginal carinæ small, minutely and densely tuberculated, excepting upon its lower third, where the tubercles disappear, and it becomes plicated or obscurely nodose; inner carina large, transversely prominently plicated; median carina small and distinct, represented at the upper half of the shell by a row of minute, closely placed tubercles, which become obsolete over the lower half of the area. Escutcheon depressed, somewhat excavated, flattened obliquely, irregularly plicated; its length is considerable, or exceeding half the length of the entire valve; no part of its surface is visible when a valve is placed horizontally and viewed from above.

The rows of tuberculated costæ are about eighteen in the fully developed form; they are narrow and elevated with about twelve or thirteen tubercles in each row, the rows are widely separated, have little curvature, their carinal extremities become nearly perpendicular and are much attenuated and imperfectly subnodose or cord-like.

The most prominent distinctive features in this large species consist in the short depressed, subtrigonal form, the narrow area, its steep slope; the widely separated narrow costæ attenuated at both of their extremities, together with the considerable angle which they form with the marginal carina.

The specimen figured, Pl. XXXIX, figs. 1 a, 2, exposes the diverging hinge processes of the right valve which are unusually lengthened, more especially the posteal one, which is upwards of eighteen lines in length, the entire border of the escutcheon having a length of thirty-two lines; the nymphal plate is also much lengthened.

For the loan of this fine *Trigonia* I am indebted to W. H. Hudleston, Esq., who states " that he obtained it in a limestone quarry at Cawklass, in the North Riding of Yorkshire, in a compact calcareous stone full of sparry shells, and having a few oolite granules. This rock belongs in all probability to the upper part of the Coralline Limestones associated with Corals, though there are no Corals in this bed."

The imperfect specimen, Pl. XXXIV, fig. 5, is from the Coral Rag of Heddington, near Oxford, and is deposited in the Oxford Museum.

The smaller specimen, Pl. XXXIV, fig. 6, has the outline nearly perfect; it is from the Elsworth Rock, Cambridgeshire, and belongs to the collection of Mr. J. F. Walker, of York.

Compared with the allied species *T. Alina*, the general figure has less convexity; the rows of costæ are smaller and less curved, or become nearly perpendicular as they approach the carina, and are much attenuated at each of their extremities; the tubercles in the rows are irregular and unequal.

TRIGONIA BRODIEI, *Lyc.* Pl. XXXV, figs. 8, 9 (CLAVELLATÆ).

TRIGONIA STRIATA, *Quenstedt.* Handbuch der Petrefactenkunde, tab. xliii, fig. 13 (not *T. striata*, Mill.), 1867.

Shell smaller than *T. striata*, more oblong, with greater convexity. Umbones large, prominent, obtuse, much recurved, placed one third the length of the shell from the anterior border, which is convex and prominently rounded; the superior or hinge-border is lengthened and concave; the area is narrow, transversely delicately striated, having a mesial furrow, and bounded by minutely tuberculated carinæ; they form together with the area a concave space at the hinge-border. The costæ, about twelve in number, are narrow and elevated; their tubercles are small, closely arranged, irregular, rounded, and attenuated anteally; the rows of costæ have considerable curvature, they are more widely separated than is usual in allied species of the same group.

As a British *Trigonia* this species appears to be rare; my knowledge of it is limited to the two specimens herewith figured; the left-hand specimen is from the collection of the Rev. P. B. Brodie, and was obtained by him in the Inferior Oolite (Northampton Sands) of Milcomb Hill, Oxon., a locality described by Mr. Beesley, in his excellent little memoir on the Geology of Banbury (Warwickshire Nat. Field Club, 1872); the numerous Inferior Oolite fossils recorded at the locality in question contain *Trigonia costata, T. signata, T. formosa,* and *T. Phillipsi,* the last named being intended for the present species; the locality, with a list of the fossils, is also mentioned by Prof. J. W. Judd, in the 'Memoirs of the Geological Survey' (Rutland, &c.), pp. 23 and 25, 1875, and in Phillips's 'Geology of Oxford,' Appendix B, p. 512. The other specimen figured agrees entirely in its aspect and matrix with the Milcomb specimen, and I have no doubt pertains to the same rock; it belongs to my collection, but unfortunately no note has been retained of the locality.

The small specimen figured by Quenstedt is slightly more lengthened, but offers no other material distinction; it possesses some differences with the *T. striata* of the same author ('Jura,' pl. xlvi, figs. 2, 3); both forms are sufficiently distinct from the specimens figured in the 'Mineral Conchology.'

TRIGONIA KEEPINGI, *Lyc.*, sp. nov. Pl. XXXV, figs. 1, 2 (CLAVELLATÆ).

Shell with the general figure much shorter and more convex than *T. ingens;* it has also much greater breadth across the pallial surface of the valve; the escutcheon is depressed, of moderate breadth, its upper border is somewhat raised. The area is comparatively narrow and slightly convex or raised; it has a well-marked median furrow; it is crossed by some irregular and unequal plications, which differ much in prominence upon the two specimens at my disposal. The marginal carina is represented by a row of large rounded, or ovate, closely arranged nodes; there is a row of minute papillary prominences at the position of the inner carina, but the area generally is irregular in its ornamentation. The other portion of the valve has the rows of costæ very numerous, regular, and closely placed; they meet the carina at a considerable angle and have only a small curvature; the nodes in the rows, sixteen or more in each, are nearly equal in size, rounded, prominent, and very closely arranged those; of the last-formed two or three rows become ovate; their longer diameters are across the rows. Upon the whole the rows diminish somewhat in size near to the marginal carina. *T. ingens,* Lyc., and *T. Keepingi,* Lyc., represent the only examples of the *Clavellatæ* known in the Cretaceous Rocks.

The name adopted for this species is that of the able Curator of the Woodwardian Museum, Cambridge, who obtained the specimens in the Middle Neocomian formation at Acre House near Tealby, and kindly brought them to my notice. For ample information respecting the geological position of the Tealby beds and their relations to Neocomian strata at other localities refer to three memoirs by Prof. J. W. Judd, 'Quart.

ADDENDA.

Jour. Geol. Soc.,' vol. xxiii, p. 227, 1867; vol. xxiv, p. 218; also vol. xxvi, p. 326, 1870.

The difficulties attendant upon the examination of Clavellated *Trigoniæ* with fragile crystalline tests preserved in a hard matrix have necessitated the use of specimens defective in their outline, but otherwise in a fine condition of preservation; they were obtained in the vicinity of Tealby, in the bed of hard limestone which has also yielded *T. Tealbyensis*, a bed distinct from the brown ferruginous pisolite of the same locality, which is the source of the specimens of *T. ingens*, Pl. XXXVI, figs. 5, 6. The originals of *T. Keepingi* are in the Woodwardian Museum, Cambridge, and had not been discovered when p. 24 was printed, where *T. ingens* is stated to be the only example of the *Clavellatæ* known in the *Cretaceous Rocks*. The general aspect of these clavellated forms is such that they would readily be mistaken for Jurassic *Trigoniæ* in the absence of all knowledge of their geological position—a fact which appears to lend some support to the views of those palæontologists who would arrange the Neocomian formation with the Upper Jurassic rather than with the Cretaceous Rocks. One distinctive feature exemplified in these delicately preserved *Trigoniæ* consists in the lines of growth having their edges minutely and densely fringed or granulated, visible only under a magnifier, and resembling a similar feature in the Cretaceous *Glabræ*, as in *T. excentrica*; a kind of surface which is absent in the Jurassic *Clavellatæ*, and is distinct from the epidermal tegument which is not preserved.

TRIGONIA WITCHELLI, *Lyc*., sp. nov. Pl. XXXVIII, figs. 8, 9. (CLAVELLATÆ.)

For examples of this small species I am indebted to Mr. Witchell, of Stroud, who discovered them in the Fuller's Earth of that locality; their condition of preservation is only indifferent, the test is not preserved, the ornaments of the surface are in relief upon a black pigment, which renders them distinct and prominent, notwithstanding their minuteness and delicacy. They possess some general resemblance to *T. imbricata*, Pl. XXXVI, figs. 9, 10, but are sufficiently distinct. Compared with that little species, they are larger, more lengthened, and pointed posteally; the umbones are more elevated and produced; the surface ornaments are more minute.

The area is narrow, having three very small plain, linear carinæ, and is traversed transversely by a few widely separated, regular costellæ. The escutcheon is narrow, its superior border is somewhat raised. Upon the sides of the valves the rows of tuberculated costæ, about 9 or 10 in number, are concentric, much curved, equal in size, small, and minutely papillated. The larger specimen has two of the costæ with their anteal portions broken, and their papillæ scattered irregularly; their figures are slightly lengthened and pointed downwards. The very diminutive size of the costæ causes the rows to appear widely separated—a feature which at once serves to separate it from *T. imbricata.*

T. Witchelli has also some affinities with a little Kimmeridge Clay species kindly communicated by Wm. Topley, Esq., and known only from a partially exposed posteal portion of the valve in black shale, brought up in the Sub-Wealden exploration from a depth of 402 feet. The general figure appears to be similar, but the area has no median carina; the rows of costæ are more closely arranged, their nodes are also larger; each of them is slightly prolonged or pointed downwards; they are distinct in the rows : apparently about a third part of the valves is exposed.

Dimensions. The length of the largest specimen of *T. Witchelli* measured upon the marginal carina is 9 lines; the opposite measurement is 6 lines; apparently the species has but little convexity.

Position and Locality. The few specimens obtained are ill preserved; they are the sole representatives of the genus hitherto known in the Fuller's Earth. The matrix is a soft pale grey marly rock; in the same bed were fragments of *Ammonites Parkinsoni*. Hitherto this *Trigonia* has been obtained only at a single locality adjacent to the town of Stroud.[1]

TRIGONIA SNAINTONENSIS, *Lyc.*, sp. nov. Plate XLI, figs. 1, 2.

Shell having a resemblance in its general aspect and the surface ornaments to *T. recticosta* (p. 16, Pl. I, figs. 4—6), but having a much more considerable convexity; the area also differs materially in its more narrow figure and in the absence of a median furrow; in the latter feature it exactly resembles *T. gemmata* (p. 15, Pl. I, fig. 7), and is similarly bounded by a plain marginal carina and a minutely papillated inner carina. The first few rows of costæ are concentric and tuberculated as in *T. recticosta;* the others, whose perpendicular direction accords with those of the latter species, are larger, more ridge-like, more elevated, and are much less distinctly and less regularly tuberculated; they are also somewhat fewer in number; the whole aspect of the surface has, therefore, much less neatness and minuteness in its ornamentation.

The imperfect examples figured are almost the only specimens known; they were obtained by Mr. W. H. Hudleston, and are communicated for the present Monograph.

[1] With much regret I announce that the two specimens so carefully figured on Pl. XXXVIII were lost on their return to Stroud through miscarriage of the post. It may be hoped that the loss is not altogether irreparable, as, although the species has been obtained only at one locality, other specimens are, I believe, in the cabinet of Mr. Witchell. I am not aware that in any other instance a loss of fossils has occurred in transmission through the post.

ADDENDA.

The larger of the two specimens has the costæ somewhat more oblique, but does not appear to differ as a species from the smaller specimen. The position and locality is the Lower Calcareous Grit of Snainton, near Scarborough.

TRIGONIA RUPELLENSIS, *D'Orb.* Page 28, Plate VIII, fig. 4; also Plate XXXVI, figs. 1, 2, 3, 4.

During many years the original of our figure Pl. VIII, fig. 4, was the only well-preserved British specimen known. Recently a considerable number, representing every stage of growth and varying greatly in figure and in surface ornaments, have been procured in shore beds of Kelloway Rock at Cayton Bay, near Scarborough. The rock, hard, subsiliceous, varying in colour and structure, has been found to contain over a small area a profusion of these *Trigoniæ*, with both separated and united valves; but, owing to the intractable and tough matrix, only a small minority have been obtained in a condition suitable for the cabinet. The additional figures, on Pl. XXXVI, exhibit considerable differences both in the figure of the shell and in the surface ornaments.

In some instances, as in Pl. VIII, fig. 4, the form is ovately trigonal and short posteally; the costæ are curved, and have not much general irregularity; more frequently the figure is ovately oblong, lengthened posteally; the costæ, or some of them, are broken mesially, angulated, directed anteally; or in other examples the rows of costæ descend obliquely in a confused and irregular manner to the pallial border. The surface ornaments generally have so much irregularity that scarcely any two specimens fully developed present any near approximation in their general aspect. The nodes in the rows also partake of the general variability; usually the larger nodes are those near to the marginal carina; they are rounded, obtuse, and depressed; the smaller nodes are more compressed, pointed, and elevated.

Specimens representing the earlier stages of growth have but little of the variability exhibited by more adult forms; they might readily be mistaken for young examples of other clavellated species, and have therefore not been figured upon our plates.

Numerous specimens have suffered compression, or have their tests only partially preserved; a few examples have the valves in contact, and in such the internal moulds have been more or less exposed. The cardinal processes are small, the hinge-border is concave, the posteal portion is depressed, the borders are rounded; the test is thin; the lines of growth are large, uniform, and conspicuous whenever the surface is well preserved.

Notwithstanding the considerable differences of figure and of surface ornaments, it does not seem possible to arrange them as distinct varieties; the additional figures illustrate sufficiently the several aspects of this species.

Compared with *T. Scarburgensis*, Pl. IV, figs. 1—4 (which is also a very variable species, almost limited stratigraphically to the bed of Cornbrash), the short example of *T. Rupellensis*, Pl. VIII, fig. 4, and other similar forms, show only remote alliance; neither does *T. Rupellensis* exhibit that considerable difference in the surface ornaments of opposite valves of the same shell, which so commonly occurs in the Cornbrash form. Upon the whole there is much affinity between the more lengthened forms of the two species; and, if we exclude from comparison the greater number of the left valves of *T. Scarburgensis* similar to the specimen, Pl. IV, fig. 3, it will occasionally be found difficult to give any definite distinction between the two forms. Generally it may be stated that the variability in the costæ, and also in the general figure of the valves, is much more considerable in the Kelloway Rock form, and that none of the latter have the figure so much lengthened and depressed posteally as the Cornbrash species; the latter also have the rows of costæ usually more horizontal, and they approach the carina at a lesser angle than in *T. Rupellensis*. The result of an ample comparison of specimens has been to confirm the propriety of retaining the two allied forms as separated both by palæontological and stratigraphical distinctions.

TRIGONIA UNDULATA, *From.*, var. ARATA. Page 77, Plate XVI, figs. 9—11; Plate XVII, figs. 5, 6.

The British examples figured Pl. XVI, figs. 9—11, and Pl. XVII, figs. 5, 6, described at p. 77, and alluded to p. 48 may be regarded as a variety of *T. undulata*, figured by Agassiz ('Trigonies,' tab. 10, figs. 14—16) from the Great Oolite of Piedmont. Other fine examples of the typical form have since been obtained in the mountain district of the Lebanon to the eastward of the town of Beyrout; specimens from the latter locality have been known under the name of *Trigonia Syriaca*. Compared with the British variety *arata*, it has somewhat greater convexity upon the middle and umbonal portions of the valves; the marginal carina has greater prominence and the siphonal border is more lengthened or more oblique, thus shortening the length of the hinge-border. As these differing features are very persistent, there can be no doubt of the propriety of separating the British fossils as a variety when compared with the typical form from Italy and from the Lebanon. All the latter specimens examined have the last-formed costæ scarcely developed. The subjoined engravings represent a Syrian specimen of full dimensions.

It may be a subject of doubt whether our British variety *arata* may not be fitly separated from the continental or typical form, and constitute a distinct species. Our Cornbrash and Great Oolite specimens possess much variability, and more than one of them which have come under my notice are separated but little from the typical form in the almost entire absence of tubercles upon the costæ.

ADDENDA. 201

The Syrian examples possess equal variability, and occasionally have the costæ not less prominently tuberculated: the typical form has the marginal carinæ always more strongly defined, and the siphonal border more lengthened, than the variety *arata*.

Trigonia undulata, From. Locality, the Lebanon east of Beyrout.

Mr. Damon, of Weymouth, who obtained the Syrian fossil, informs me that specimens are found at three localities near Abich: one in the village itself in a loose earth south of the village, 2400 feet above the sea; the best locality is two miles distant to the south-west at an elevation of 2000 feet; the third locality is directly east.

TRIGONIA SPINOSA, *Park.*, var. SUBOVATA. Page 136, Plate XXIII, fig. 10; Plate XXVIII, figs. 1, 2.

In the Upper Greensand of the Isle of Wight, *T. spinosa* occurs in the defective condition common to the Testacea of that formation; occasionally the test is partially preserved, and the costæ with their obtuse spines are more or less distinct. Usually the species is represented by a variety which also occurs not uncommonly in the Upper Greensand of Wiltshire, where it has undergone compression, and the moulds of external casts have the surface ornaments only obscurely visible. The Isle of Wight specimens sometimes have the surface better preserved, and the moulds are more free from compression. Pl. XXIII, fig. 10, represents a small specimen, the only one having the test preserved and uncompressed with which I am acquainted. Compared with the typical form (Pl. XXIV, figs. 8, 9), the convexity near the umbo is more considerable; the pallial costæ are somewhat more straight and are directed more towards the lower border; the costellæ upon the area are much decussated by the lines of growth, the costellæ also are directed more posteally or towards the siphonal border; both the costæ and costellæ are therefore less radiating, and have somewhat less curvature; the general outline of the shell is less orbicular, or is more lengthened in the direction of the divisional angle of the valve. The moulds have the costæ less conspicuous; their edges are almost smooth, having only slight indications of the obtuse spines which ornament the test. Upon the whole this variety from the Upper Greensand is readily recognised when compared with the typical form from the Blackdown Greensand; the more lengthened form appears to require a varietal designation (*subovata*), as it is readily recognised irrespective of the lesser differences alluded to.

TRIGONIA FORMOSA, *Lyc.*, var. LATA. Page 35, Pl. XXIX, figs. 11, 12; Pl. XXXV, fig. 7.

The specimens of *T. formosa* figured Pl. V, figs. 4—6, also Pl. XI, fig. 2, and an additional specimen, Pl. XXXVII, fig. 10, are characteristic forms of the species as it occurs in the Cotteswold Hills. In Somersetshire a shell occurs, recognised generally as *T. formosa*, which I regard as a variety (*lata*) of the Cotteswold form; these examples are figured, Pl. XXIX, figs. 11, 12; and Pl. XXXV, fig. 7.

Compared with the Cotteswold or typical form, the shell has somewhat less convexity, more especially upon the anal portion, which is more flattened and expanded; the area has greater breadth; the siphonal border is more lengthened and oblique, forming a smaller angle with the escutcheon, which is shorter and more horizontal; the transverse striations, which are small and regular in the Cotteswold form, become larger and fewer near to the apex in the variety; the costæ are variable in the specimens examined, but do not present any clear distinction.

At Bradford Abbas this variety is not uncommon; it occurs also at Haselbury, Somerset, accompanied by *Trigonia costata* and by the Conchifera usually met with at the former locality, but the state of preservation is much inferior. At Milcomb Hill, Oxon., it has been recorded by Mr. Beesley in the Northampton Sands.

TRIGONIA ARCHIACIANA, *D'Orb.* Page 140, Pl. XXIII, fig. 7.

In reference to this species I have been favoured with the following remarks by Mr. C. J. A. Meÿer, whose researches in connection with associated species of the Upper Greensands and Chloritic Marls of the southern counties of England entitle his opinions to the highest consideration. I fully agree with the following conclusions :

"It appears likely that two or three nearly allied species have long been included under the name of *Archiaciana* on account of the great similarity of their surface markings and the unusually indifferent condition of the specimens examined. And supposing that there are three species which have been included under the one name, the following might, I think, be a safe provisional arrangement :

"1. *T. Archiaciana*, D'Orb.
  Syn. *T. Archiaciana*, Pictet and Renevier.
} Horizon : Aptian.
  Loc. Varennes (Meuse), Perte du Rhône.

"2. *Trigonia*, a small species which would include
  Pl. XXV, fig. 10.
  Syn. *T. spinosa*, Ag. 'Trig.,' Pl. VII, fig. 6.
  ? *T. pumila*, Nilsson.
} Horizon : Upper Greensand, Chloritic Marl, Grèsvert.
  Loc. Dunscomb Cliffs, Isle of Wight, and Warminster.

"3. *T. Vicaryana*, Lyc.   Pl. XXV, figs. 8, 9; Pl. XXIII, fig. 7; Pl. XL, figs. 3, 4; Pl. XXVIII, figs. 4, 4 *a*.
  Syn. *T. Archiaciana*, Pictet and Roux.
  —   —   'Morris Catal.,' 1854.
  — *T. spinosa*, D'Orb.
} Horizon : Upper Greensand and Chloritic Marl.
  Loc. Chardstock, Axmouth, Dunscomb Cliffs, Great Haldon.

"No. 2 seems to occur sparingly in Dunscomb Cliffs, in company with *T. Vicaryana*, from which it appears to differ in being smaller, more convex, and less elongated. It *appears* (?) also to want the closely set series of small, oblique, supplementary costellæ on the upper half of the pedal border of the shell, which are very conspicuous in well-preserved specimens of *T. Vicaryana*.

"The third species (your *T. Vicaryana*) seems to be sufficiently distinguishable from others of the *spinosa* group by its large size and (usually) more oblique outline.

"It seems probable that the *T. Archiaciana*, D'Orb., may have to be given up as a British species, in so far at least as the Dunscomb and Great Haldon examples are concerned."

TRIGONIA VICARYANA, *Lyc.*   Page 141, Pl. XXIII, fig. 7; Pl. XXV, figs. 8, 9; Pl. XXVIII, figs. 4, 4 *a*; Pl. XL, figs. 3, 4.

Recent researches of Mr. Meÿer have shown that this species is abundant in the Chloritic Marl of Dunscombe Cliffs, more especially in the beds 10 and 12 of his classified section. See 'Quart. Journ. Geol. Soc.,' vol. xxx, p. 371.

Mr. Vicary has also obtained specimens in the pebble-bed which overlies the Greensand at Great Haldon, where the species, although more rare, has the surface ornaments preserved with great delicacy and beauty; the little perpendicular pillars forming the sides of the costæ have great uniformity and prominence, but are scarcely depicted with sufficient distinctness upon the magnified figure, Pl. XXVIII, fig. 4 *a*; their upper extremities are obtuse, forming upon each row a high narrow ridge bordering upon the channelled base of each succeeding costa, features which are altogether distinct from the plain step-like rows of costæ depicted by D'Orbigny upon the magnified surface of *T. Archiaciana*. The specimen, Pl. XXIII, fig. 7, tabulated *T. Archiaciana*, proves to belong to *T. Vicaryana*. Other small examples from the highest bed of the Haldon

Greensand possess similar features and reduce the supposed examples of *T. Archiaciana* to the little moulds exemplified, Pl. XXV, fig. 16, from the Upper Greensand of the Isle of Wight and of Warminster; these are, however, altogether ill preserved and doubtful as examples of that species. See also p. 141.

I am also inclined to regard our specimens of *T. Vicaryana* as identical with the *T. spinosa* of D'Orbigny, which that author mistook for the *T. spinosa* of Parkinson and of Sowerby; no figure has been given of *T. Pyrrha*, D'Orb., but the few words of description agree with the *T. spinosa* of British authorities.

TRIGONIA SIGNATA, *Ag.* Page 29, Plate II, figs. 1, 2, 3.

In the description of this species, p. 29, no allusion was made to the figure of *T. clavellata* in the work of Knorr ('Verst.,' vol. ii, pl. B, fig. 1 *a*; 1775), which was referred to by Agassiz as one of the types of his *T. signata*; this omission resulted from a lack of confidence in an engraving of one of the *Clavellatæ* in a work of such considerable antiquity.

The figures of *T. clavellata* in Zieten's 'Die Versteinerungen Würtembergs' are also quoted by Agassiz as one of his types of *T. signata*; and, as they are free from the objection above referred to, and also agree generally with British Inferior Oolite specimens within slight limits of variation, and as the number of specimens collected within the last thirty years from Yorkshire, Oxfordshire, and Gloucestershire are very considerable and all approach nearly to Zieten's type, I preferred to regard the latter as the species intended by Agassiz.

In this arrangement we should regard the figure given by Agassiz, 'Trig.' pl. ix, fig. 5, as a variety, excluding his pl. iii, fig. 8, which represents a specimen very defective in condition and doubtful as a species. The example given in 'Trigonies,' pl. ix, fig. 5, is apparently founded upon Knorr's figure, and differs as a variety from the figures by Zieten; it is remarkable for the much greater upward curvature given to the posteal portions of the costæ, which are also more attenuated; the same feature equally characterises the imperfect specimen figured by Dewalque and Chapuis, 'Pal. Luxemb.,' p. 172, pl. xxvi, fig. 1. As the two figures (Knorr's and Agassiz') above mentioned differ from all the known British specimens, and the latter have a general unity of aspect and accordance with Zieten's figures, I have adopted the last for the type of *T. signata*.

The description at p. 29 sufficiently records the differing positions in the Inferior Oolite in which the species has occurred; it may, however, be mentioned that the Upper Trigonia-grit of the Cotteswolds has supplied the specimens having the growth most fully developed, and that in such the rows of costæ anteally sometimes become irregular and confused; in specimens from other positions in Oxfordshire and Yorkshire the rows of costæ are remarkable for their regularity and uniformity.

ADDENDA. 205

TRIGONIA COSTIGERA, *Lyc.* Plate XLI, fig. 17 (Mould).

I am indebted to Mr. C. J. A. Meÿer, F.G.S., for information respecting this imperfectly known species discovered by that gentlemen in the Chloritic Marl rocks of the South Devon Coast, near to Beer Head; the bed is No. 10 of Mr. Meÿer's classified section. Several examples have been observed larger than our Trigonia on Plate XLI; these, however, consisted only of impressions of the costæ. The small mould herewith figured is the only one hitherto obtained; it has some small portion of the shell attached, including three short, horizontal costæ upon the anteal portion of the valve; these have some traces of crenulations. The figure is ovately subtrigonal, moderately convex posteally and depressed anteally; there are some obscure indications of a marginal carina; the umbones are submesial, prominent, and pointed; the unusual shortness of the form is remarkable, the length and height being nearly equal; the abruptness of the posterior slope chiefly contributes to this peculiarity. These few features are insufficient to characterise the species, and its sectional position is somewhat doubtful. I am inclined to arrange it with that group of the *Scabræ* which includes *T. sulcataria*, *T. pennata*, *T. Meÿeri*, and *T. Nereis*, which have the posteal portions of the costæ small, sometimes ill defined, and bent upwards perpendicularly or at right angles to their anteal portions. None of these features are preserved upon our specimen, but in the absence of any well-grounded expectation that better specimens will be obtained I have ventured to figure this very defective shell.

TRIGONIA BLAKEI, *Lyc.*, sp. nov. Plate XLI, fig. 4.

Shell with the general figure ovately oblong; umbones prominent and pointed, placed at the boundary of the anteal third of the valve; posteal slope straight and lengthened; borders of the valves elliptically curved. Area moderately wide, somewhat concave, distinctly bipartite, with three slightly developed tuberculated carinæ. Escutcheon small and depressed. Costæ about thirteen; the first formed eight or nine are regular, closely arranged, and concentrically or elliptically curved, narrow, high-ridged, and imperfectly tuberculated; the last formed four or five costæ are more widely separated, and are more nearly horizontal, excepting their posteal portions, which have the tubercles, to the number of three or four, larger and more curved upwards, approaching the carina at a considerable angle.

Length 14 lines; height 11 lines; diameter through the single valve 4 lines.

The general figure and ornamentation approaches to *T. concentrica* Ag., but the rows of costæ are much less regularly and less distinctly tuberculated, the last formed rows more especially are very narrow, less moniliform, and more horizontal, excepting their posteal portions; these differences apply equally to the figure of *T. concentrica* given by De Loriol and Pellat, 'Portl. de Boulogne,' pl. 8, fig. 2. The convexity of the last-named figure is also much more considerable than our British Calcareous Grit species.

Stratigraphical Position and Locality.—The passage-beds over the Lower Calcareous grit of Snainton, Yorkshire: accompanied by numerous Conchifera, including *Trigonia Snaintonensis* and *T. clavellata*. It appears to be rare, and only single valves have been collected. Obtained by W. H. Huddleston, Esq. The name is from the Rev. J. F. Blake, whose important contributions to the Geology and Palæontology of the Northern Counties of England are so well known.

TRIGONIA CONCENTRICA, *Ag.*

Since the notice of this species at p. 52 was written, no additional materials to illustrate it have come under my observation. The specimens first examined and provisionally assigned to it in 1870 were very imperfect, fragmentary, and altogether insufficient to characterise the species. The hope that more satisfactory and less doubtful specimens would be obtained not having been realised, it becomes necessary to remove *Trigonia concentrica* from the ascertained list of British Trigoniæ.

A nearly allied species, obtained by Mr. Huddleston in the passage-beds over the Lower Calcareous Grit of Snainton, may possibly be identical as a species with the fragmentary specimens alluded to. A good example figured upon Pl. XLI, fig 4, under the name of *T. Blakei*, has enabled me to correct the error upon page 52, and to describe a species of the Lower Calcareous Grit distinct from the *T. concentrica* of Agassiz and of De Loriol.

TRIGONIA PAUCICOSTA. Plate XI, figs. 8, 9; Plate XVI, fig. 7; Plate XXXVII, fig. 3, p. 57.

TRIGONIA ANGULATA. Pl. xiv, figs. 5, 6; pl. xxxvii, figs. 7, 8, 9, p. 54.

Supplementary to the comparisons between these species at page 58, the number of specimens of *T. paucicosta* since obtained have been so considerable as to illustrate the distinctive differences with much certainty. With a species so variable in its surface-ornaments numerous examples are necessary to exemplify its aspects. I retain sixteen specimens, and have examined probably not less than a hundred; these, however, form but an inconsiderable portion of specimens destroyed in endeavouring to separate them from the hard Kelloway Rock at Cayton Bay. These numbers, collected over a very small surface area, evince the gregarious habits of the species. Unlike the Scarborough shell, the Inferior Oolite *T. angulata* was not gregarious; it has occurred at various localities always very sparingly, and apparently is not limited to a single bed or horizon of that formation. Collected during the last half century by geologists and local observers it remains a somewhat rare species, and is absent in collections of Inferior Oolite fossils which are unconnected with the Cotteswolds.

Distinctive Differences.—*T. paucicosta* is the smaller of the two species; its general convexity is greater, the anterior side is shorter, giving to the umbones a more anteal position; the marginal carina in its upper portion has a row of well separated and rounded tubercles; these do not occur in *T. angulata*. The rows of costæ have much variability in both species, but more especially in *T. paucicosta*; usually these terminate posteally with two or three large nodes in each row, or these are sometimes united and become a single varix.

In *T. angulata* the posteal portions of the rows of costæ have much greater uniformity; in common with the *Undulatæ* generally they are subtuberculated, become attenuated, and curve upwards to the carina with a graceful undulation (see Pl. XIV, fig. 6; Pl. XXXVII, figs. 7, 8, 9). These differences indicate the propriety of a zoological not less than of a stratigraphical separation.

As a correction to p. 59, line 4, read, " few examples of *Trigonia paucicosta* have occurred at that locality."

TRIGONIA INGENS, *Lyc.* Plate VIII, figs. 1, 2, 3; Plate XXXVI, figs. 5, 6, p. 24.

Trigonia ingens of the Middle Neocomian formation, compared with *T. signata* of the Inferior Oolite, Zieten's variety.

Subsequent to the publication of the figures and descriptions of *Trigonia ingens*, numerous fine examples, with the test preserved and representing every stage of growth, have been obtained by Mr. Keeping, of the Woodwardian Museum, Cambridge, in the Middle Neocomian formation at Acre House, near Tealby; the bed is a brown ferruginous pisolite; a portion of the rock worked for iron-ore at that locality is described by Professor J. W. Judd, 'Quart. Journ. Geol. Soc.,' vol. xxiii, p. 227. Two additional figures of small specimens from that locality will be found, Plate XXXVI, figs. 5, 6; the general aspect is altogether that of the Jurassic *Clavellatæ*, and bears so considerable a resemblance to British specimens of *T. signata* from the Inferior Oolite that without care it might be mistaken for that species.

Compared with the Jurassic shell the general figure has much greater convexity, or is more ovately oblong; the area is more narrow, steep, convex, and less expanded; its transverse plications are more prominent, rugose, and irregular; its bounding carinæ are less distinct, and sometimes disappear, or degenerate into plications; the position of the median carina is occupied by a groove; the umbones are smaller, more pointed, and anterior; the rows of costæ have less curvature, they are more nearly transverse or approach to the horizontal figure, their tubercles are very irregular and unequal, always much elevated, and sometimes pointed or spinose. All the Lincolnshire specimens are remarkable for the regularity and uniformity of the rows of costæ; each row ends posteriorly with a tubercle, which is one of the largest, affording a marked contrast with the postcal extremities of the costæ in the Inferior Oolite species, in which they become attenuated, and curve upwards to the carina at a more considerable angle. An examination of very numerous Tealby specimens proves that the large Norfolk specimen (Plate VIII, fig. 1) represents the ultimate stage of growth, and that the little accessory costæ near the pallial border is altogether an exceptional feature, and is not represented in Tealby specimens.

TRIGONIA RADIATA, *Ben.* Page 73.

The remarks upon the French example of this species figured by Messrs. De Loriol and Pellat, 'Mon. Paléont. de l'étage Portlandien de Boulogne,' pl. 8, fig. 1, forming the concluding sentences of p. 73, require the following emendation. My friend Dr. Wright, who has compared the original specimen in the possession of M. De Loriol with the figure in the work of that author, informs me that the anteal portion of the specimen retains the test, which is therefore altogether devoid of ornamentation. The only undoubted British specimen known continues to be the one figured by Miss Benett.

TRIGONIA PRODUCTA, *Lyc.* Pl. XIII, figs. 1, 2, 3, 4 ; Pl. XXXVII, figs. 1, 2, p. 60.

It having been objected to the figs. of this species on Plate XIII that they do not represent sufficiently the usual aspect of the Cotteswold forms, the two figures on Plate XXXVII are added, as they exhibit the more frequent condition in which the species occurs, together with the partial effacement of the surface ornaments over the middle portion of the valves. Plate XIII, fig. 3, represents the hinge-processes of the specimen, Plate XXXVII, fig. 1. Plate XIII, fig. 2, represents a specimen in an unusually fine condition of preservation, having the tubercles both of the costæ and carinæ, more than usually prominent for one of the *Undulatæ*, over the whole of the specimen.

In the Cotteswolds it is a rare Trigonia, and is limited in position to the hard whitish limestone of the Upper Trigonia beds, or to the sandy grits by which it is replaced. In Oxfordshire it has occurred less rarely, and has been collected by the Officers of the Geological Survey in the sandy beds at Hook Norton, associated with *Trigonia signata*. As a correction to p. 62, line 7, erase Northamptonshire and substitute Oxfordshire.

ADDENDA. 209

TRIGONIA IMBRICATA, *Sow*. Plate VI, fig. 5; Plate XXXVI, figs. 9, 10; also Plate XLI, figs. 10, 11, 12, p. 33.

The figures on Plates VI and XXXVI represent specimens obtained in the Great Oolite of Ancliff, and do not sufficiently express the little perpendicular pillars or elongations of the tubercles downwards in each, features which characterise the species. The figures on Plate XLI are drawings of Fullers Earth specimens, deprived of the test; they nevertheless expose the little characters indicated, and it is hoped therefore that they will aid in illustrating this species. The specimens have been procured by Mr. Witchell, who has kindly forwarded them to me to be used in this Monograph. Compared with an allied species, *T. tuberculosa*, Plate V, figs. 9, 10, the latter has the rows of tubercles much more closely arranged, and the area has transverse striations in lieu of the widely separated costellæ of *T. imbricata*.

Locality.—The Fullers Earth of Stroud, associated with *Trigonia Witchelli, Posidonomya opalina, Sowerbya triangularis*, and other Conchifera.

TRIGONIA BRONNII, *Ag*.

At p. 23 is a description of this species founded upon examples from the Coral Rag of Glos, Normandy. Professor Hébert, in his 'Memoir on certain Clavellated Trigoniæ of the Oxford Clay and Coral Rag,' there quoted, refers to four British specimens of *T. Bronnii* obtained in the Calcareous Grit of Weymouth. On examination some small examples of clavellated forms from the latter locality, in which the rows of costæ are nearly horizontal, appeared to coincide with some French examples of *T. Bronnii*, a species which has considerable variability even when obtained from a single locality. Subsequent comparisons and examinations of various Weymouth and French specimens have convinced me of the fallible character of this single distinctive feature, and of the necessity of merging all such Weymouth specimens in *T. clavellata*, to which species, therefore, fig. 8, Plate IV, should be referred; and *T. Bronnii* should be removed from the list of British species.

The subjoined figures represent a common or medium-sized example of *T. Bronnii* from Glos.

Trigonia Bronnii, Ag.

TRIGONIA CONOCARDIIFORMIS, *Krauss.*

Of this remarkable species, so abundant in certain districts in Southern Africa, the single imperfect example in the British Museum is herewith figured. I have deemed it expedient to give the subjoined figures partly to correct an error induced by the reduced and inadequate figures given by Krauss, which appeared to me to represent one of the *Clavellatæ;* the Museum specimen undoubtedly associates it with the crenulated examples of the *Scabræ* (see pp. 120, 121).

The general figure is unusually lengthened; the numerous curved, slightly crenulated costæ, widely separated anteally, are much smaller and more closely arranged posteally; they all disappear upon the upper surface of the valve; the spaces representing the area and escutcheon are separated and apparently plain. The specimen, which is the only one known to me, is imperfect posteally, and would be slightly more lengthened when entire. The interior exhibits the hinge-processes and sulcations of the left valve, massive and spreading, but partially destroyed.

This gigantic species is not without a certain resemblance, both in the general figure and arrangement of the costæ, to the Belgian *T. Elisæ* of the Whetstones of Bracquegnies, but the larger anteal costæ are without the rounded papillary prominences of that species. The smaller posteal costæ are nearly straight and directed retrally, as in the smaller Belgian form.

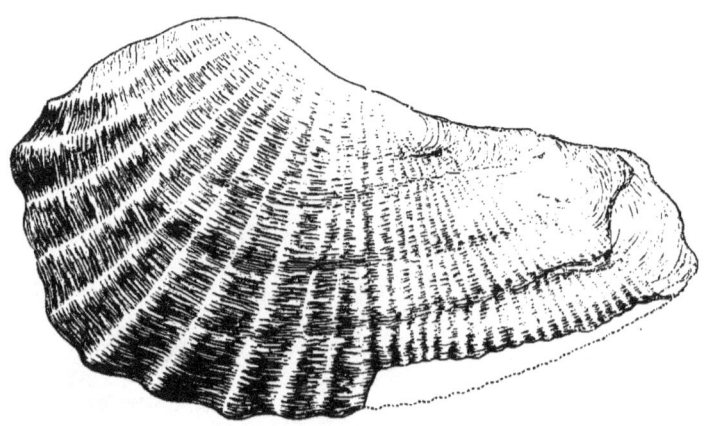

Trigonia conocardiiformis, Krauss. South Africa.

ADDENDA.

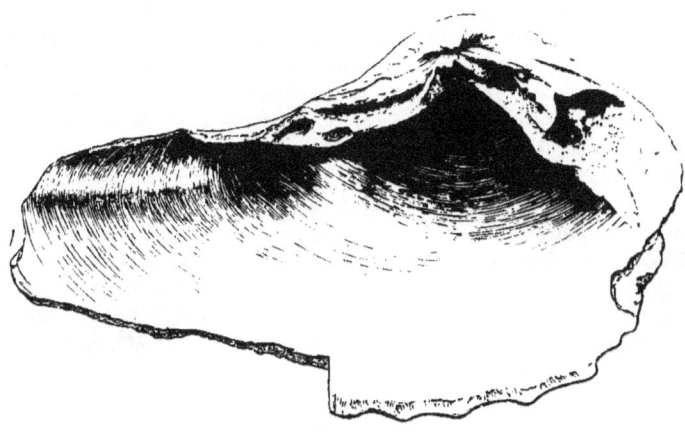

Trigonia conocardiiformis, Krauss.   Inner surface of valve.

Comparison of the Trigoniæ of the Blackdown Beds with those of Bracquegnies.—The Whetstones (meule) of Bracquegnies, Belgium, are upon the same horizon, and are identical lithologically, for the most part, with the Blackdown Whetstones (Cornet and Briart, "Description de la Meule de Bracquegnies," 'Mémoires Couronnes et Mém. des Savants Étrangers,' Acad. Royale de Belgique, t. xxxiv, 1868). Like to the British deposits, they are characterised by the prevalence of *Trigonia dædalea*, Park., Cornet and Briart, pl. 6, figs. 1, 3. The numerous specimens have individual peculiarities, but they all differ from the usual Blackdown form, of which good illustrative examples are given upon our Plate XXIII, figs. 2 and 3, which represent specimens of full dimensions; the Belgian specimens are identical with our large variety *confusa* (Plate XXIII, fig. 1, p. 102). In Britain this variety is comparatively rare, and is found a little higher in stratigraphical position. In Devonshire it has occurred only at Little Haldon, at the base of the Upper Greensands, in a pebbly bed special to that region. The British Museum has upon its tablets a fine and varied series of adult forms of this large variety, which are exhibited as examples of *T. dædalea*. I have placed it as a variety, but possibly other observers may be inclined to regard it as a distinct species. The most prominent varietal features consist in the unusually large, confused, rounded tubercles, which cover and crowd the larger or posteal portion of the shell, and in the large tuberculated escutcheon, rendering the carinal nodes indistinct or only well defined near to the umbones.

A series of the Bracquegnies specimens, kindly forwarded to me by Dr. C. Barrois, of Lille, illustrates every stage of growth in the variety *confusa*. The few first-formed rows of costæ are plain and angulated or nearly destitute of nodes or tubercles, excepting at the boundary of the escutcheon, where they form a carinal angularity; the escutcheon is well defined, and its surface equally, as in the area, is closely and profusely tuberculated. The ligamental cavity is larger and more lengthened than in the other form. Usually the rows of carinal nodes cannot be distinguished over the middle and posteal thirds of the valve, the entire surface of which is occupied by the large, crowded tubercles. This appears to be the only Belgian variety of *T. dædalea*; the convexity of the valves is greater than in the typical form, resulting from the greater breadth of the escutcheon.

With the foregoing species is associated another abundant and equally characteristic form, which in the Belgian beds seems to replace the *aliformis* group of Blackdown; this is the *Trigonia Elisæ* of Cornet and Briart. I am indebted to the liberality of Dr. Charles Barrois, of the Faculty of Sciences, Lille, for the gift of a series of each of these *Trigoniæ*. He refers them to the zone of *Am. inflatus*, Sow. *T. Elisæ* is a

Trigonia Elisæ, Cornet and Briart.

much ornamented and characteristic example of the *Scabræ*, allied to the *aliformis* group, and in common with others of its allies remarkable for the great length of the hinge-border and the shortness of the siphonal border; it is moderately convex anteally, produced and attenuated posteally. The rows of costæ covering the sides of the valves are very numerous and closely arranged, with rounded, depressed nodes; they are concentrically or obliquely arranged anteally; the rows rapidly diminish in size posteally, where their ornamentation becomes obscure. All the costæ have their posteal portions much attenuated, straight, perpendicular, or inclined retrally. The escutcheon is narrow and concave; it has delicate, closely placed, transverse costellæ, which also pass across the upper half of the narrow and flattened area, the lower half of which is smooth; there are no distinct carinal elevations, but the boundaries of the area and escutcheon are well defined.

TRIGONIA? MODESTA, *Tate*. Plate XLI, figs. 13, 13 *a*.

   TRIGONIA MODESTA, *Tate and Blake*. 'The Yorkshire Lias,' *Lamellibranchiata*, by
     R. *Tate*, p. 386, pl. xiv, fig. 4, 1876.

ADDENDA.

The extension of the genus *Trigonia* to the *Armatus*-zone of the Lower Lias in Britain could only be accepted upon the clearest evidence, which unfortunately in the present instance is wanting. Entertaining doubts of the propriety of describing either of the two following minute specimens as examples of *Trigonia*, I have omitted to figure one of them, which in my opinion probably and apparently pertains to another genus of *Lamellibranchiata*.

The figure and description of *T. modesta* from the Yorkshire Lias as a British species compel the present notice.

The materials upon which it was endeavoured to found this species consisted of two very small, incomplete, and imperfect specimens, altogether insufficient for the purpose, obtained by Professor Tate in the *Ammonites-armatus*-zone of the Lower Lias of Warter and Robin Hood's Bays, North Yorkshire, and unfortunately no subsequent discovery of the shell has taken place. I am indebted to the courtesy of the Rev. J. F. Blake for the loan of these specimens.

It appears scarcely possible to convey information by any fair delineation of the little imperfect object upon our Plate; it may represent the very young condition of one of the *Trigoniæ Costatæ*, but even the genus is doubtful.

This ill-defined little shell, about $\frac{2}{10}$ths of an inch across the valve (the left), is in a soft shaly matrix; the test is ill preserved, and the apex has disappeared; the area, of which only the portion adjacent to the carina remains, is plain and somewhat concave; the marginal carina or posteal angle is distinct, plain, and slightly curved; the dorsal costæ are small, numerous, indistinct, and nearly horizontal; this apparently was the specimen which induced Professor Tate to give the name of *Trigonia modesta*. The small portion of the surface preserved posterior to the carinal angle is plain, a feature which militates against it being one of the *Trigoniæ Costatæ*.

Another specimen, of nearly similar dimensions, and attributed to the same species, appears to me to be distinct; it is in a fragment of hard, dark-coloured, shelly limestone from the same zone and locality; the test has disappeared, but the specimen generally is better preserved than the other; it represents an ovately oblong, moderately convex shell, of which the lower and posterior borders are not exposed, and the exact figure is therefore doubtful; the umbones have not much elevation, they are curved forwards, and placed anterior to the middle of the valve; the anterior border is produced and rounded; the posterior border is imperfect, it has a slightly defined carinal angle, and a narrow umbonal, smooth space posteal to it; the upper portion of the valve has a numerous series (about 12) of small dorsal costæ, which pass forwards from the posteal divisional angle and carina horizontally, and become evanescent about the middle of the valve; the anteal half of the valve is therefore without ornamentation. The anteal direction of the umbones appears to be sufficient to remove it from *Trigonia*; it must therefore remain one of the doubtful examples of the *Lamellibranchiata*; possibly it may be a *Corbula*. The posteal extremities of the few last formed costæ extend nearly to

2 S

the border of the valve, which there exhibits no appearance of a carinal area or escutcheon.

Genus—MYOPHORIA. *Bronn*, 1835.

Before concluding the present Monograph I was tempted to depart so far from its scope and intention as to refer to a genus, and single British species, allied to *Trigonia*, and constituting its immediate precursor. *Myophoria*, a genus of *Conchifera* special to the Trias, established by Bronn in 1835, has been amply illustrated in his 'Lethæa Geognostica,' and by Goldfuss in his 'Petrefacta Germaniæ,' including several species of *Myophoria* from the Muschelkalk, which the latter author assigned to *Trigonia* (*Lyrodon*). In Germany and the Tyrol the *Myophoriæ* occur in the Hallstadt, the St.-Cassian, and the Kössen or Rhætic rocks, important fossiliferous formations, which in Britain are represented only by the Penarth beds, the highest stage of the Trias, and reduced considerably in thickness. At the base of the Lower Lias are certain brown sandstones, grey or greenish marls, black clays, and shales, with occasional limestone bands, having a thickness of from 30 to 100 feet, containing a very characteristic series of Rhætic fossils, more or less exposed at numerous localities, in the long course of the Lias between Somerset and North Yorkshire, probably extending uninterruptedly, but chiefly concealed in its course through the intervening counties.

The specimens of *Myophoria* herewith figured represent selected examples of the only recorded British species of that genus, obtained by my friend, Mr. C. Moore, in a bed of hard limestone one foot in thickness, called the "flinty bed of Bere Crowcombe," disclosed by a section made by a canal tunnel at an obscure locality near the town of Ilminster; the blocks of limestone also contained a considerable series of *Gasteropoda* and *Conchifera* included in Mr. Moore's collection of Rhætic fossils from the County of Somerset, described and figured by him in the Memoir subsequently cited.

Myophoria differs as a genus from *Trigonia* chiefly in the absence of transverse sulcations upon the diverging hinge-processes, and not less universally by the direction of the umbones, which, unlike those in *Trigoniæ*, are turned forwards as in the *Conchifera* generally.

Limited stratigraphically to the Trias, this genus of small Conchifers exhibits only a portion of that diversity of aspect, both in groups and species, found in its more important analogue *Trigonia*; the species may be arranged into three sectional divisions, succinctly described as follows :

The first group, trigonal in figure, has one, two, or three large costæ or varices diverging from the umbones, which for the most part disappear before reaching the lower border. Examples : *Myophoria vulgaris*, Bronn, 'Lethæa,' tab. xi, fig. 6 ; *Myophoria pes-anseris*, Bronn, ibid., tab. xi, fig. 8; *Lyrodon Kefersteinii*, Münster, Goldf., 'Petrefacta,' tab. cxxxvi, fig. 2. There is nothing analogous to this section in the genus *Trigonia*.

The second group, for the most part also subtrigonal in figure, has the surface destitute of ornamentation, or has only longitudinal plications; it is without any clearly defined posteal area, or has an imperfect marginal angle. Examples : *Lyrodon ovatum*, Goldfuss, 'Petrefacta,' tab. cxxxv, fig. 11; *Lyrodon lævigatum*, Goldf., ibid., tab. cxxxv, fig. 12; also *Lyrodon simplex*, Goldf., ibid., tab. cxxxv, fig. 14. This group in its surface characters has affinities with some of the *Trigoniæ glabræ*, and more especially with *Trigonia Lingonensis*, Dum.

The third group has longitudinal costæ and a posteal area, which is separated from the other portion of the valve by a marginal carina; the area has also oblique costellæ; it thus approaches to the *Trigoniæ Costatæ* in various features, and more especially by the differences exhibited by the opposite valves of the same species; differences, however, which are wholly distinct from those exhibited by the Jurassic *Trigoniæ* of the allied section. The single British species *Myophoria postera* belongs to this third group. Other examples are *Lyrodon lineatum*, Münster, 'Petrefacta,' tab. cxxxvi, fig. 4; and *Myophoria Goldfussii*, Bronn, 'Lethæa,' tab. xi, fig. 7. This group is allied to the *Trigoniæ costatæ*, yet possesses an unerring distinctive feature; the Jurassic *Trigoniæ* have the marginal carina of the right valve larger than that of the other, and is never divided, as in the *Myophoriæ*, into two or three smaller carina, which cross the dorsal portion of the valve.

In venturing to propose the foregoing sectional divisions for *Myophoria*, I would offer them only as exemplifying the present knowledge of that genus, at the same time avowing the possibility that future discoveries in Triassic palæontology may tend to modify very materially the divisional groups here proposed.

MYOPHORIA POSTERA, *Quenst.*, sp. Pl. XLI, figs. 6, 7, 8, 9; 6 *a*, 7 *a*, 8 *a*, 9 *a*.

TRIGONIA POSTERA, *Quenstedt*. Der Jura, tab. i, figs. 3—6, p. 28, 1857.
MYOPHORIA — *Moore*. Quart. Journ. Geol. Soc., p. 507, pl. xvi, figs. 8, 9, 10, 1861.

Diagnostic Characters.—Shell very convex at the divisional angle of the valve, subtrigonal; umbones rather depressed, directed anteriorly, forming a buccal excavation anteally; the lower margin is lengthened and curved elliptically. The left valve has an elevated marginal carina, which is denticulated more or less prominently upon its lower half; it is bounded anteally by a strongly defined ante-carinal groove; there is no distinct median or inner carina. The area, which is steep and wide, has a series of large, acutely ridged plications; they are somewhat irregular and unequal in their prominence, the surface becoming somewhat depressed at the position of the escutcheon, where the plications are smaller and less distinct; about eight or nine plications are visible upon the left valve; the hinge-border slopes obliquely downwards, its length exceeding that of the siphonal border, which is almost perpendicular with the lower border. The dorsal portion of the surface is covered by a series of linear, depressed, horizontal costæ; their size is unequal in different specimens and irregular, sometimes even upon the same valve; they do not enlarge posteally, and disappear in the ante-carinal groove; nearly fifty of these linear costæ may be distinguished. The right valve has its area somewhat more excavated; its plications are very irregular in prominence and unequal in size, but smaller and more numerous than those of the left valve; the marginal carina is smaller, it is plicated more or less distinctly upon its lower portion; its upper portion, which is smooth, divides into two carinæ, which continue separate and distinct to the lower border. There is also a third dorsal carina at a little distance anteally to the second carina, defined chiefly by the greater depression of the surface occupied by these three carinæ when compared with the general dorsal linear costæ of the right valve, which do not differ materially from those of the other valve.

The shell is of moderate thickness, even near the lower border. The mould does not exhibit any portion of the external ornamentation. Only single valves have been obtained, and the hinge-characters have not been exposed.

Height of the largest specimen 5 lines; length $5\frac{1}{2}$ lines.

The Bere specimens, in common with some Jurassic *Conchifera* having a hard limestone matrix, appear to have had their surfaces covered and their ornaments concealed by the infiltration of a layer of carbonate of lime between the outer surface of the fossil and its matrix. Small portions of this white film-like layer are still visible upon some of the specimens. I did not consider this exterior surface as having formed any portion of the test.

CONCLUDING SYNOPTICAL OBSERVATIONS.

During the period occupied by the publication of the earlier portions of this Monograph it was suggested to me upon more than one occasion by a palæontologist, since deceased, whose varied and extensive knowledge entitled his opinions upon such a subject to high consideration, that I should reconstruct the *Trigoniæ* by arranging them into a family, separating its species into from five to ten genera. The consideration of a similar proposal has probably occurred to other naturalists, and had, in fact, been present to my own mind during many previous years, and had induced me to bestow more than usual attention upon the various aspects assumed by the genus. The results of these observations had, however, tended in a direction the opposite of that proposed to me; they had led to the perception of a general resemblance between the several groups of species in features of sufficient importance to induce me to regard them as forming only so many portions of one great whole,—as so many allied forms greatly varied, which, whether viewed separately or in combination, constituted only a single generic idea, the subordinate features of which were elaborated, some synchronously, others in a certain order of geological succession, never occurring all together or in a single stratigraphical position. It therefore appeared to me that to regard certain differences between such groups as of generic value would be an attempt to dissociate forms which are by natural affinity in close relationship,—to do violence to the chain of life disclosed by an important genus through the great geological periods to which it belonged,—and not less an endeavour to dissolve the association which exists between the more ancient Mesozoic forms and the Tertiary and living portion of the genus which still remain to us.

As examples of such proposed reconstruction grounded upon differences of figures and of surface-ornaments, it would apparently become necessary to divide the extensive section of the *Scabræ* into three genera, one type form of which would have its representative in *T. spinosa*, Park., Plate XXIV; a second in *T. pennata*, Sow., Plate XXIV; the third in *T. aliformis*, Park., Plate XXV; species remarkably distinguished when separated from the *Trigoniæ* generally and brought together for comparison, in their general forms, their ornaments, and even to some extent in their internal hinge-characters, but which will be found to form a gradual approximation when they are compared through the connecting links of other examples of the *Scabræ*,—forms from which they can only possibly be as so many species.

The *Quadratæ* also, although they are sufficiently distinct in the more short or quadrate forms as *T. quadrata*, Ag., p. 105, and *T. spectabilis*, Sow., Plate XXVI, present in other examples approaches both to the groups *spinosa* and *aliformis* of the *Scabræ*.

Perhaps no sectional forms are usually more clearly separated than the *Glabræ* and the *Clavellatæ*, but certain Upper Jurassic species of the former section (see Plates XVIII, XIX, XXI) occasionally acquire much of the exterior ornaments of the *Clavellatæ*; this latter section and the *Scaphoideæ* can sometimes only be separated rather doubtfully, and a similar remark will also not unfrequently apply to the separation between the *Clavellatæ* and the *Undulatæ*. For illustrations of the *Undulatæ* see the figures in Plates XIII, XIV, XV, XVI, and XVII. Doubtless the group of the *Costatæ* possess the most strongly defined sectional characters, more especially in the posteal slope of the area and escutcheon, with the peculiarities of their carinæ and costellæ, together with the differences which they present in opposite valves of the same species; but even these features, so important in their combination, become in some instances modified or only slightly defined.

The proposed division of *Trigonia* into several separate genera is based upon the high value attributed to the exterior form and ornaments as examples of generic character, the internal and chiefly the hinge-characters being regarded as constituting features pertaining to a great natural family, embracing all the groups of species. The separation of the genus adopted in this Monograph is based upon the opposite principle,—that the internal characters are the only features which can be relied upon as affording decided distinctions more important than those of species or of subgenera, and that the modifications which embrace all the features connected with the external figure and surface ornaments are only of subordinate or sectional value, more or less linked together, and are chiefly of interest and importance in comparing the stratigraphic value or succession in geological time of these several features, and of affording separation between the several series of forms of which such groups are composed. The generic distinction based chiefly upon the hinge characters has the further advantage in the genus *Trigonia* of its great convenience in legislating upon the fossil internal moulds of *Conchifera*, which afford usually so few features supplying distinctive characters. It becomes of great importance to have upon the moulds indentations of any undoubted hinge characters distinct from all others, and such are supplied by the transverse sulcations upon both sides of the diverging hinge-processes in *Trigonia*, a feature which can always be recognised in well-preserved moulds and is free from ambiguity, and by which also its hinge is clearly separated from that of its allied genus *Myophoria*.

Upon contemplating the changes exhibited by the *Trigoniæ* during the long succession of Molluscan life disclosed by the two great Jurassic and Cretaceous portions of the Mesozoic æra, as developed both in Britain and over the continents generally, we observe a long, partially broken, imperfect chain of life, even the earlier portions of which seem, but as continuations of previously existing groups of allied organic forms; of these the more immediate precursor is the *Myophoria* of Bronn, an important genus of small Conchifers represented by numerous species, all of which are more or less allied to *Trigonia* and one group more especially to the section of the *Costatæ*; *Myophoria* is special to the

Trias, its species occur in France, Alsace, Hanover, Brandenburg, Saxony, Swabia, Wurtemberg, and the Tyrol. In Britain it is represented by a single species obtained only in one bed of the highest or Rhætic stage of the Trias, exemplified by *Myophoria postera*, Quenst., Jura., tab. i, figs. 3—6, discovered by Mr. Moore at an obscure locality in the county of Somerset. The anteal direction of the umbones serves effectually to separate it from the *Trigonia costatæ*, irrespective of the distinctions afforded by the hinge-processes.

The exterior ornaments in *Myophoria postera*, p. 215, are analogous to, but are altogether distinct from, the Jurassic *costatæ*; the opposite valves more especially offer important differences.

Throughout the several zones of Molluscan life disclosed by the Lower Lias the genus *Trigonia* is known only by one or two minute, imperfectly defined specimens, doubtful generically, obtained by Professor Tate in the *Armatus-zone*, and named by him *Trigonia modesta*, 'Yorkshire Lias,' p. 386, pl. xiv, fig. 4, depicted also of the natural size upon our Plate XLI, fig. 13: but in the absence of more satisfactory specimens the presence of this genus in the Lower Lias remains doubtful.

The Middle Lias of France has produced rather abundantly, and that of Britain very rarely, and apparently localised in each country, a single species of the *Trigoniæ glabræ*, *T. Lingonensis*, Dum. (Plate XXII), p. 98, an abnormal form when compared with the genus generally, almost devoid of surface ornaments, and in that feature approaching to a group of the more ancient Triassic *Myophoria*, but even in this exceptional species the figure agrees with that of the genus subsequently developed; the umbones, unlike those of *Myophoriæ*, are recurved or directed backwards, thus departing both from the figure of that genus and from the *Conchiferæ* generally, but agreeing with the *Trigoniæ*, and imparting a degree of concavity both to the area and escutcheon. *T. Lingonensis* is limited to the main ironstone band of the Zone of *Ammonites spinatus*, both in France and in Yorkshire.

Since the notice of the discovery of this species in Yorkshire by Professor Tate in 1872, English geologists working upon the long course of the Middle Lias have searched for this remarkable *Trigonia* in the midland and southern counties of England without success with one exception, in which several specimens were obtained in the *Spinatus*-zone in the vicinity of Banbury by Mr. E. A. Walford of that place during the past summer of 1877; the same gentleman has also forwarded to me an internal mould of *Trigonia* obtained by him in the Zone of *Ammonites Henleyi* in the same vicinity; it is not certain to what species it belongs, apparently it does not differ materially from moulds of *T. Lingonensis*; in either case it establishes the presence of the genus Trigonia in a lower zone of the Middle Lias.

In the Cleveland Ironstones this species continues to be one of its most rare forms, notwithstanding that the bed has been extensively worked and its fossils diligently collected at numerous localities over wide areas and during many years.

In the Upper Lias and Supra-liassic Sands of Britain the usual Jurassic sectional forms of *Trigonia*, consisting of the *Costatæ*, *Clavellatæ*, and *Undulatæ*, already acquired some of the importance and variety by which they were subsequently characterised.

In the *Costatæ* are surface ornaments and some other external features observed in the *Myophoriæ*, but the umbones are never directed forwards as in that genus, and the hinge-processes present differences, both in their figures and in their deeply sculptured sulcations, which are not limited to one side only of each process; the right valve presents the more important differences, the marginal carina is not divided into two or three costellæ as in *Myophoria*, but forms a single carina larger than that of the left valve; the costellæ upon the area are also fewer and larger than those of the other valve. The Lower Oolitic rocks abound with this important sectional form, which in Britain is represented by upwards of twelve species besides varieties; abundant and dwarfed in the oolitic limestones of the Inferior and Great Oolite, they acquire large dimensions in the clays and argillaceous shales. The stratigraphical range of the section *Costatæ* is not considerable, ten of the British species occur only in the Lower Oolites; one of these, *T. hemisphærica*, and its variety named *gregaria*, Plates XXXI and XXXIII, although distinct generically from *Myophoria*, yet has some external resemblance to *Myophoria lineata*, Munst., and to *M. postera*, Quenst.; the variable and minute longitudinal costæ are more especially analogous. The British Upper Oolites have only two additional species of the *Costatæ* special to those stages. These are *T. Meriani*, Ag., and *T. monilifera*, Ag., so that we have no example higher than the Kimmeridge Clay; and even *T. monilifera* (Plate XXXI), locally so abundant in the lower beds of that stage, is but a continuation of the same form which occurs more rarely in the beds of Upper Calcareous Grit. The *Costatæ* occur apparently in greater variety in the Upper Jurassic rocks of France, Germany, and Switzerland. Only a minority of these have been figured and described: probably more than twice the number of ascertained species would be required to make up the sum of the European *Costatæ*. All of them disappear with the lower portion of the Portland formation.

Two abnormal costated forms occur in the Neocomian period if, indeed, we should include with that section *T. carinata*, Ag., and *T. peninsularis*, Coq., species which were deprived of the surface ornaments ere they attained adult growth, when one, and perhaps each of them, acquired a byssal aperture. *T. carinata* is always recognised by the great obliquity of the costæ; *T. peninsularis* has the costæ very irregular, imperfect in the rows, subangulated, and almost evanescent. In the distinctions between the opposite valves of the *Costatæ* described at p. 9, we see reproduced, modified, and less strongly defined some of the generic features of the *Myophoriæ* fading out, it may be, and modified upon the *Trigoniæ*. The variability in the number and size of the longitudinal costæ described as marking *varieties* of species in this Monograph were also produced as a variable feature in the more ancient *Myophoria postera*, of the Upper Trias; it is equally present in the Neocomian *T. carinata*, where the variability is so considerable as to induce Agassiz to separate it into two species, the one with the few and large costæ constituting his *T. sulcata* ('Trigoniæ,' p. 44). Possibly *T. peninsularis* may be equally variable, but unfortunately Coquand has given only a single figure and of the left valve only ('Terr. Aptien d'Espagn.,' pl. xxiii, fig. 3).

However clearly separated are the *Costatæ* as a section, they also approach a very different section (that of the *Glabræ*) in certain Swabian species which have the middle portion of the valve entirely destitute of surface ornaments, as in *T. zonata*, Ag., *T. interlævigata*, Quenst., and *T. triangularis*, Goldf. They also approach to the *Clavellatæ* in *T. hybrida*, Roem., and in *T. geographica*, Ag., and in two less known forms alluded to at p. 161 under the names of *T. fimbriata*, Lyc., and *T. granigera*, Cont., which have the longitudinal costæ and also the plications upon the marginal carina minutely clavellated.

The *Clavellatæ* are peculiarly varied in species and locally abundant in the Middle and Upper Oolitic Rocks of Britain, numbering upwards of thirty-three species; probably they are not less abundant in France, Switzerland, and Germany, where, in the Oxfordian and Kimmeridgian stages, they appear to attain their maximum of numbers, and then suddenly disappear.

The *Undulatæ* are exclusively Jurassic, and little less varied than the *Clavellatæ*. Britain has twenty species. They occur in the Upper Lias, in the Supra-liassic Sands, and more abundantly in the Lower Oolites; locally they are gregarious. In the Middle and Upper Oolites they are much less conspicuous, and are represented by four species only. *T. paucicosta* alone can be said to be even locally abundant. They disappear altogether in the lower portion of the Portland formation.

The *Scaphoideæ*, comparatively a small section, occurs only locally, and of few species, chiefly in the Upper Lias and Inferior Oolite, after which nothing more is known of the section until it reappears in the Lower Calcareous Grit of Yorkshire, and, after another stratigraphical interval, widely separated, in the Middle Neocomian beds of Norfolk, where it is represented by two species, one of which, *T. exaltata*, nob., Plate XXXVIII, is of gigantic dimensions. In the higher assemblages of Cretaceous fossils the *Scaphoideæ* are unknown.

The *Glabræ* in the circumstances under which they occur offer a marked contrast to the leading sectional forms of the genus both Jurassic and Cretaceous; belonging to the whole of the Mesozoic Epoch, it is only in the Portlandian beds of Britain that they become predominating forms; their occurrence as a single species dates from the earliest record of the genus in the Middle Lias above referred to. In the Inferior Oolite they are represented by the rare, anomalous, and in some respects almost unique, *T. Beesleyana*, after which nothing more is known of the section until its reappearance in the Portland formation, represented in Britain by five species, two of which, *T. gibbosa* and *T. Damoniana*, occur in great abundance. All of these Portlandian *Glabræ* have a considerable family resemblance in their short, subglobose, or ovately rounded forms; their surface ornaments consisting for the most part of small, longitudinal, subtuberculated costæ; the plain antecarinal space is always well developed, and is usually the only smooth portion of the shell. The variability assumed by the three species *T. gibbosa, T. Damoniana,* and *T. Manseli,* in their surface-ornaments, are so considerable (exemplified on our Plates XVIII, XIX, and XXI) as to offer a remarkable contrast to other examples of the *Glabræ,* and surpass in diversity, perhaps, any other of the more ornamented examples of the genus. *T. gibbosa,* more especially, which I have arranged in three varieties, becomes in one of them a shell almost devoid of ornament, and is chiefly remarkable for the prominence of its zonal sulcations; in other forms its surface is crowded by its excentric costæ, and roughened by their small prominent tubercles.

The Isle of Portland, Tisbury, Warminster, Brill, and Swindon are the localities at which the *Trigoniæ Glabræ* are the predominating forms, the other sectional examples of the genus having altogether disappeared. The European continental *Glabræ* of the same period are *T. Michellotti,* De Lor. and Pellat, 'Boulogne,' pl. vii, fig. 6; *T. variegata,* Credner, 'Ob. Jura,' pl. viii, fig. 22, also a variety of the latter from Boulogne figured by De Loriol and Pellat, pl. xi, fig. 9. The latter species has the posteal portions of its costæ broken, forming lengthened oblong nodes, unlike the British species which are all tuberculated; *T. Boulogniensis,* De Lor. and Pel. 'Boulogne,' pl. vii, fig. 10, has irregular plicated costæ, destitute of tubercles, and has no zonal sulcations.

Of the seventy-eight species of Jurassic *Trigoniæ,* exclusive of varieties, found in Britain, and recorded in this Monograph, twenty-eight are also obtained at various European continental localities, but chiefly in France and Switzerland; six of these also occur in Southern Germany, and one species (*T. triquetra,* Sub.), Plates VI and XXXVI, in Northern Germany.

The general European assemblage of Jurassic *Trigoniæ* contained in the British Museum are so considerable, varied, and abundant, and their state of preservation so exceptionally fine, that it becomes a subject of regret so small a portion of them should come under the observation of the public; eventually this defect will be removed and a series of *Trigoniæ* disclosed, so remarkable and attractive that it may confidently be predicted they will engage the special attention of future palæontologists. The prevailing forms belong to the *Clavellatæ, Undulatæ,* and *Costatæ,* only a minority of which can be identified with recorded species.

Other European Jurassic Trigoniæ, not British, figured and described, are:
Trigonia costata, Chap. et Duv. Foss. de Luxemb., tab. 25, figs. 6, 7.
— *costata,* var. *triangularis,* Goldf. Petref., tab. 137, fig. 3.
— *similis,* Ag. Trig., tab. 2, figs. 18—21.
— *navis,* Lam. Ag. Trig., tab. 1, 2, figs. 21—24.
— *costellata,* Ag. Trig., tab. 2, figs. 1—12.
— *Bronnii,* Ag. Trig., tab. 5, fig. 19.
— *maxima,* Ag. Trig., tab. 4, figs. 6—9.
— *Golfussii,* Ag. Trig., p. 31.

CONCLUDING SYNOPTICAL OBSERVATIONS.

Trigonia notata, Ag. Trig., tab. 5, figs, 1—3.
— *reticulata*, Ag. Trig., tab. 11, fig. 10.
— *parvula*, Ag. Trig., tab. 11, fig. 8.
— *rostrum*, Ag. Trig., tab. 9, fig. 1, tab. 5, fig. 10.
— *concentrica*, Ag. Trig., tab. 6, fig. 10.
— *picta*, Ag. Trig., tab. 6, fig. 11.
— *Parkinsoni*, Ag. Trig., tab. 10, fig. 6.
— *suprajurensis*, Ag. Trig., tab. 5. figs. 1—6.
— *truncata*, Ag. Trig., tab. 5, figs. 7—9.
— *clathrata*, Ag. Trig., tab. 9, fig. 9.
— *sexcostata*, Roemer, Nord. Ool., tab. 6, fig. 1.
— *inflata*, Roemer, Nord. Ool., tab. 19, fig. 22.
— *hybrida*, Roemer, Nord. Ool., tab. 6, fig. 1.
— *cardissa*, Ag. Trig., tab. 11, fig. 4.
— *concinna*, Roemer, Nord. Ool., tab. 19, fig. 21.
— *Bouchardi*, Oppel, Juraform., p. 486.
— *aspera*, Lam. Héb. Jour. de Conchyl., tab. 7, fig. 3.
— *inter-lævigata*, Quenst. Jura, tab. 67, figs. 7, 8.
— *Barrensis*, Buvig. Meuse Atlas, tab. 16, figs. 30, 32.
— *Arduennea*, Buvig. Meuse Atlas, tab. 4, figs. 10—14.
— *Kurri*, Opp. Juraform., p. 485.
— *variegata*, Credner, Ob. Jura, tab. 8, fig. 22.
— *verrucosa*, Cred., Ob. Jura, tab. 8, fig. 23.
— *clivosa*, Cred., Ob. Jura, tab. 9, fig. 24.
— *Boulogniensis*, D'Lor., Kim. of Boulogne, tab. 7, fig. 10.
— *Rigauxiana*, Mun. Chal., Bull. Soc. Lin. de Normandie, vol. 9, tab. IV, fig. 2.
— *Baylei*, Dolf. Kimmer., tab. 10, fig. 4.
— *granigera*, Contejean, Ét. Kimmer. de Montbeliard, tab. 14, fig. 4.
— *pseudo-cyprina*, Cont. Montbeliard, tab. 14, fig. 7.
— *trigona*, Waagen, Beitr., tab. 29, fig. 3.
— *Etalloni*, de Loriol, Royer et Tombeck, Ét. Jur. Sup. Haut. Marne, tab. 17, figs. 7—10.
— *spinifera*, De L., R., et T., tab. 18, fig. 2.
— *Cottaldi*, Mun., De L., R., et T., tab. 17, fig. 3.
— *Tombecki*, De L., R., et T. tab. 18, fig. 21.
— *Matronensis*, De L., R., et T., tab. 18, fig. 24.
— *Carmontensis*, De L., R., et T., tab. 17, figs 10, 11.
— *Moutierensis*, Lyc. Monogr. Trigoniæ, p. 36.
— *Michellotti*, D'Lor. et Pellat, Boulogne, tab. 7, fig. 6.

Our notices of the Jurassic *Trigoniæ* of the other continents must be understood as relating chiefly to their natural affinities as species, and not to their stratigraphical positions, which, in some instances, are only doubtfully and imperfectly known.

We are absolutely without information of the presence of a single Jurassic *Trigonia* in the continents of America. In Asia one of the *Undulatæ* occurs in the mountain district of the Lebanon, to the eastward of the town of Beyroot; apparently it is identical with *T. undulata*, Fromherz, from the Piedmontese flanks of the Alps (p. 77) and nearly allied to a *Trigonia* of the Cornbrash and Great Oolite in Britain (p. 201).

In Cutch three of the *Costatæ* have been obtained, two of which nearly resemble *T. costata*, Sow., and *T. pullus*, Sow., British species of the Middle and Lower Oolites, and are so named 'Geol. Trans.,' 2nd ser., vol. v, pls. 21—23 ; the third, a large, lengthened, and oblong form, distinct from all others, is named by Sowerby *T. Smeeii*, 'Geol. Trans.,' 2nd ser., vol. V, pl. 61. The Ammonites associated with these species are, for the most part, identical with British Kelloway Rock forms.

In Northern India the Spiti Pass of the Himalayas affords a fossil Jurassic fauna, which Dr. Stoliczka has assigned to certain *Conchifera* of the European Rhætic beds, Lias, and Lower Oolites, *Trigonia costata* is one of these, ' Quart. Jour. Geol. Soc.,' 1868, p. 506, vol. xxiv.

The only record we possess of Jurassic *Trigoniæ* in Australia is the memoir on Australian Mesozoic Geology and Palæontology, 'Quart. Jour. Geol. Soc.,' May, 1870, by Mr. C, Moore, who has therein figured and described two considerable series of fossils from Queensland, and from Western Australia upon the opposite sides of Australia, separated by upwards of 38° of longitude, and by the great central Sahara. The West Australian series contains Ammonites, Conchifera, and Brachiopoda, some of which cannot be distinguished from British species of the Middle Lias, Upper Lias, and Lower Oolites, including *Trigonia Moorei*, Lyc. (p. 151), which occurs in some abundance, a single block having contained fifteen specimens. The Queensland fossils from Wollumbilla and other localities are almost entirely distinct from known European forms : they are preserved in loose blocks evidently derived from beds more ancient than the surrounding strata; numerous in species, two only are believed to be identical with European shells : one is *Avicula Braamburiensis*, a species which has a considerable stratigraphical range in Britain, and is so nearly allied to other forms of the same genus that great care and an excellent condition of preservation are necessary in its discrimination ; the other is *Lingula ovalis*, which differs little from the *L. subovalis* of Neocomian strata. Mr. Moore has described and figured *Trigonia lineata*, pl. 13, fig. 12, an ill-preserved fossil, one of the *Glabræ*, a short, gibbose form, with numerous concentric regular lines upon the anteal and middle portion of the shell, but the posteal or anal portion is imperfect ; it has some affinities with the Portlandian group of *T. gibbosa*, and is apparently yet more nearly allied to an Indian Cretaceous species, *T. orientalis* or *suborbicularis*, Forbes, from Southern India (see page 121); the latter has the concentric

CONCLUDING SYNOPTICAL OBSERVATIONS.

lines much more prominent than in *T. lineata ;* the posteal slope is destitute of bounding carinæ. The presence of a gigantic *Crioceras* at the same locality is also strongly indicative of the Cretaceous rocks.

It is evident that further explorations and collections of fossils, and more especially of *T. lineata*, will be required ere the stratigraphical position of the Queensland beds can be determined without ambiguity.

On passing from the Jurassic to the Cretaceous rocks the genus *Trigonia* presents some very remarkable changes; the large and conspicuous section of the *Undulatæ* is found to have altogether disappeared, neither is there known a single continental example of the *Clavellatæ*.

In the Neocomian period generally the great sectional series of *Trigoniæ*, constituting the *Scabræ* and *Quadratæ*, first appear, replacing the lost *Costatæ, Clavellatæ,* and *Undulatæ*. Varied in species, considerable in numbers, and conspicuous beyond all other sectional forms in their ornaments and in their individualisation, they continue prominent and generally predominating in every considerable assemblage of *Conchifera* throughout the varied beds of the Cretaceous rocks in whatever country they are discovered, so that *a clavellated Trigonia having an ornamented escutcheon becomes an infallible indication of the presence of the Cretaceous rocks*. The converse of this statement holds good as to the clavellated species with smooth escutcheons; they are all Jurassic. The two examples figured in this Monograph, *T. ingens*, Lyc., Pl. VIII, figs. 1—3, and Pl. XXXVI, figs. 5, 6; and *F. Keepingi*, Lyc., Pl. XXXV, figs. 1, 2, are the only known Cretaceous *Clavellatæ*. They have been obtained only within the very limited area occupied by the Middle Neocomian formation in Britain.

It is in the Cretaceous *Trigoniæ* also that we find the genus represented by the greatest variety of figure, ranging from the short, suborbicular, or subquadrate examples of the *Quadratæ* to the subcrescentic attenuated forms of a portion of the *Scabræ* or group of *T. aliformis*, the latter having their siphonal borders so short that the incurrent and excurrent respiratory orifices are brought into near proximity. They differ likewise from all other of the Mesozoic *Trigoniæ* in having the lower borders of the valves toothed, a feature which is reproduced in the living Australian section of the *Pectinidæ*. These differences of figure are so considerable that when they are found to apply to the vital organs it can scarcely be supposed that the Mollusca so differently constituted could have had the same habitats in depths of waters, or that their characteristic habits were similar. Nevertheless, throughout the Trigonia-bearing beds of the Cretaceous rocks, from the Middle Neocomian beds even to the highest beds of Chloritic Marls, we find these forms so dissimilar associated, their valves sometimes together, but more frequently separated, leading to the inference that the dead Testacea were in some instances drifted to the places where they are found in such considerable numbers.

The Cretaceous *Quadratæ* also possess a peculiar feature which must have a connection with the general economy of the Mollusk. The hinge-processes, even in the largest examples of that section, are remarkably small, but the valves have internally near the middle of the lower border, and external to the pallial scar, a few closely placed oblong pits, apparently intended to afford attachment to an accessory ligamental appendage, enabling the Mollusk to hold the valves closed with greater power. As these pits existed at several stages of growth, it is evident that they were reproduced periodically, and were obliterated at each period by the growth of new shell-substance (for examples of this feature see Pl. XXIV). The *Quadratæ*, also, had their hinge-plates strengthened by additional ligamentary support at the posteal extremity of the plate, which has several oblique grooves, a feature which also obtains in the *aliformis* group of the *Scabræ*, but is not seen in any other section of the genus.

The Blackdown and Haldon Greensands are represented in Belgium by the *Meule de Bracquegnies*, in which those beds reappear both lithologically and palæontologically; the *Trigoniæ* are, however, for the most part distinct in species. The Belgian *T. dædalea* differs from the well-known form of Blackdown, but is identical with our variety *confusa*, Pl. XXIII, fig. 1, which in Britain occurs rarely, at Little Haldon, much to the westward of the typical form; it is abundant at Bracquegnies, as is also *T. Elisæ*, one of the *aliformis* group special to that locality (Cornet and Briart, 'Acad. Roy. de Belgique,' t. xxxiv).

The *Trigoniæ*, so abundant in the Cretaceous glauconitic sands and marls, disappear suddenly and entirely with the advent of the Chalk. Apparently this change is not an exceptional feature as regards the *Trigoniæ*, but is connected with a similar loss of genera in other Dimyarian *Conchifera*, a circumstance which becomes remarkable when compared with the general abundance and variety of Monomyarian forms in the same deposits; a fact which has long been observed, but which has not hitherto received a satisfactory explanation.

The only record we have of the genus *Trigonia* in the White Chalk consists of some impressions, ill preserved, named by d'Orbigny *Trigonia inornata*, 'Pal. Fran. Terr. Crét.,' pl. 297, from Royan, Charente Inférieure, a moderately convex, subovate shell, having apparently a plain area and escutcheon, and a numerous closely arranged series of obliquely curved costæ (about forty) covering the other portion of the shell, passing from the angle of the valve to the anteal and lower borders, but so faintly traced that the entire surface appears almost devoid of ornamentation; thus approximating or apparently intermediate to, the *Scabræ* and *Glabræ*.

The Cretaceous *Trigoniæ* are, for the most part, localised; of the thirty-one species yielded by the British rocks fourteen only have been identified at continental localities, and limited to neighbouring countries, as France, Belgium, Switzerland, Southern Germany, and the Spanish Peninsula. A considerable proportion of the French Cretaceous *Trigoniæ* appear also to be special to that country.

The Aptian beds of Spain, Province of Teruel (Coquand, 'Monogr. de l'étage Aptien de l'Espagne'), appear to have been deposited in a portion of a Mediterranean

CONCLUDING SYNOPTICAL OBSERVATIONS. 227

basin of the Lower Cretaceous period ; their *Trigoniæ* are for the most part special to them ; *T. peninsularis*, Coq., is an abnormal example of the *Costatæ*; *T. Picteti*, Coq., and *T. abrupta*, Coq., belong the *Scabræ*; *T. Hondeana*, Coq., to the *Quadratæ*. Neither of the two latter forms will allow of any close comparison with the South American species from the equatorial region, to which the same names had previously been given by Lea and by Von Buch. Their localisation is as distinct as that of the British Upper Neocomian *Trigoniæ*.

The Spanish *T. abrupta* is nearly allied to a small British *Trigoniæ* from the Chloritic Marls of South Devon (see *T. Meyeri*, p. 125, Pl. XXIII ; also Pl. XLI, figs. 15, 16). The two very dissimilar figures given of the American species in the works of Von Buch and D'Orbigny will, perhaps, account for this supposed identity between the American and Spanish species.

From near the western coast of the same peninsula, in the vicinity of Torres Vedras, a series of Testacea has been described and figured by Mr. D. Sharpe ('Quart. Jour. Geol. Soc.,' vol. vi, pls. 20—24, and referred to the Suberetaceous, or lower portion of the Cretaceous rocks. Among them is *Trigoniæ Lusitanica*, Sharpe, a characteristic example of the *Scabræ*, previously figured by Goldfuss, together with three other *Trigoniæ*, all of which the latter author erroneously referred to the *T. literata* of Young and Bird and of Phillips. The Neocomian *T. caudata*, Ag., is also associated with *T. Lusitanica*.

Other Cretaceous European *Trigoniæ*, not British, already figured, are :

Trigonia disparilis, d'Orb. Terr. Crét., 3, pl. 299, fig. 2.
— *limbata*, d'Orb. Terr. Crét., 3, pl. 298.
— *Lamarkii*, Matheron. Catal., pl. 24, figs. 5—7.
— *tenuisulcata*, Dujard. Mém. Soc. Géol. Fr., t. 2, pl. 15, fig. 14.
— *excentrica*, Goldf. Petref., 3, pl. 137.
— *pulchella*, Reuss. Bohem., pl. 41, fig. 3.
— *scabra*, Lam., d'Orb. Terr. Crét., pl. 296.
— *Coquandiana*, d'Orb. Terr. Crét., 3, pl. 294.
— *crenulata*, Lam., d'Orb. Terr. Crét., 3, pl. 295.
— *divaricata*, d'Orb. Terr. Crét., 3, pl. 288, figs. 2—4.
— *longa*, Ag. Trigon., pl. 8, fig. 1.
— *Robinaldina*, d'Orb. Terr. Crét., 3, pl. 299, figs. 1, 2.
— *paradoxa*, Ag. Trigon., pl. 10, figs. 12, 13.
— *Delafossei*, Leymerie. Mém. Soc. Géol. Fr., 2 ser., t. 4, pl. 8.
— *Constantii*, d'Orb. Terr. Crét., pl. 291, figs. 3, 4, vol. 3.
— *palmata*, Desh. Mém. Soc. Géol. Fr., vol. 5, pl. 8, fig. 5.
— *nodosa*, Pictet and Roux. Grés Verts, pl. 35, fig. 5.
— *sanctæ-crucis*, Pictet. Paléont. Suisse, pl. 128, figs. 2—5.
— *nodosa*, Pictet. Paléont. Suisse, pl. 12, figs. 1, 2.
— *Elisæ*, Cornet and Briart. Meule de Brecq., pl. 6, figs. 4, 5.

In Britain the Cretaceous *Trigoniæ* constitute portions of four well-marked zoological assemblages, separated in stratigraphical position; they assist materially in imparting a characteristic *facies* to each series. Of the Middle Neocomian *Trigoniæ*, six in number, one only of the *Quadratæ*, *T. nodosa*, constitutes a variety of a species which in the Upper Neocomian beds is represented by other forms.

Of the others, two belong to the *Clavellatæ*, and have a Jurassic aspect so remarkable that one of them, *T. ingens*, Lyc. (Pl. VIII, figs. 1, 2, 3; Pl. XXXVI, figs. 5, 6), might readily be mistaken for the *T. signata* of the Inferior Oolite; the other species, *T. Keepingi*, Lyc. (Pl. XXXV, figs. 1, 2), with a shorter figure, has an aspect equally Jurassic. In the absence of all knowledge of their position and associated fossils they would undoubtedly have been assigned to the more ancient period. The Upper Neocomian formation is represented by seven *Trigoniæ*, of which one (*T. nodusa*, Sow.) is a variety of the Middle Neocomian form; a second (*T. carinata*, Ag.) passes upwards into a higher stage; the other five species are all *Scabræ*, and are special to that stage. The third, or stage of the Blackdown and Haldon Greensands and Gault (identical also with the Belgian *Meule de Bracquenies*), has nine ascertained species, including two of the *Quadratæ*, viz. *T. dædalea*, Park., and *T. spectabilis*, Sow., two of the *Glabræ*, *T. excentrica*, Park., and *T. læviuscula*, Lyc., and five of the *Scabræ*. There is also an internal mould in the Red Chalk or Gault of Hunstanton not sufficiently characterised. The fourth or highest stage, consisting of Upper Greensand and Chloritic Marls, has upwards of sixteen species; and three others have been observed by Mr. Meÿer in the hard rocks and marly beds of the South Devon Coast, which hitherto he has not been able to add to his collection. Of the sixteen, three are varieties of forms met with in the third series, as *T. aliformis*, *T. spinosa*, and *T. Vicaryana*; three others which have passed upwards without apparent change, as *T. scabricula*, *T. læviuscula*, *T. spectabilis*, and *T. carinata*; leaving ten other species apparently special to the highest stage.

D'Orbigny assigned the maximum development of the genus *Trigonia* to the highest fossiliferous beds of the Cretaceous rocks (*Cours élément. de paléont.*, Tableau 8). In Britain our stratigraphical table records the greatest number of species in the Lower Mesozoic rocks or Inferior Oolite, and in the Upper Greensand and Chloritic Marls of Wilts and of the Isle of Wight. The hard rocks in the cliffs of the South Devon Coast have also produced *Trigoniæ*, for the most part in a very ill state of preservation; and great difficulty is experienced in procuring useful and reliable specimens. Enough, however, is ascertained to assure us that the genus was represented in Britain at this the period of its final disappearance in a manner both ample and varied. For these the reader is referred to the stratigraphical table, also to the figures on Plates 23, 24, 25, 26, 28, 35, 37, 40, and 41.

The Cretaceous *Trigoniæ* obtained in the American continent, although not numerous in species, are not less remarkable and well characterised. With the exception of a single lengthened species of the *Glabræ* from Columbia (*T. Lajoyei*, d'Orb.), all pertain to to the *Scabræ* and *Quadratæ*, *T. Thoracica*, Mor. ('Synopsis,' pl. xv, fig. 13), from the State of New Jersey and from Alabama, is allied to the *Aliformis* group. To the same group is also allied a large and abundant Mexican species, *T. plicato-costata*, figured and described by Nyst and Galeotti ('Bull. l'Acad. de Bruxelles,' tom. vii, No. 10), from the great principal Cordillera of Anahuac, Mexico, several thousands of feet above the sea. This species, which was erroneously referred by the authors to the Jurassic rocks, has a near ally in our *T. scabricola*. *T. Humboldtii*, von Buch ('Petref. Americ.,' figs. 29, 30), another of the *Scabræ*, has radiating costellæ passing from the umbones retrally over the upper and siphonal half of the shell; its locality is San Felipe, Central America.

In the elevated region of equatorial South America, in New Granada and Columbia, the same groups are represented by several unusually large and remarkable species. *T. abrupta*, von Buch ('Petref. Amer.,' fig. 21), an ovately oblong form, with numerous delicate, almost evanescent, straight, oblique, retral or nearly perpendicular, minutely crenulated costæ; the area and escutcheon, which have considerable breadth, are almost plain. Also a large species named *T. aliformis*, by von Buch ('Petref. Amer.,' fig. 10), which may be a more fully developed example of *T. thoracica*, Mor. Another is *T. subcrenulata*, d'Orb. ('Coq. Foss. de Colomb.,' pl. iv, figs. 7, 8), a remarkably inflated subcrescentic shell, allied to *T. crenulata*, Lam., in the general features of its ornamentation, distinguished by its more inflated and lengthened form, by the small and deep concavity formed by the indistinctly separated area and escutcheon, by the zigzag costellæ of their transverse ornaments, and by the small, perpendicular, widely separated crenulated rows of costæ. Another is a gigantic example of the *Quadratæ* from Bogota, *T. Hondeana*, Lea, *T. Boussignaultii* (d'Orbigny, 'Coquilles fossiles de Colombie,' pl. iv, figs. 1, 2), distinguished by the gigantic size, by the extreme shortness of the general figure, by the few perpendicular rows of small, widely separated, crenulated costæ, and by the great breadth of the area and escutcheon, whose transverse, curved costellæ agree with the ornaments upon the other portion of the shell; the umbones are obtuse, and the borders of the valves are rounded.

Not less distinct and well characterised are the prevailing Trigoniæ of the South African provinces to the eastward of the Cape of Good Hope; these belong to two very different stages of the Cretaceous formations. That of the province of Uitenhage has been investigated and illustrated by Dr. Krauss, of Stuttgart, by Dr. Rubidge, and Dr. Atherstone, who assigned them to the Cretaceous rocks. The fossils from the same beds have been examined by Mr. D. Sharpe and Professor R. Tate, and the opinion of these two palæontologists, founded upon the analogies of the fossils generally, was that they were Jurassic and should be referred to the Lower Oolites. In offering an opinion adverse to the latter conclusion, I would admit the Jurassic aspect of some of the Conchifera, which, in common with certain European forms, indicate that the *Jurassic facies* did not disappear suddenly and entirely with the close of the Jurassic period, but was in some instances continued partially into the Molluscan fauna of the lower portion of the Cretaceous rocks. Notwithstanding, however, certain specific resemblances, I doubt the absolute identity of any one of these African species with European allied forms. The occurrence of *T. Herzogii*, Haussman, of *T. conocardiiformis*, Krauss (p. 210), and of *T. ventricosa*, Krauss (p. 119), in such profuse numbers and distributed over so wide a region, the first a member of the *Quadratæ*, the others of the *Scabræ*, may, in the absence of reliable and guiding Ammonites, be regarded as affording strong, and to my mind, decisive evidence of the Cretaceous character of the series. The resemblance (perhaps even identity) of *T. ventricosa* with the Indian Cretaceous *T. tuberculifera*, Stol. ('Mem. Geol. Surv. India,' vol. iii, pl 15), tends materially to support the same conclusion.

Two other Trigoniæ, mentioned in Professor Tate's memoir, require notice. Pl. 7, fig. 6, of that memoir represents the magnified figure of a very young Trigonia, which is attributed to *T. Goldfussi*, Ag. This, in common with other very young shells, might possibly pertain to one of the *Quadratæ*, and is even allied to certain young specimens of *T. dædalea*, Park. The other Trigonia mentioned is a single valve of one of the *Costatæ*, believed to represent a young specimen of *T. Cassiope*, d'Orb. (see Pl. XXXII). There can be no doubt of the importance which attaches to the presence of the *Costatæ* when the age of the stage is a question of doubt, but the presence of a single specimen cannot be accepted as affording any decisive proof in such a question.

The second, and apparently newer series of Cretaceous fossils, collected in the region of the Umptafuna and Umzanbani rivers, contains *T. elegans*, Baily, a small and much ornamented example of the *Scabræ* ('Quart. Jour. Geol. Soc.,' vol. xi, plate 13, figs. 3 *a, b*).

In Asia the Cretaceous Trigoniæ are scarcely known beyond the limits of the British Indian Empire. In the southern region, near to Pondicherry, two short subglobose species of the *Glabræ* have been figured and described by Forbes ('Geol. Trans.,' 1846, vol. vii, p. 150, pl. 18) under the names of *T. semiculta*, *T. orientalis*, and *T. suborbicularis*. The first of these species has the longitudinal costæ interrupted about the middle of the shell by the usual smooth antecarinal space; the other two, which I can only regard as varieties of one species, have the costæ, which are unusually prominent, continued without interruption across the whole surface; the area and escutcheon are only slightly developed. The last-named form has the costæ less prominent. They have some general resemblance to the *Glabræ* of the Jurassic Portland group, but differing in having their costæ entirely devoid of tubercles.

A small species of the *Scabræ*, *T. Forbesii* (p. 122), is also distinct from European forms. The Geological Survey of India has added but few additional Trigoniæ. *T. tuberculifera*, Stol., an inflated example of the *Aliformis* group, presents in its surface-ornaments varied aspects in its numerous specimens, which are altogether analogous to those assumed by its near ally the *T. ventricosa*, Kr., of Southern Africa, a considerable and instructive series of which are in the British Museum. Another small form of the *Glabræ*, *T. indica*,

CONCLUDING SYNOPTICAL OBSERVATIONS.

Stol., and two others of the *Scabræ*, equally insignificant, *T. crenifera*, Stol., and *T. minuta*, Stol., complete the list of Indian Cretaceous Trigoniæ.

The section *Pectinidæ* was established by Agassiz in 1840 ('Trigoniæ,' pp. 10 and 48) upon the *Trigonia pectinata* of Lamarck, at that time the only known species of the section, so named from the external resemblance which their ornamentation bears to the *Pectines and Limæ*; the species, both living and Tertiary, are exclusively Australian. They present in their suborbicular Cardium-like forms, in their crenulated costæ radiating from the umbones, in their areas destitute of bounding carinæ or of divisional sulcations, in the absence of any clear separation between the dorsal and the anal or siphonal portions of the surface, a remarkable contrast to the *Trigoniæ* of the other continents; differences which are rendered the more remarkable when we examine the hinge features, which present little or no modification of the older Mesozoic forms of the genus; even the changes observable in the interiors of certain of the Cretaceous *Scabræ* and *Quadratæ* have also disappeared, and in the *Pectinidæ* we find reproduced unchanged the more ancient hinge features of the Mesozoic *Glabræ* in all their original prominence.

The European and American Tertiary formations, although occupying such extensive tracts of country and presenting every gradation of molluscan life, from extinct to living forms, are altogether destitute of the genus: it is only in the Tertiary Australian deposits of Victoria and of South Australia, associated with other existent generic forms, that we again discover *Trigoniæ*, represented solely by the group of the *Pectinidæ*, and more or less nearly allied to the few forms of the genus which inhabit the seas and tidal waters of the same region. This great hiatus in the chain of *Trigonia* life may possibly be eventually filled up by discoveries in some unknown series of the Tertiary formations. Widely, indeed, as the *Pectinidæ* are separated from the usual Mesozoic sectional forms, we discover some resemblance to a portion of the Cretaceous *Scabræ* both in their surface ornaments, and in the not less important absence of carinæ upon the superior and siphonal portions of the shell. There may also be observed in an Australian Tertiary species (*T. Howitii*, McCoy) a tendency to effacement of the ornaments over the middle portion of the valves, various examples of which occur in the Mesozoic *Undulatæ* and *Glabræ*. Perhaps some modification of this statement may be deduced from an examination of some known Cretaceous *Trigoniæ*. In *T. disparilis*, d'Orb. ('Ter. Crét.,' pl. 229), one of the *Scabræ* of the Terrain Sénonien, or highest chloritic marls of Tours, some approximation may be seen in the numerous crenulated costæ which radiate from the umbones over the middle portion of the valves; also in the minute *T. pulchella*, Reuss (Böhm., tab. 41), from a similar position in Bohemia; if, indeed, the latter species be not the very young condition of the former, which I am inclined to believe.

The American *T. Humboldtii*, von Buch, which has likewise costellæ radiating from the umbones over the upper, the anal, and the median portions of the shell, may also be regarded as a transitive or connecting species; these, however, are rare and exceptional forms in the great section of the *Scabræ*.

Within the last few years the Tertiary deposits of Australia have yielded several forms of *Trigonia;* these all belong to the section of the *Pectinidæ*, more or less allied to living forms. One small one, more especially allied to the species of Sydney Harbour, is named by Jenkins *T. Lamarckii* (' Geol. Mag.,' vol. iii, 1866, pl. 10, figs 3, 7), a name previously given by Matheron to a Cretaceous species (see p. 138). This Tertiary species is named by McCoy *T. acuticosta* (' Geol. Mag.,' 1866, p. 481), from the beds of Mordialloc in Hobson's Bay. Other Tertiary species are *T. semiundulata*, McCoy, from Bird Rock Bluff, and *T. Howitii*, McCoy, from near the entrance to the Gippsland lakes. The latter in size nearly equals some of the larger Mesozoic forms, the length is twenty-six lines, the height twenty lines; it is remarkable for the great length of the hinge-border compared with the short perpendicular siphonal border; the middle portion of the shell has a tendency to the effacement of surface ornaments (' Ann. and Mag. Nat. Hist.,' ser. 4, vol. xv, pl. 18).

The distinctive differences of *T. acuticosta* as compared with the species of Sydney consist in the far more numerous costæ, their angulated forms, the lesser convexity of the valves, their more inequilateral figure, the greater breadth of the posteal slope, and greater length of the siphonal border : differences of small importance when viewed separately, but in the aggregate appearing to justify the separation claimed by McCoy for the Tertiary forms. They are, however, for the most part such as have been regarded as *varieties* among the Mesozoic *Trigoniæ* figured in this Monograph; and, should it eventually be decided to separate this small Tertiary species from the neighbouring living forms, it would undoubtedly be necessary to erect into species certain Mesozoic forms here tabulated as *varieties*.

T. semiundulata, McCoy, considered to approximate to the Jurassic *Trigoniæ costatæ* in its apparent concentric ribbing, is, nevertheless, one of the *Pectinidæ*, having the anteal and mesial costæ only slightly defined, crossed by undulating concentric ridges or lines of growth; in no other feature does it approach to Mesozoic forms.

The Tertiary Australian *Trigoniæ* deposited in the Melbourne National Museum have been assigned to the Miocene and Pliocene stages from a comparison of the percentage of living forms with which they are associated; but the question whether such a rule is applicable to Australian geology, and whether it affords a criterion whereby it may be measured with European Tertiary deposits, has yet to be determined.

The foregoing three Tertiary forms are so distinct from each other that there can scarcely remain a doubt of the propriety of regarding them as separate species; the small *T. acuticosta*, in its approach to the living Australian forms, will require further investigation, which will include the question of the separation of the living *Pectinidæ* into species and varieties, concerning which naturalists are much divided in opinion. Our present very insufficient knowledge upon this subject may be greatly augmented by the results of future dredging operations. A more precise estimate will be thus obtained of the hydrographical limits and habitats of forms, at present tabulated as species, which are distributed in the shallow Coralline seas girding the eastern coasts of Australia, throughout more than thirty degrees of south latitude, from Cape York to Tasmania. Known, in one instance only, as a species of the sea, in other instances as denizens of land-locked waters, or of brackish waters in tidal rivers, these *Trigoniæ* do not appear to form varieties at any one locality. The differences between these species or varieties, chiefly founded upon their exterior forms and ornaments, will have to be considered and determined in connection with their aspects over the entire marine area occupied by each form. Eventually it may in this manner be possible to ground our knowledge of this section upon the living, in comparison with the Australian Tertiary, forms of the genus, and thus to legislate, with greater authority, upon the questions of species and varieties of the *Pectinidæ*.

In making comparisons of the species or varieties of the living *Pectinidæ*, we may select two Australian forms nearly allied, which have, of late years, become well known from their abundance; one of these is the *Trigonia* of that land-locked, fine expanse of water constituting Sydney Harbour, and the mud of its tributary, the Paramatta river, it is the *T. Lamarckii* of Jenkins and the *T. Jukesii* of Adams. It has been regarded by some naturalists as a variety only of a larger and nearly allied Tasmanian form, to which Lamarck's name, *Pectinata*, is now exclusively applied. The latter is abundant, buried in the black mud of the Launceston river, or in the tidal portion of it, the brackish water of which extends up the course of the channel for many miles. Separated from the habitat of the other shell by eight degrees of latitude, the difference of form, although only inconsiderable, is very persistent at each locality, and is instantly detected in the adult stage of growth. *T. pectinata* has the lesser convexity; the umbones are smaller and more oblique; the length of the siphonal border is greater; the costæ over the valve generally are less numerous; they are closely arranged anteally, but become widely separated over the middle of the valve, the spaces between them increasing towards the posteal slope; the costellæ upon the slope are small and inconspicuous, the costæ near to the border and throughout the circumference of the valve degenerate in their crenulations into closely placed imbricated lamellæ of growth, which are obscured by the greater development of the epidermal tegument; a similar feature is seen in various Mesozoic species, and notably in the Neocomian *Trigonia nodosa*.

The Sydney species has a similar kind of ornamentation; but, having larger and less oblique umbones, it is more globose; the hinge-area, corresponding with the escutcheon in the Mesozoic forms, is larger, and slightly excavated, and its ornamentation is more prominent than in *T. pectinata*; the costæ over the shell generally have smaller interstitial spaces, their crenulations become more square or flat-topped over the middle of the valves and near to the lower border. The siphonal border is always shorter and more perpendicular, forming a more considerable angle with the hinge-border; this feature from its prominence would alone be sufficient to separate the two forms. The young shell is usually more orbicular, having the siphonal portion less developed, it is therefore sometimes not sufficiently illustrative of the fully developed growth of the species.[1]

The Sydney *T. Jukesii* is, however, nearly allied to another form which has been obtained rarely in the Coralline sea at Cape York, the most northernly point of Australia, separated by twenty-two degrees of latitude, and figured by Gray under the name of *Trigonia uniophora* ('Voyage of the Fly,' 1847, Appendix, pl. 2, fig. 5). The only essential difference between the latter and the species of Sydney appears to consist in the greater breadth of the postcal slope, and the greater length of the siphonal border in Gray's species—features which, unaccompanied by other distinctions, can only be regarded as constituting a varietal character.

To the foregoing living Australian forms of the *Pectinidæ* must also be added a single unnamed *Trigonia* upon the tablets of the British Museum, remarkable for the bizarre and anomalous character of the external ornaments, and especially for the characters of the costæ.

With our experience of the last few years it is easy to foresee that the missing connecting links between the Mesozoic and living *Trigoniæ* may be expected to be found in the Tertiary formations of Australia.

[1] Specimens with individual peculiarities occur; one of the Sydney Harbour *T. Jukesii*, Adams, in my possession, of adult growth, has in each valve an arrest of growth near to the pedal border; the left valve has three additional narrow interstitial costæ, the right valve having one such. An inordinate secretion of shell substance internally causes the valves to gape at the hinge-border, exposing the hinge processes with their transverse sulcations. A single, small, interstitial costa not unfrequently occurs in this species.

GEOLOGICAL DISTRIBUTION OF SPECIES AND VARIETIES OF BRITISH TRIGONIÆ, WITH REFERENCES TO THE FIGURES AND DESCRIPTIONS IN THE PRESENT MONOGRAPH.

TRIASSIC SYSTEM.
Rhætic Formation: allied genus, Myophoria.

		PAGE
Myophoria postera, *Quenst.* Pl. xli, figs. 6, 6 *a*, 7, 7 *a*, 8, 8 *a*, 9, 9 *a*		212

JURASSIC SYSTEM.
LOWER LIAS.
COSTATÆ..............? Trigonia modesta, *Tate.* Pl. xli, figs. 13, 13 *a* . . 212

MIDDLE LIAS.
GLABRÆ..............Trigonia Lingonensis, *Dum.* Pl. xxii, figs. 1, 1 *a*, 2, 3, 4   98

UPPER LIAS.
SCAPHOIDEÆ ....T. pulchella, *Ag.* Pl. xxxviii, figs. 10, 11, 12, 12 *a*   185
UNDULATÆ............T. literata, *Young & Bird.* Pl. xiv, figs. 1, 1 *a*, 2, 3, 4 .   64

MIDFORD OR SUPRA-LIASSIC SANDS.
UNDULATÆ ....T. Leckenbyi, *Lyc.* Pl. xvi, figs. 1, 2   71
CLAVELLATÆ...... { T. formosa, *Lyc.* Pl. v, figs. 4, 5, 6 .   35
              { T. Ramsayi, *Wright.* Pl. vi, fig. 6 .   49
COSTATÆ..............T. denticulata, *Ag.* Pl. xxix, figs. 1, 2, 3, 4   152

INFERIOR OOLITE.
SCAPHOIDEÆ...... { T. Bathonica, *Lyc.* Pl. i, fig. 3 .   17
              { T. duplicata, *Sow.* Pl. i, figs. 8, 9, 10   14
              { T. gemmata, *Lyc.* Pl. i, fig. 7 .   15
              { — (*var.* bifera), *Lyc.* Footnote   239
              { T. recticosta, *Lyc.* Pl. i, figs. 4, 5, 6   16, 198

GEOLOGICAL DISTRIBUTION.

		PAGE
CLAVELLATÆ	T. Brodiei, *Lyc.* Pl. xxxv, figs. 8, 9	195
	T. formosa, *Lyc.* Pl. v, figs. 4, 5, 6; pl. xi, fig. 2; pl. xxxvii, fig. 10	35, 202
	T. — (*var.* lata). Pl. xxix, figs. 11, 12; pl. xxxv, fig. 7	35, 202
	T. parcinoda, *Lyc.* Pl. xxxvi, fig. 8	46
	T. Phillipsi, *Mor. & Lyc.* Pl. vi, figs. 3, 4	38
	T. signata, *Ag.* (Zieten's var.). Pl. ii, figs. 1, 2, 3	29, 204, 207
	T. spinulosa, *Young & Bird.* Pl. iii, figs. 4, 5 *a*, 5 *b*, 6	44
	T. striata, *Mil.* Pl. v, figs. 6′, 7, 8	36
	T. tuberculosa, *Lyc.* Pl. v, figs. 9, 10	33
	T. Witchelli, *Lyc.* Pl. xxxviii, figs. 8, 9 (Fuller's Earth)	197
UNDULATÆ	T. angulata, *Sow.* Pl. xiv, figs. 5, 6; pl. xxxvii, figs. 7, 8, 9	54, 206
	T. compta, *Lyc.* Pl. xv, figs. 5, 6, 7; pl. xxxviii, fig. 4	70
	T. conjungens, *Phil.* Pl. x, figs. 5, 7, 8; pl. xiii, fig. 6	62
	T. costatula, *Lyc.* Pl. xv, figs. 8, 9, 10; pl. xii, fig. 6, 6 *a*	81
	T. producta, *Lyc.* Pl. xiii, figs. 1, 2, 3, 4; pl. xxxvii, figs. 1, 2	60, 208
	T. Sharpiana, *Lyc.* Pl. xv, fig. 11; pl. xvi, figs. 3, 4, 5, 6	79
	T. subglobosa, *Lyc.* Pl. xii, figs. 8, 9, 10	68
	T. v.-costata, *Lyc.* Pl. xiii, fig. 5; pl. xv, figs. 1, 2, 3, 4	66
GLABRÆ	T. Beesleyana, *Lyc.* Pl. xvii, figs. 2, 3, 4	91
COSTATÆ	T. bella, *Lyc.* Pl. xxxii, figs. 6, 7, 8, 8 *a*	126
	T. costata, *Sow.* Pl. xxix, figs. 5, 6, 7, 8	147
	T. — (*var.* lata). Pl. xxix, figs. 9, 10	149
	T. Culleni, *Lyc.* Pl. xxxi, figs. 9, 9 *a*	173
	T. denticulata, *Ag.* Pl. xxix, figs. 1, 2, 3, 4	152
	T. hemisphærica, *Lyc.* Pl. xxxi, figs. 4, 5, 6, 7, 8	174
	T. — (*var.* gregaria). Pl. xxxiii, figs. 4, 5, 6	176
	T. pullus, *Sow.* (small var.). Pl. xxxiv, figs. 7, 7 *a*, 8, 9	164
	T. sculpta, *Lyc.* Pl. xxxiv, figs. 1, 2, 2 *a*	157
	T. — (*var.* Cheltensis). Pl. xxxiv, fig. 3	159
	T. tenuicosta, *Lyc.* Pl. xxxiii, figs. 7, 8, 9, 9 *a*	160

GREAT OOLITE, STONESFIELD SLATE, FOREST-MARBLE, AND CORNBRASH.

CLAVELLATÆ	T. Griesbachi, *Lyc.* Pl. iii, figs. 10, 10 *a*, *b*; pl. xxxvi, figs. 11, 11 *a*	34
	T. imbricata, *Sow.* Pl. vi, fig. 5; pl. xxxvi, figs. 9, 10. Fuller's Earth, pl. xli, figs. 10, 11, 12	33, 209
	T. impressa, *Sow.* Pl. vii, figs. 4, 5	46
	T. Morstoni, *Mor. & Lyc.* Pl. ii, figs. 4, 5, 7, 8; pl. iv, fig. 6	47
	T. Scarburgensis, *Lyc.* Pl. iv, figs. 1, 2, 3, 4	31, 200
UNDULATÆ	T. arata, *Lyc.* (*var.* of T. undulata, *From.*). Pl. xvi, figs. 9, 10, 11; pl. xvii, figs. 5, 6	77, 200
	T. Clytia, *d'Orb.* Pl. xi, figs. 4, 5; pl. xvii, fig. 7	76
	T. detrita, *Terq.* Pl. x, figs. 3, 3 *a*, 4	75
	T. flecta, *Lyc.* Pl. xiv, figs. 7, 8, 9, 10	55
	T. Painei, *Lyc.* Pl. xii, figs. 2, 3, 4, 5	59
	T. tripartita, *Forbes.* Pl. xii, fig. 7	74

GEOLOGICAL DISTRIBUTION OF BRITISH TRIGONIÆ. 237

		PAGE
COSTATÆ	T. Cassiope, *d'Orb.* Pl. xxxii, figs. 1, 2, 4, 5 .	170
	T. costata (Cornbr.). Pl. xxix, figs. 5, 6, 7, 8 .	147
	T. elongata (*var.* angustata). Pl. xxx, figs. 1, 1 *a*, 2	154
	T. — (*var.* lata). Pl. xxx, figs. 4, 5 .	154
	T. sculpta, *Lyc.* (*var.* Rolandi). Pl. xxxiv, fig. 4	157
	T. pullus, *Sow.* Pl. xxxiv, figs. 7, 7 *a*, 8, 9 .	164

KELLOWAY ROCK, OXFORD CLAY, CORALLINE OOLITE, INCLUDING CALC. GRIT AND CORAL RAG.

SCAPHOIDEÆ	T. Snaintonensis, *Lyc.* Pl. xli, figs. 1, 1 *a*, 2, 3	. 198
CLAVELLATÆ	T. Blakei, *Lyc.* Pl. xli, fig. 4 . . .	. 205
	T. clavellata, *Sow.* Pl. i, figs. 1, 1 *a*, 2 ; pl. iv, fig. 8 .	18, 209
	T. complanata, *Lyc.* Pl. vii, fig. 3 . . .	. 49
	T. corallina, *d'Orb.* Pl. viii, fig. 5 ; pl. iii, figs. 7, 8, 9, 11	. 45
	T. Hudlestoni, *Lyc.* Pl. xxxiv, figs. 5, 6 ; pl. xxxix, figs. 1 *a*, 2 .	. 194
	T. irregularis, *Seeb.* Pl. v, figs. 1 *a*, 1 *b*, 2 ; pl. vii, fig. 6 ; pl. xxxix, fig. 3	39
	T. perlata, *Ag.* Pl. iii, figs. 1, 2, 3 ; pl. xi, fig. 3 .	. 22
	T. Rupellensis, *d'Orb.* Pl. viii, fig. 4 ; pl. xxxvi, figs. 1, 2, 3, 4	28, 199
	T. triquetra, *Seeb.* Pl. vi, figs. 1, 1 *a*, 2 ; pl. xxxvi, fig. 7	. 26
	T. Williamsoni, *Lyc.* Pl. xvi, fig. 8 ; pl. xxxviii, fig. 7 .	. 53
UNDULATÆ	T. geographica, *Ag.* Pl. x, fig. 6 ; pl. xxxii, fig. 9 .	69
	T. Joassi, *Lyc.* Pl. xx, figs. 2, 3, 4 . .	. 82
	T. paucicosta, *Lyc.* Pl. xi, figs. 8, 9 ; pl. xvi, fig. 7 ; pl. xxxvii, fig. 3	57, 206
COSTATÆ	T. elongata, *Sow.* Pl. xxx, figs. 3, 3 *a*, 3 *b*, 6 .	. 154
	T. Meriani, *Ag.* Pl. xxxiii, figs. 1, 2, 3 .	. 167

KIMMERIDGE CLAY AND PORTLAND OOLITE.

CLAVELLATÆ	T. Alina (Cont.). Pl. xxxviii, fig. 3 ; pl. ix, fig. 2 .	193
	T. cymba (Cont.). Pl. xxxviii, fig. 1 . .	. 192
	T. incurva, *Ben.* Pl. ix, figs. 2, 3, 4, 5, 6 .	42
	T. irregularis, *Seeb.* (*var.*). Pl. xxxix, fig. 3 . .	39
	T. Juddiana, *Lyc.* Pl. ii, figs. 6, 6 *a*, 6 *b* ; pl. iv, figs. 5, 7 .	25
	T. muricata, *Goldf.* Pl. ix, fig. 1 . .	50
	T. Pellati, *Mun. Chal.* Pl. vii, figs. 1, 2, 2 *a* ; pl. xi, fig. 1 ; pl. xxxix, fig. 1 .	41
	T. Voltzii, *Ag.* Pl. x, figs. 1, 2 (T. Thurmanni Cont.) .	. 20
	T. Woodwardi, *Lyc.* Pl. xvii, fig. 1 . .	40
UNDULATÆ	T. Carrei, *Mun. Chal.* Pl. xii, fig. 1 . .	72
COSTATÆ	T. monilifera, *Ag.* Pl. xxxi, figs. 1, 1 *a*, 1 *b*, 2, 2 *a*, 3 .	. 165
GLABRÆ	T. Damoniana, *De Lor.* Pl. xviii, fig. 3 ; pl. xix, figs. 1, 1 b, 1 c ; pl. xxi, figs. 2, 2 *a*, 2 *b*, 3, 4, 5 .	88
	T. gibbosa, *Sow.* Pl. xviii, figs. 1, 2, 2 *a*, 4, 5, 6 ; pl xix, fig. 2 ; pl. xxi, fig. 1	84
	T. Manseli, *Lyc.* Pl. xix, figs. 3, 4, 4 *a*, 4 *b* .	. 86
	T. Michellotti, *de Lor.* Pl. xx, fig. 7 (*variety*) .	92
	T. tenuitexta, *Lyc.* Pl. xx, figs. 1, 1 *a*, 1 *b* .	90

238   GEOLOGICAL DISTRIBUTION OF BRITISH TRIGONIÆ.

CRETACEOUS SYSTEM.

MIDDLE NEOCOMIAN FORMATION.

		PAGE
CLAVELLATÆ......	T. ingens, *Lyc.* Pl. viii, figs. 1, 2 a, 2 b, 3; pl. xxxvi, figs. 5, 6 (Tealby and W. Norfolk)	24, 207
	T. Keepingi, *Lyc.* Pl. xxxv, figs. 1, 2 (Tealby and W. Norfolk)	196
QUADRATÆ	T. nodosa, *Sow.* (*var.*) Pl. xxv, fig. 2 (Tealby)	106
	T. Tealbyensis, *Lyc.* Pl. xxviii, fig. 7 (Tealby)	114
SCAPHOIDEÆ......	T. exaltata, *Lyc.* Pl. xxxviii, fig. 2 (W. Norfolk)	184
	T. scapha, *Ag.* Pl. xxxviii, fig. 6 (W. Norfolk)	183

UPPER NEOCOMIAN FORMATION.

QUADRATÆ............	T. nodosa, *Sow.* Pl. xxiv, figs. 1, 1 a, 2, 3 (*var.* Orbignyana); pl. xxxvii, figs. 5, 5 a, 6	106
	T. caudata, *Ag.* Pl. xxvi, figs. 5, 6, 6 a, 6 b, 7	129
	T. Etheridgei, *Lyc.* Pl. xxvii, figs. 1, 1 a, 1 b, 2, 3, 3 a	127
SCABRÆ............	T. ornata, *d'Orb.* Pl. xxiv, figs. 6, 7	139
	T. Upwarensis, *Lyc.* Pl. xxiii, figs. 8, 9; pl. xxxix, fig. 4 (Upware)	143
	T. Vectiana, *Lyc.* Pl. xxiv, figs. 10, 10 a, 10 b, 11; pl. xxv, fig. 7	123
BYSSIFERÆ	T. carinata, *Ag.* Pl. xxxv, figs. 4, 4 a, 5, 5 a, 6, 6 a	179

GREEN SANDS OF BLACKDOWN AND HALDON.

QUADRATÆ	T. dædalea, *Park* (Blackdown). Pl. xxiii, figs. 2, 3; pl. xxii, figs. 7, 8; pl. xxviii, fig. 8	100
	— var. confusa, *Lyc.* (Little Haldon). Pl. xxiii, fig. 1	102, 211
	T. spectabilis, *Sow.* Pl. xxvi, figs. 1, 2, 3, 4	112
SCABRÆ............	T. aliformis, *Park.* Pl. xxv, figs. 3, 3 a, 4, 4 a; pl. xxviii, figs. 5, 5 a	116
	T. Fittoni, *Desh.* Pl. xxiii, figs. 4, 4 a, 4 b, 5 (Gault)	132
	T. scabricola, *Lyc.* Pl. xxvii, figs. 4, 5, 5 a, 5 b	130
	T. spinosa, *Park.* P. xxiv, figs. 8, 9	136
	T. Vicaryana, *Lyc.* Pl. xxiii, fig. 7	141, 203
GLABRÆ............	T. excentrica, *Park.* Pl. xx, figs. 5, 6; pl. xxii, figs. 5, 5 a; pl. xxviii, figs. 6, 6 a, 9, 10	94
	T. læviuscula, *Lyc.* Pl. xxii, fig. 6	96

UPPER GREEN SANDS AND CHLORITIC MARLS.

QUADRATÆ............	Species doubtful. Dunscombe Cliffs (*Meyer*).	
	T. aliformis (*var.* attenuata), *Lyc.* Pl. xxv, figs. 5, 6	118
	T. Archiaciana, *d'Orb.?* Pl. xxv, fig, 10	140, 202
	T. costigera, *Lyc.* Pl. xl, fig. 17	205
	T. crenulifera, *Lyc.* Pl. xl, figs. 1, 1 a, 1 b, 7, 9 a, 9 b	189
	T. Cunningtoni, *Lyc.* Pl. xxiii, fig. 4	146
	T. Meyeri, *Lyc.* Pl. xxiii, fig. 6; pl. xli, figs. 15, 16	125
SCABRÆ............	T. pennata, *Sow.* Pl. xxiv, figs. 4, 5; pl. xxxvii, figs. 4, 4 a	133
	T. scabricola, *Lyc.* Pl. xxvii, figs. 4, 5, 5 a, 5 b	130
	T. spinosa, *Park.* Pl. xxiii, fig. 10 (*var.* suborata); pl. xxviii, figs. 1, 2	201

CORRIGENDA.

		PAGE
	T. sulcataria, *Lam.* Pl. xxvi, fig. 8; pl. xxviii, fig. 3	135
	T. Vicaryana, *Lyc.* Pl. xxv, fig. 9; pl. xxvii. figs. 4, 4 *a*; pl. xl, figs. 3, 4	141, 203
	— (*var.*) Pl. xxv, fig. 8	141
	T. affinis, *Mill.* Pl. xl, fig. 2; pl. xxi, fig. 7	187
GLABRÆ............	T. debilis, *Lyc.* Pl. xl, figs. 8, 8 *a*; pl. xli, fig. 5	189
	T. Dunscombensis, *Lyc.* Pl. xl, figs. 5, 6; pl. xli, fig. 41	188
	T. læviuscula, *Lyc.* Pl. xxii, fig. 6	96
BYSSIFERÆ............	T. carinata, *Ag.* Pl. xxxv, fig. 3 (Ventnor)	179

NOTE.—Subsequently to the completion of this Monograph Mr. Witchell kindly forwarded to me a variety of *T. gemmata*, Pl. I, fig. 7, p. 15, which I propose to designate by the varietal name *bifera*. It has seven rows of concentric costæ, which occupy more than half the height of the valve. The oblique or perpendicular costæ, nine in number, have three only which originate at the marginal carina; all the others proceed from the last-formed concentric costa to the pallial border; the general figure is somewhat shorter than the typical form. Both varieties have occurred very rarely in a freestone bed of the Upper Trigonia Grit of the Inferior Oolite in the vicinity of Stroud.

CORRIGENDA.

PAGE

11 (Introduction). *For* the general sketch of the distribution of British Trigoniæ commencing at this page *substitute* the revised stratigraphical table of their distribution at the end of the Monograph, p. 235.

23, line 4. *Erase T. Bronnii*, Ag., pl. iv, fig. 8, and *substitute T. clavellata*, Ag., young specimen. See also p. 209 for description and figures of *T. Bronnii*, Ag.

42. Title to *T. incurva*, Ben., *alter* fig. 2 to *T. Alina*, Cont., var. *Alter* also explanation, fig. 2, facing pl. ix; *make it T. Alina*, Cont. It is also corrected upon the stratigraphical table, p. 237. Refer these to *T. incurva* and to *T. Alina*.

43. *Erase* line 26, commencing "No. 2 has suffered," to the word "pointed" at end of the sentence.

52. *For T. concentrica*, Ag., *see* p. 206.

59, line 4. *Read* few examples of *Trigonia paucicosta* have occurred at that locality.

62, line 7. *For* Northamptonshire *read* Oxfordshire at Hook Norton; see Pl. XXXVII, figs. 1, 2.

69. *T. geographica*, Ag.; see also Pl. XXXII, fig. 9.

74. *Erase* the concluding sentence of the description of *T. Carrei*, which refers to an unsatisfactory and doubtful specimen not figured.

77. *T. undulata*, From., var. *arata*, Pls. XVI and XVII; see also p. 201 for figures of the typical form of *T. undulata*.

84. *T. gibbosa*, Sow. *Erase* Pl. XVIII, fig. 3; also Pl. XIX, figs. 1, 1 *a*, 1 *b*. The references facing the plates are correct.

88. *T. Damoniana*, de Lor. *Erase* Pl. XX, figs. 1, 2, 2 *a*, 2 *b*, *substitute* Pl. XXI, figs. 2, 3, 4, 5, 2 *a*, 2 *b*. The references facing the plates are correct.

91. *T. Beesleyana*, Lyc. Additional well-preserved specimens exhibit a narrow, lengthened, depressed space upon the superior border, representing the escutcheon; the postenl slope, therefore, represents the area with its delicate transverse costellæ. There are no carinal elevations.

CORRIGENDA.

PAGE

94. *T. excentrica*, Park. *Alter* this name to *T. affinis*, Mill. and Sow.; see p. 187, Pl. XXI, fig. 7, and Pl. XXII, figs. 5, 5 *a*; also Pl. XL, fig. 2.

96, line 23. *Erase* the sentence commencing "The Chloritic Marls."

115, *T. Agassizii*, line 23. *Erase* this name and *substitute T. Upwarensis*, p. 143.

126, line 7 from the bottom. *For T. divaricata* read *T. disparilis*.

138, line 12. *For* Petref. Sulc. *read* Petref. Suecc.

140. Pl. XXIII, fig. 7. *Alter* to *T. Vicaryana*, Lyc.

159, line 2 from the top. *For* Shamford *read* Stamford.

174. *Trigonia hemisphærica*, var. *gregaria*, Pl. XXXIII, figs. 4, 5, 6. The examples of the small variety here represented have been incautiously selected for their good condition of preservation. The ribbing is unusually large, and does not exemplify the more common, smaller, and less clearly defined examples with minute or variable ribbing.

211, Addenda. *Erase* line 13, commencing "at the base of the Upper Greensands," and *substitute* "in one of the lower beds of Greensand at that locality." The pebble bed here alluded to is at the base of the fourth or highest stage of the Trigonia-bearing beds. For its species, &c., see the stratigraphical table.

EXPLANATION OF PLATES.

Pl. IV, fig. 8. *Erase* the name *T. Bronnii* and *substitute T. clavellata*, Sow., young specimen, see pp. 23 and 209.

Pl. IX, fig. 1. *Erase* Wilts and *substitute* St. Adhelm's Head, Dorset.

Pl. IX, fig. 2. *Alter T. incurva* to *T. Alina*, Cont. *Alter* page 42 to page 193.

Pl. IX, fig. 3. *Add T. incurva*.

Pl. IX, fig. 4. *Alter* the locality to St. Adhelm's Head, Dorset.

Pl. XIX, figs. 4 *a*, 4 *b*. The encircling costæ upon the umbones of *T. Manseli* are not sufficiently numerous and minute.

Pl. XXI, figs. 2 *a*, 2 *b*. The encircling costæ upon the umbones are not sufficiently linear and minute; this feature alone is sufficient to separate the form from *T. gibbosa*.

Pl. XXI, fig. 7. *Add* the words *T. affinis*. See page 187.

Pl. XXIII, fig. 7. *Alter T. Archiaciana*, D'Orb., to *T. Vicaryana*, Lyc. *Alter* page 140 to page 203.

Pl. XXXIV, fig. 1. *T. sculpta*. The figure of the right valve appears to represent an ante-carinal groove adjacent to the marginal carina. This is owing to the person who cleared the fossil from its matrix having heedlessly removed the posteal extremities of the costa, which should extend to the carina, as in other examples of the *Costatæ*.

The author desires to record his obligations to the artists engaged upon the plates of this Monograph for the general care and fidelity to nature which they evince—the first nine plates by the late Mr. Lackerbauer, the succeeding plates by M. Karmansky, the last plate by a son of the late Mr. Lackerbauer. The wood engravings up to p. 122 are by Mr. Dewilde; the subsequent ones, by Mr. G. Shayler, are carefully drawn exemplifications of foreign Trigoniæ.

ALPHABETICAL INDEX OF SPECIES AND VARIETIES OF TRIGONIA REFERRED TO IN THIS MONOGRAPH.

Species known under other names and synonyms are printed in *italics*. The letters F. S. following species indicate that the form has not been found in Britain.

	PAGE
Myophoria postera, *Quenstedt*	215, 219
— *varigeræ, glabræ, costatæ*, F. S.	214
Trigonia abrupta, Coquand, F. S.	126, 227
— — De Buch, F. S.	126, 227, 229
— *acuticosta*, McCoy, F. S.	232
— affinis, *Miller*	94, 187
— aliformis, *Parkinson*	116, 123
— — var. attenuata, *Lycett*	118
— — *Fitton*; see T. Vectiana	123
— *aliformis*, Forbes; see T. Vectiana	116, 122
— — Pictet and Renevier	116, 123
— — De Buch, F. S.	119, 229
— — d'Orbigny, F. S.	122
— — Mantell; see T. Vectiana	123
— — Ibbetson and Forbes; see T. Vectiana	123
— — Judd; see T. Vectiana	123
— — (*Lyrodon aliforme*, Goldfuss), F. S.	116, 122, 131
— Alina, *Contejean*	193
— angulata, *Sowerby*	54, 56, 58, 63, 67, 71, 206
— *angulata*, Phillips; see T. V. costata	66
— — Oppel; see T. flecta	55
— *angulosa*, Agassiz; see T. angulata	54
— *arata*, *Lycett*; see T. undulata, Fromherz	48, 77, 200
— Archiaciana, *d'Orbigny*	140, 202

	PAGE
Trigonia Arduennea, Buvignier, F. S.	60, 223
— aspera, Lamark, F. S.	223
— *Barrensis*, Buvignier	223
— Bathonica, *Lycett*	17
— *Baylei*, Dolfuss, F.S.	223
— Beesleyana, *Lycett*	99, 221
— bella, *Lycett*	162
— Blakei, *Lycett*	205
— *Bouchardi*, Oppel, F. S.	18, 223
— *Boulogniensis*, De Loriol, F. S.	223
— *Boussignaultii*, d'Orbigny, F. S.	223
— Brodiei, *Lycett*	195
— *Bronnii*, Agassiz, F. S.	23, 209, 222
— *cardissa*, Agassiz, F. S.	157, 223
— carinata, *Agassiz*	179, 220
— Carmontensis, De Loriol, Royer, and Tombeck, F. S.	223
— Carrei, *Munier-Chalmas*	72
— Cassiope, *d'Orbigny*	170, 172
— caudata, *Agassiz*	127, 129
— cincta, Agassiz; see T. *nodosa*	106, 111
— clapensis, Terquem and Jourdy; see T. Moretoni	47
— clathrata, Agassiz, F. S.	223
— clavellata, *Sowerby*	18, 20, 23, 53, 63
— *clavellata*, Knorr; see T. signata, F. S.	204
— — *major*, Luid	18
— — var. *jurensis*, Grewingk; see T. corallina	45

INDEX.

Trigonia *clavellata*, Ziethen ; see T. signata 204
— clavocostata, *Lycett* 29
— *clavulosa*, Rigaux and Sauvage, F. S. 33
— *clivosa*, Credner, F. S. 26
— clytia, *d'Orbigny* 76
— complanata, *Lycett* 49
— compta, *Lycett* 63, 70, 80
— *concentrica*, Agassiz, F. S.
45, 52, 206, 223
— *concinna*, Roemer, F. S. 223
— conjungens, *Phillips* 17, 62, 68, 71
— *conocardiiformis*, Krauss, F.S.
120, 210, 230
— *Constantii*, d'Orbigny, F.S....... 115, 227
— *Coquandiana*, d'Orbigny, F. S.... 97, 227
— corallina, *d'Orbigny*........... 45, 49, 52
— costata, *Sowerby* 147
— — var. lata, *Lycett* 149
— *costata*, Knorr ; see T. sculpta 150
— — Encycl. Method. ; see T.
sculpta 150
— — Parkinson (doubtful) 150
— — Smith ; see var. Rolandi ... 150
— — Bronn ; see T. sculpta 150
— — Sowerby, in Grant's Mem. ;
see T. elongata 150
— — Ziethen ; see T. denticulata 150
— — Young and Bird ; see T.
Meriani 150
— — Pusch. ; see T. zonata, F. S. 151
— — var. *triangularis*, Goldfuss,
F. S. 151
— — var. *silicea*, Quenstedt, F. S. 167
— — Chapuis and Dewalque,
F. S. 151
— costatula, *Lycett*........... 69, 70, 81
— *costellata*, Agassiz, F. S......... 45, 66, 222
— costigera, *Lycett*................. 205
— *Cottaldi*, Munier-Chalmas, F. S. 192, 223
— *crenifera*, Stoliczka, F. S. 143, 231
— *crenulata*, Lamark, F. S.......... 190, 227
— crenulifera, *Lycett* 189
— Culleni, *Lycett* 173
— Cunningtoni, *Lycett* 146
— cuspidata, *Sowerby*................ 59
— cymba, *Contejean* 192
— Dæmoniana, *De Loriol*......... 86, 88, 91

Trigonia debilis, *Lycett*............... 189
— decorata, Lycett ; see T. signata ... 29
— dædalea, *Parkinson* 100
— — var. confusa, *Lycett*... 102, 211
— *dædalea*, Agassiz, F. S. 104
— — d'Orbigny, F. S. 105
— — Pictet and Renevier ; see
T. nodosa, var............. 106
— *Delafossei*, Leymerie, F. S. ... 120, 227
— denticulata, *Agassiz* 45, 152
— detrita, *Terq*. and *Jour*. 75
— divaricata, d'Orbigny, F. S...... 126, 227
— Dunscombensis, *Lycett* 188
— duplicata *Sowerby* 14, 16
— *elegans*, Bailey, F. S. 115, 230
— *Elisæ*, Cornet and Briart, F. S. 212, 227
— elongata, *Sowerby* 154
— — var. *angustata*, Lycett ... 156
— — var. *lata*, Lycett............ 156
— *Etalloni*, De Loriol, F. S. 223
— Etheridgei, *Lycett* 127
— exaltata, *Lycett* 184, 221
— excentrica, *Parkinson*............ 92, 94
— excentricum (Lyrodon), *Goldfuss* ;
see T. Michellotti 92, 227
— *exigua*, Lycett ; see T. costatula ... 81
— *Ferryi*, Munier-Chalmas ; see T.
radiata............................ 73
— *fimbriata*, Lycett, F. S. 221
— Fittoni, *Deshayes* 132
— flecta, *Lycett* 55, 60
— Forbesii, Lycett, F. S................. 230
— formosa, *Lycett* 35, 38, 44
— — var. lata, *Lycett* 202
— gemmata, *Lycett*........... 15, 16, 17
— — var. bifera, *Lycett* 239
— geographica, *Agassiz* 69
— gibbosa, *Sowerby* 84
— — vars. *a, b*, *Lycett*............ 85
— — var. *c*, *Lycett* 85
— *gibbosa*, Seebach, F. S. 86
— *Goldfussii*, Morris and Lycett, F. S.
59, 222, 230
— *granigera*, Contejean, F. S.
162, 221, 223
— Griesbachi, *Lycett* 33, 34
— harpa, *Deshayes* ; see T. carinata ... 179

INDEX. 243

Trigonia Heberti, Munier-Chalmas; see T.
incurva 42
— hemisphærica, *Lycett* 174
— — var. gregaria, *Lycett* 11, 176
— *Herzogii*, Hausmann, F. S. ... 120, 230
— *Hondeana*, Lea, F. S. 229
— — Coquand, F. S. 92, 227
— *Howittii*, McCoy, F. S. 231, 232
— *Hudlestoni*, *Lycett* 194
— *Humboldtii*, De Buch, F. S.
126, 229, 231
— *Hunstantonensis*, Seeley............... 183
— *hybrida*, Roemer, F. S. 221, 223
— imbricata, *Sowerby* 33, 209
— impressa, *Sowerby* 46, 70
— incurva, *Benett* 42
— *inflata*, Roemer, F. S................. 223
— ingens, *Lycett*.............. 24, 207, 228
— inornata, d'Orbigny, F.S. 226
— *interlævigata*, Quenstedt, F. S.
151, 221, 223
— irregularis, *Seebach*................ 32, 39
— Joassi, *Lycett* 82
— *Juddiana*, *Lycett* 25
— *Jukesii*, Adams, F. S. 233
— Keepingi, *Lycett* 196, 228
— *Kurri*, Oppel, F. S. 223
— *Lajoeci*, d'Orbigny, F. S. 228
— *Lamarkii*, Matheron, F. S. 142, 227
— — Jenkins, F. S............... 232
— læviuscula, *Lycett* 96
— Leckenbyi, *Lycett* 71
— *limbata*, d'Orbigny, F. S. 122, 227
— *lineata*, Moore, F. S. 224
— lineolata, *Agassiz*; see T. costata ... 147
— Lingonensis, *Dumortier* 98, 219
— *longa*, Agassiz, F. S. 97, 227
— literata, *Young & Bird* 64
— *literatum* (Lyrodon), Goldfuss, F. S.
59, 61, 65, 76
— litterata, *Tate*; see T. literata 64
— *Lusitanica*, Sharpe, F. S.......... 115, 227
— lyrata, *d'Orbigny*; see literata 64
— major, d'Orbigny, F. S. 19
— Manseli, *Lycett* 86
— *Matronensis*, De Loriol, Royer, and Tombeck, F. S. 223

Trigonia maxima, *Agassiz*, F. S. 222
— Meriani, *Agassiz* 167
— Meyeri, *Lycett* 125, 227
— *Michelloti*, De Loriol, F. S.
87, 92, 222, 223
— — var., *Lycett* 92, 93
— *minuta*, Stoliczka, F. S. 231
— modesta, *Tate* 212
— monilifera, *Agassiz* 46, 165
— *monilifera*, Quenstedt, F. S. ... 167, 220
— *Montierensis*, *Lycett*, F. S. 35, 36, 223
— *Moorei*, *Lycett*, F. S.............. 151, 224
— Moretoni, *Morris & Lycett*
47, 59, 63, 70, 78
— *Munieri*, Hébert; see T. Michellotti 92
— muricata, *Goldfuss* 20, 50
— *navis*, Lamark, F. S............... 16, 222
— *Nereis*, d'Orbigny, F. S......... 136, 205
— nodosa, *Sowerby* 102, 106, 111
— — var. Orbignyana, *Lycett*
106, 107, 111
— — *Pictet & Renevier*, F. S.
111, 227
— *nodosa*, Pictet & Roux, F. S. ... 111, 227
— *notata*, Agassiz, F. S. 223
— *orientalis*, Forbes, F. S.... 121, 224, 230
— ornata, *d'Orbigny* 139
— Painei, *Lycett* 56, 59, 60, 65, 68
— *palmata*, Deshayes, F. S.
100, 103, 111, 227
— papillata, *Agassiz*; see T. monilifera 167
— *paradoxa*, Agassiz, F. S........... 114, 227
— parcinoda, *Lycett* 46
— *Parkinsoni*, Agnasiz, F. S. 105, 223
— *parvula*, Agassiz, F. S. 223
— paucicosta, *Lycett* ... 29, 56, 57, 71, 206
— pectinata, *Lamark*, F. S.............. 233
— Pellati, *Munier-Chalmas*........... 41, 72
— *peninsularis*, Coquand, F. S.
169, 182, 220
— pennata, *Sowerby* 133
— perlata, *Agassiz* 22, 45
— Phillipsii, *Morris & Lycett*
37, 38, 69, 80
— picta, Agassiz, F. S................ 87, 223
— *Picteti*, Coquand, F. S. 227
— *plicata*, Agassiz, F. S. 123

INDEX.

Trigonia plicato-costata, Nyst & Galeotti, F. S. 131, 229
— producta, *Lycett* 60, 67, 208
— producta, Terquem & Jourdy, F. S. 62
— Proserpina, *d'Orbigny*; see T. duplicata 14
— *pseudo-cyprina*, Contejean, F. S. ... 223
— pulchella, *Agassiz* 45, 66, 80, 185
— *pulchella*, Reuss, F. S. ... 186, 227, 231
— pullus, *Sowerby* 164
— *pumila*, Nilsson, F. S. 138
— Pyrrha, *d'Orbigny*; see T. spinosa 137, 204
— quadrata, Sowerby; see T. dædalea 100, 104
— quadrata, *Agassiz*, F. S. 104
— radiata, *Benett* 73, 208
— Ramsayi, *Wright* 49, 71
— recticosta, *Lycett* 16
— reticulata, Agassiz, F. S. 167, 223
— *Rigauxiana*, Munier-Chalmas, F. S. 26, 223
— *Robinaldina*, d'Orbigny, F. S. 185, 227
— *rostrum*, Agassiz, F. S. 223
— rudis, *Parkinson*; see T. dædalea 100, 109
— — *d'Orbigny*; see T. nodosa ... 106
— Rupellensis, *d'Orbigny* 28, 53, 58, 199
— *sanctæ-crucis*, Valang, F. S. 96, 227
— scabra, Lamark, F. S. 130, 227
— scabricola, *Lycett* 130
— scapha, *Agassiz* 183
— Scarburgensis, *Lycett* 31
— sculpta, *Lycett* 157
— — var. Cheltensis, *Lycett* 157
— — var. Rolandi, *Cross* 157
— *semiculta*, Forbes, F. S. 96, 121, 230
— *sexcostata*, Roemer, F. S. 223
— Sharpiana, *Lycett* 79
— signata, *Agassiz*; Zieten's var. 19, 29, 50, 61, 204
— — Knorr's var., F. S. 204
— *simile* (Lyrodon), Bronn ... 66, 159, 222
— sinuata, *Parkinson* 94, 136
— *sinuata*, Agassiz; see T. sulcataria... 136
— *Smeeii*, Sowerby, F. S. 224
— Snaintonensis, *Lycett* 198

Trigonia spectabilis, *Sowerby* 102, 111, 112
— *spinifera*, d'Orbigny, F. S. 223
— spinosa, *Parkinson* 136
— — var. subovata, *Lycett* 201
— *spinosa*, Agassiz; see T. Archiaciana 137
— — d'Orbigny; see T. Vicaryana 136
— spinulosa, *Young & Bird* 37, 44, 66
— striata, *Miller* 35, 36
— *striata*, Phillips; see T. spinulosa... 44
— — d'Orbigny; see T. spinulosa 44
— — Quenstedt; see T. Brodiei 35, 195
— *subconcentrica*, Etalon, F. S. 52
— *subcrenulata*, d'Orbigny, F. S. ... 190, 229
— subglobosa, *Morris & Lycett* 65, 68
— *suborbicularis*, Forbes, F. S. 121, 224, 230
— *subundulata*, McCoy, F. S. 232
— *sulcata*, Agassiz; see T. carinata 179, 181
— *sulcatum* (Lyrodon), Goldfuss; see T. sulcataria 16, 135
— sulcataria, *Lam.* 135
— *suprajurensis*, Agassiz, F. S. 167, 223
— *Syriaca*; see T. undulata, Fromherz 200
— Tealbyensis, *Lycett* 114
— tenuicosta, *Lycett* 160, 163
— *tenuisulcata*, Dujardin, F. S. 92, 142, 227
— tenuitexta, *Lycett* 90
— *Thoracica*, Morton, F. S. 118, 229
— Thurmanni, *Contejean*; (see pl. X, figs. 1, 2) 237
— *Tombecki*, De Loriol, Royer, and Tombeck, F. S. 223
— *trigona*, Waagen, F. S. 223
— tripartita, *Forbes* 67, 74
— triquetra, *Seebach* 20, 26, 53
— truncata, Agassiz, F. S. 223
— tuberculata, Agassiz; see T. spinulosa 37, 44, 66
— *tuberculifera*, Stoliczka, F. S.... 120, 230
— tuberculosa, *Lycett* 33, 34
— *undulata*, Fromherz, F. S. 55, 58, 74, 76, 77
— — var. arata, *Lycett* 200
— *uniophora*, Gray, F. S. 234
— Upwarensis, *Lycett* 143

INDEX.

	PAGE
Trigonia variegata, Credner, F. S.	90, 223
— v. costata, *Lycett*	45, 61, 65, 66, 71, 74, 78
— Vectiana, *Lycett*	118, 122, 123
— ventricosa, Krauss, F. S.	119, 230
— Vicaryana, *Lycett*	141, 203
— — var. angustata, *Lycett*	142

	PAGE
Trigonia Voltzii, *Agassiz*	20, 24, 32, 42, 49
— Williamsoni, *Lycett*	53
— Witchelli, *Lycett*	197
— Woodwardi, *Lycett*	40
— Zonata, Agassiz, F. S.	151, 221

PLATE I.

Fig.
1, a, b. *Trigonia clavellata*, Sow. Specimen of adult growth from the Lower Calcareous Grit of Weymouth. (Page 18.)

2. ,, ,, A smaller example from the same locality.

3. ,, *Bathonica*, Lyc. Great Oolite Box, near Bath. (Page 17.)

4, 5, 6. ,, *recticosta*, Lyc. Inferior Oolite, Cloughton, near Scarborough. (Page 16.)

7. ,, *gemmata*, Lyc. Inferior Oolite, Cheltenham. (Page 15.)

8 and 10. ,, *duplicata*, Sow. Inferior Oolite, near Stroud. (Page 14.)

9. ,, ,, Inferior Oolite, near Yeovil. (Page 14.)

The small dimensions of British examples of the *Scaphoideæ* (figs. 3 to 9 inclusive) have induced me to subjoin a wood engraving of *Trigonia navis*, Lam., which is the typical species of that section, and which exemplifies its peculiar features much more prominently than is seen in the British species. Our

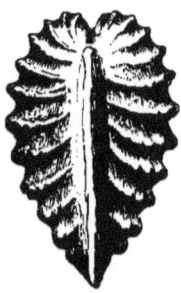

specimen, which is of adult growth, but not of the largest dimensions, is from the thick deposit of dark clays at Gundershofen (Haut Rhin), which Professor Quenstedt has shown to belong to the lowest zone of the Inferior Oolite in Southern Germany. Other localities for this species are Metz (Mozelle), Günsberg (Solothurn). For numerous figures, see Agassiz, 'Trigonies,' tab. i; also, Quenstedt, 'Der Jura,' tab. xliv, fig. 13.

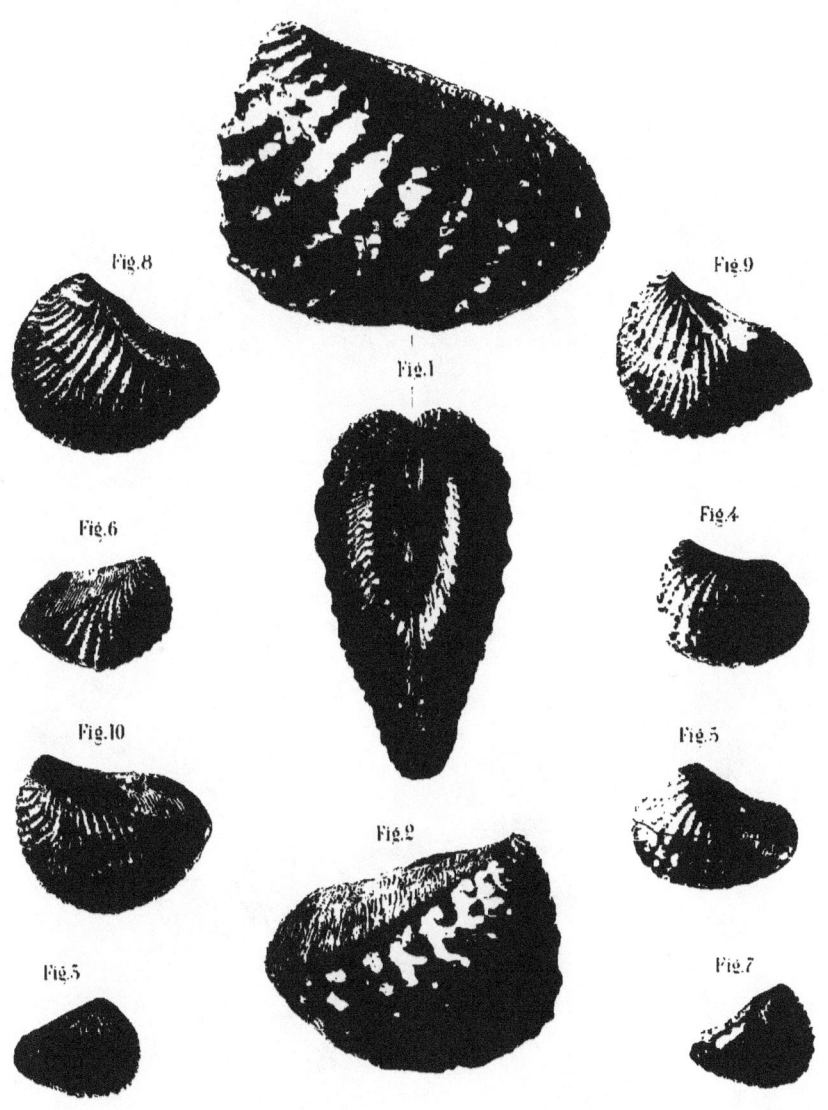

Pl. I

PLATE II.

FIG.
1. *Trigonia signata*, Ag.   Inferior Oolite, Rodborough Hill, near Stroud.   (Page 29.)
2.   ,,   ,,   A small specimen with large costæ. Inferior Oolite, Cold Comfort, near Cheltenham.
3.   ,,   ,,   Inferior Oolite (grey limestone), Cloughton, near Scarborough.
4, 5.   ,,   *Moretoni*, Mor. and Lyc.   Cornbrash, Appleby, Lincolnshire.   (Page 47.)
7.   ,,   ,,   Young example from the same locality.
8.   ,,   ,,   Young example, Great Oolite, Bisley Common, near Stroud.
6, *a, b, c*.   *Juddiana*, Lyc.   Adult specimen from the Kimmeridge Clay of Market Rasen, Lincolnshire.   (Page 25.)

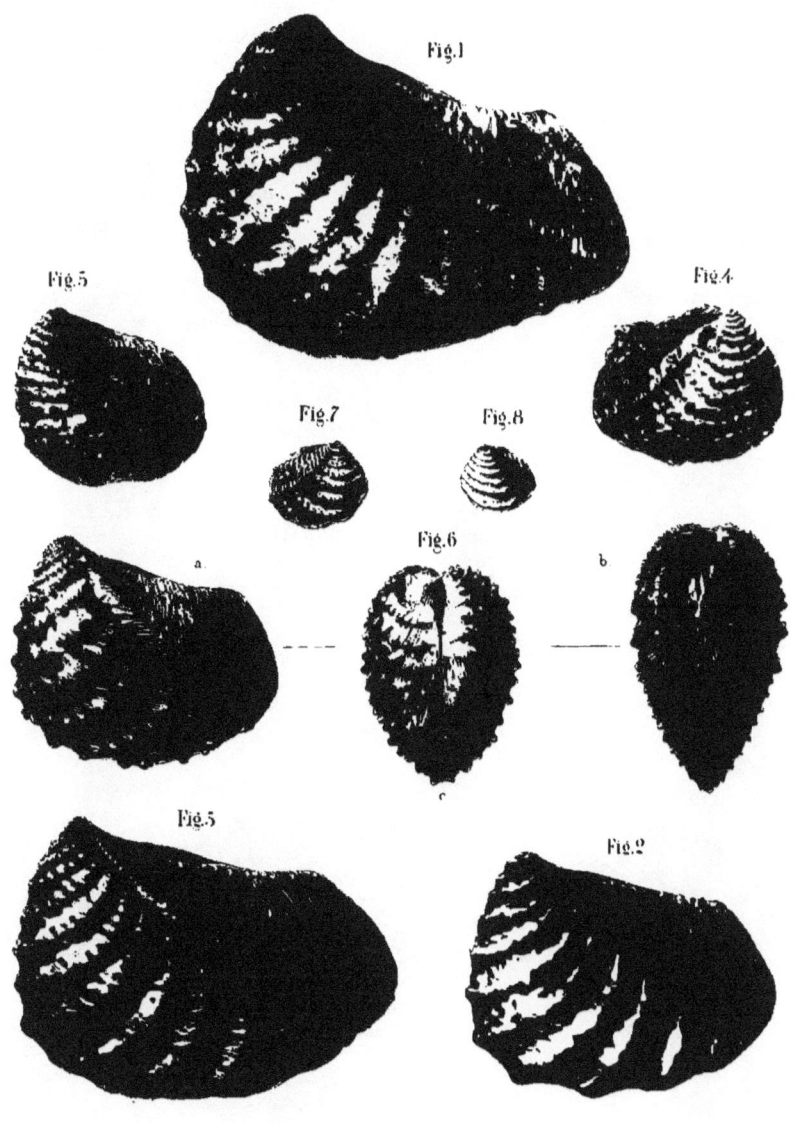

PLATE III.

Fig.
1. *Trigonia perlata*, Ag. Coral Rag, Pickering; the rows of costæ are fewer than is usual. (Page 22.)

2. ,, ,, The same formation and locality; the largest specimen known.

3. ,, ,, Young specimen from the same locality.

4, 5 a, 5 b, 6. *spinulosa*, Y. and B. Inferior Oolite (Dogger), Blue Wyke, Yorkshire. (Page 44.)

7. ,, *corallina*, D'Orb. Coralline Oolite, Pickering. (Page 45.)

8, 9, 11. ,, Young specimens, Coral Rag, Steeple Ashton, Wilts.

10, a. ,, *Griesbachi*, Lyc. Cornbrash, Rushden, Northamptonshire; natural size. (Page 34.)

10, b. ,, ,, The same specimen magnified.

Pl. III

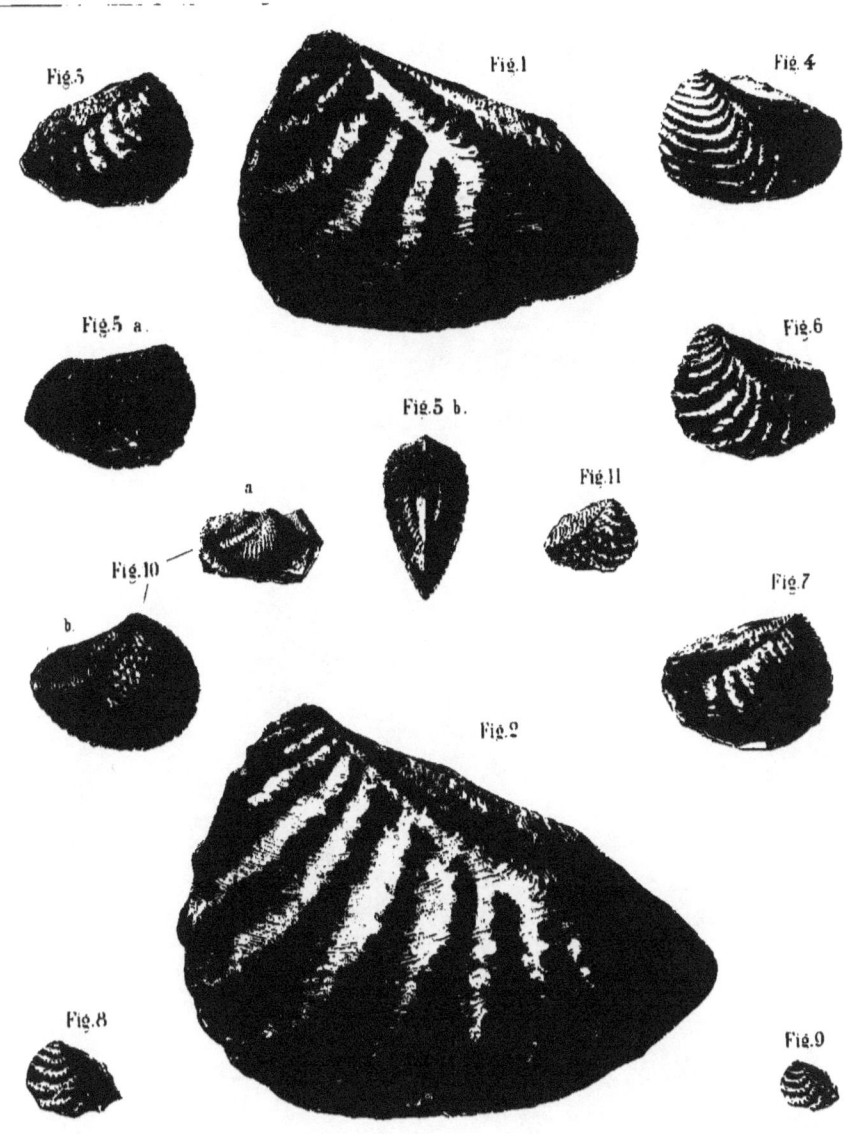

PLATE IV.

Fig.
1. *Trigonia Scarburgensis*, Lyc.   Left valve, Cornbrash, Scarborough.   (Page 31.)

2, 3.   ,,   ,,   Opposite valves of the same specimen.

4.   ,,   ,,   Young example.

5.   ,,   *Juddiana*, Lyc.   Kimmeridge Clay, Market Rasen; example with few costæ and large tubercles.   (Page 25.)

7.   ,,   ,,   Example with more numerous costæ and smaller tubercles.

6.   ,,   *Moretoni*, Mor. and Lyc.   Cornbrash, Appleby, Lincolnshire; adult specimen.   (Page 47.)

8.   ,,   *Bronnii*, Ag.   A variety from the Lower Calcareous Grit of Weymouth.   (Page 23.)

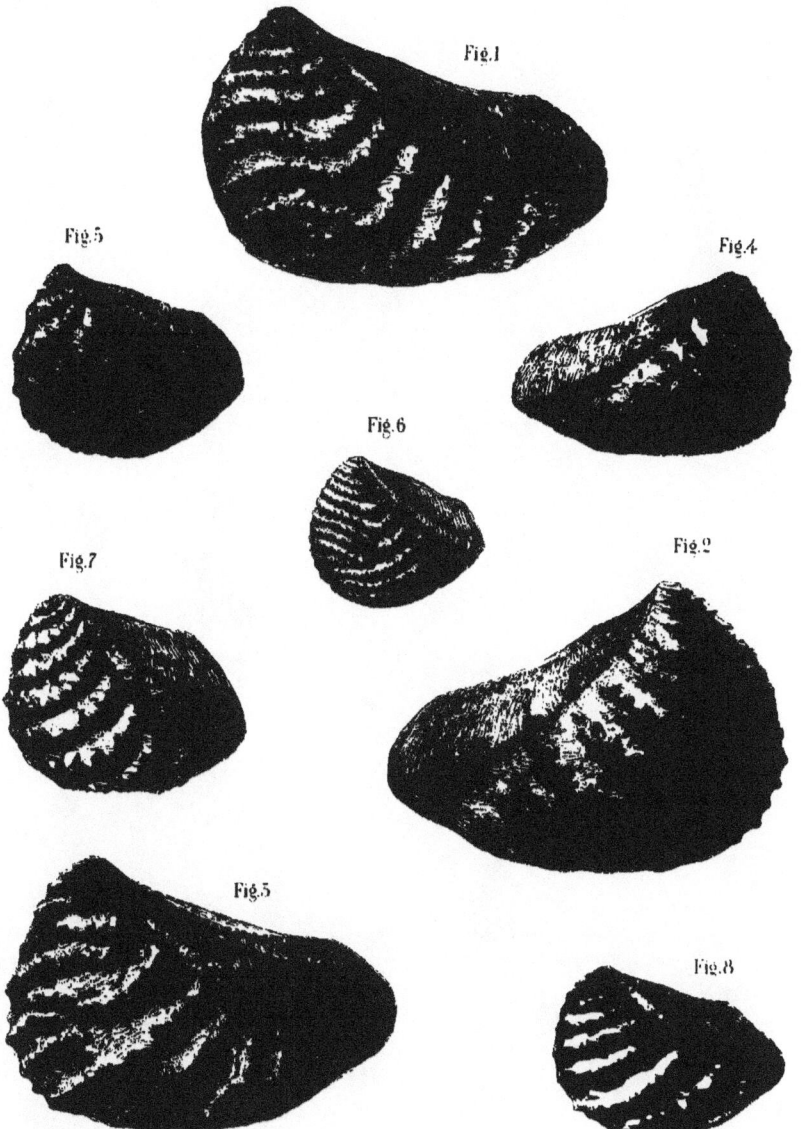

PLATE V.

Fig.

1, *a, b*. *Trigonia irregularis*, Seebach. Specimen of adult growth, Oxford Clay, Weymouth. (Page 39.)

2. ,, ,, Imperfect specimen with the rows of costæ more than usually numerous and irregular.

4. ,, *formosa*, Lyc. Young example, Inferior Oolite, Cold Comfort, near Cheltenham. (Page 35.)

5. ,, ,, Young example, Rodborough Hill, near Stroud.

6. ,, ,, Specimen of adult growth, Rodborough Hill, near Stroud.

3 and 7. ,, *striata*, Miller. Adult examples, Inferior Oolite, Bradford Abbas, Dorset. (Page 36.)

8. ,, ,, Young example from the same locality.

9, 10. ,, *tuberculosa*, Lyc. Inferior Oolite, Leckhampton Hill, near Cheltenham. (Page 33.) The perpendicular continuations downwards of the little tubercles are not sufficiently distinct.

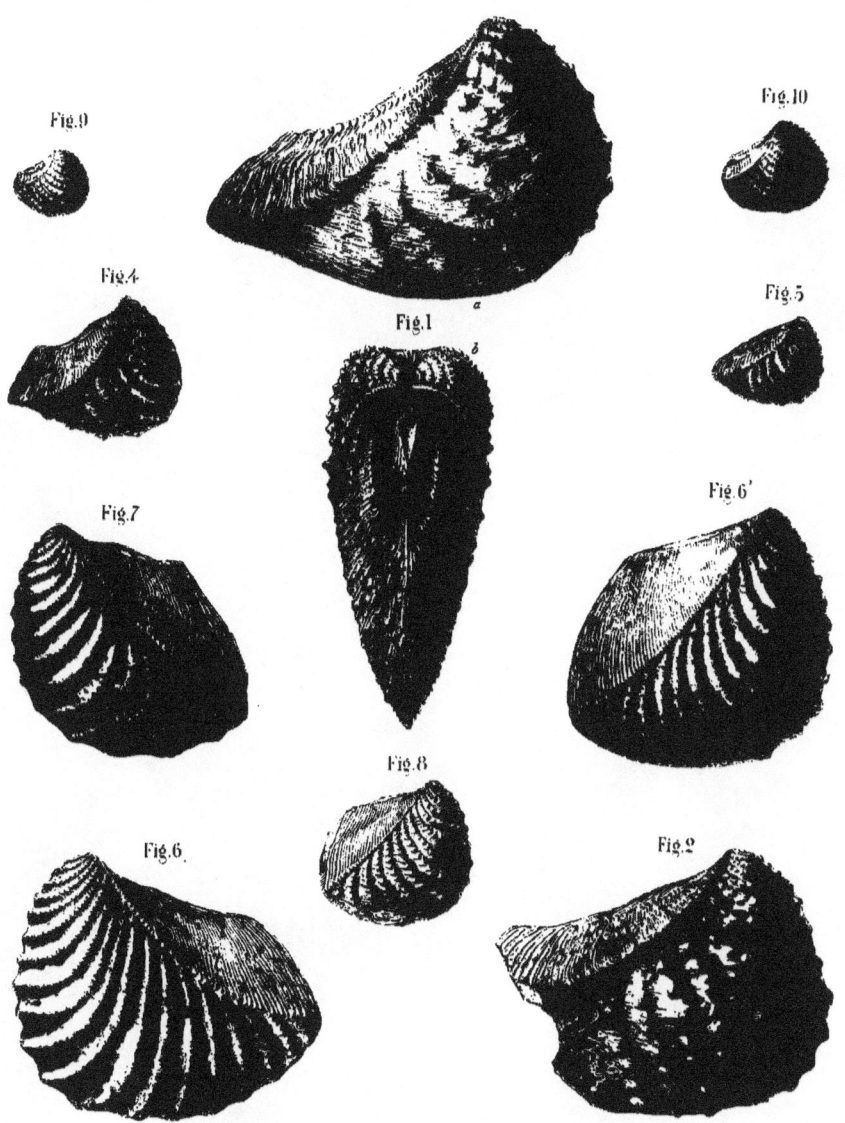

PLATE VI.

Fig.
1, a, b. *Trigonia triquetra*, Seebach. Calcareous Grit, Cumnor, Oxfordshire. (Page 26.)
2.   ,,   ,, Lower Calcareous Grit, Filey Point, Yorkshire. Specimen with fewer and larger tuberculated varices.
3, 4.   ,, *Phillipsii*, Mor. and Lyc. Inferior Oolite, Appleby, Lincolnshire. (Page 38.)
5 a.   ,, *imbricata*, Sow. Great Oolite, Bath. (Page 33.)
5 b.   ,,   ,, The same specimen magnified.
6.   ,, *Ramsayi*, Wright. Supra-Liassic Sands, Frocester Hill, Gloucestershire. (Page 49.)

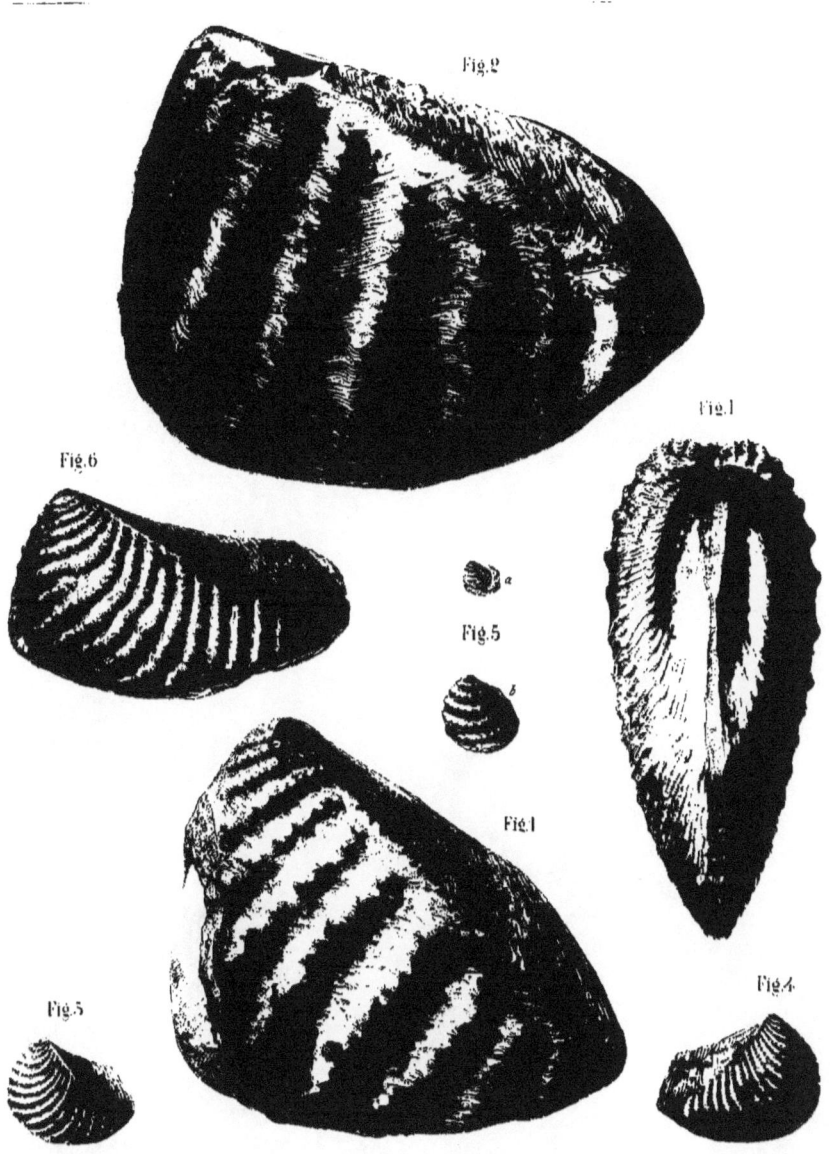

PLATE VII.

FIG.
1. *Trigonia Pellati*, Mun. Ch. Adult example, Osmington Bay, near Weymouth. Kimmeridge Clay. (Page 41.)
2. ,, ,, Another example from the same locality and formation.
3. ,, *complanata*, Lyc. Killoway Rock, Cayton Bay, near Scarborough. (Page 49.)
4, 5. ,, *impressa*, Sow. Slate, Stonesfield, Oxfordshire. (Page 46.)
6. ,, *irregularis*, Seeb. Young specimen, Oxford Clay, Weymouth. (Page 39.)

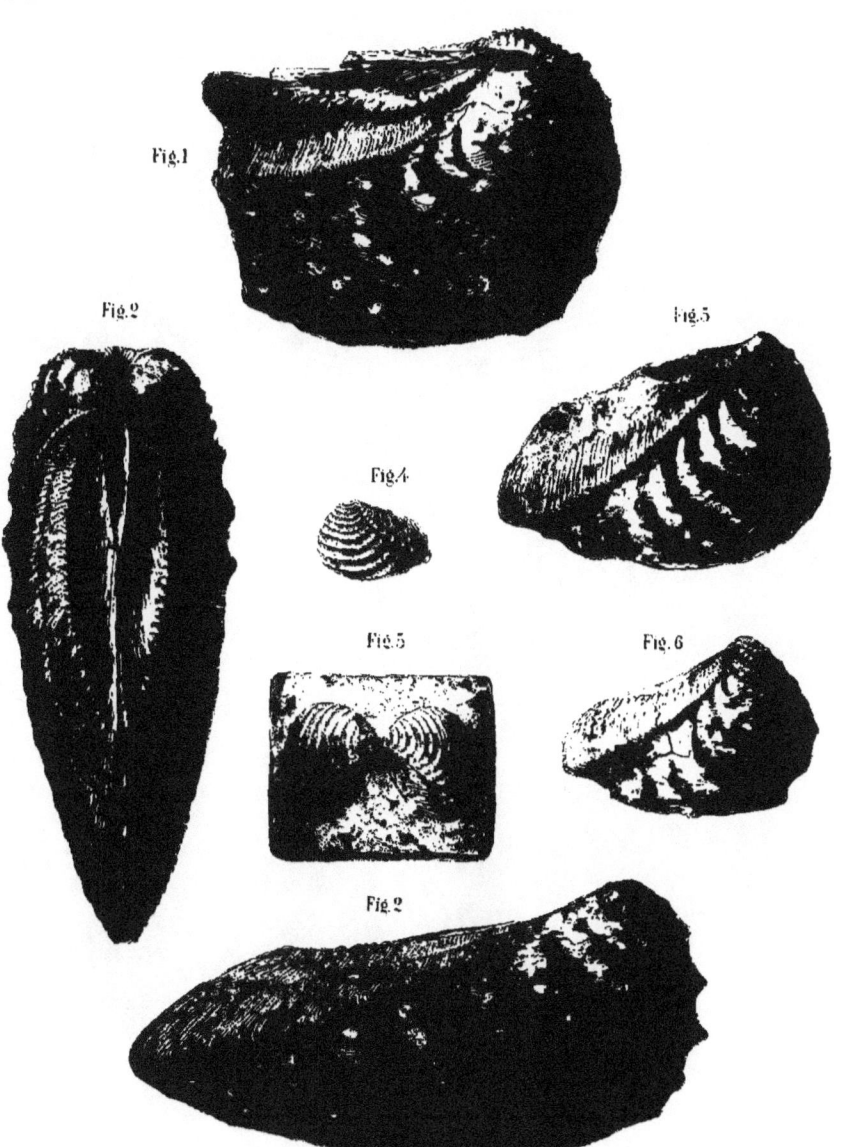

Pl.VII

PLATE VIII.

FIG.
1. *Trigonia ingens*, Lyc. Neocomian formation, Downham, Norfolk. Adult example; impression from an external cast, in coarse sandstone. (Page 24.)

2, a, b. ,, ,, Internal mould, size somewhat reduced, Downham, Norfolk.

3. ,, ,, Young example, with the test preserved, Downham, Norfolk.

4. ,, *Rupellensis*, D'Orb. Kelloway Rock, Cayton Bay, near Scarborough. (Page 28.)

5. ,, *corallina*, D'Orb. Adult example, Coral Rag, Pickering. (Page 45.)

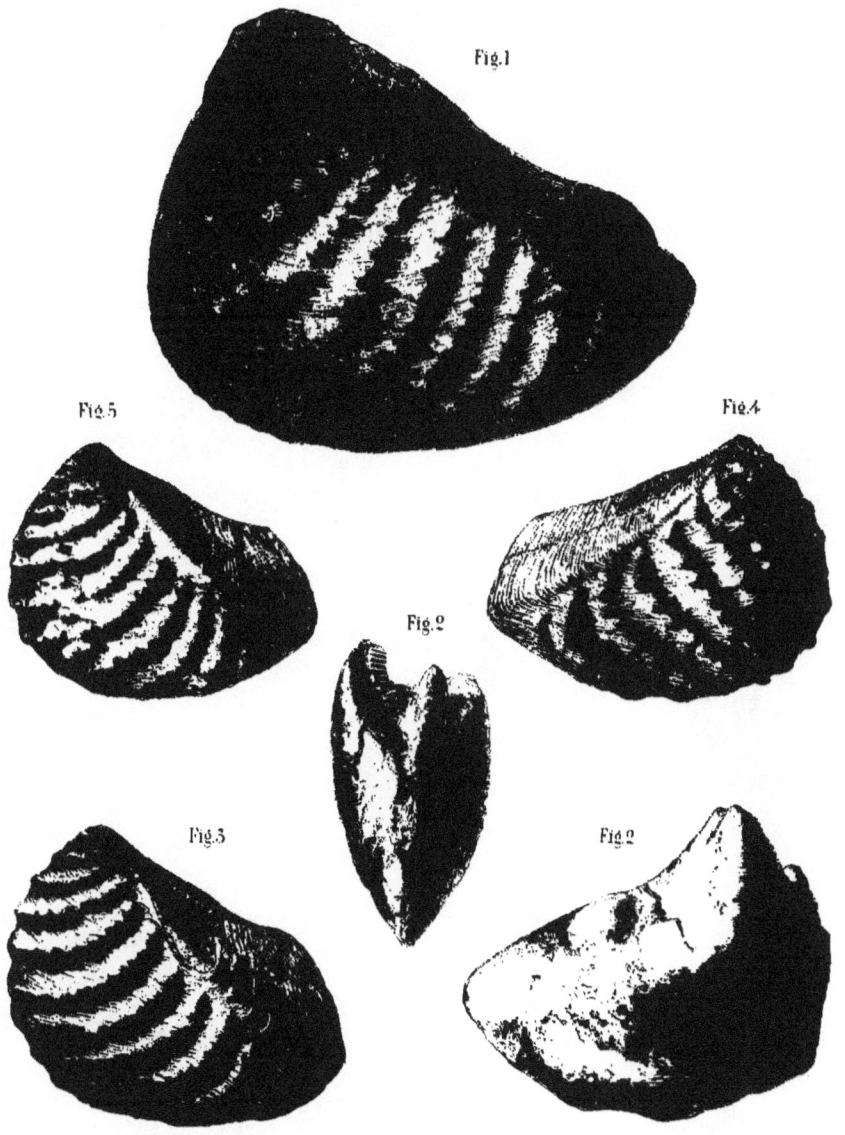

Pl. VIII

PLATE IX.

FIG.
1. *Trigonia muricata*, Goldf. Portland Limestone, Wilts. (Page 50.) The test has disappeared.
2. ,, *incurva*, Benett (a variety). Kimmeridge Clay, Dorsetshire. (Page 42.) It is somewhat flattened from vertical pressure.
3. ,, ,, Internal mould, Portland Oolite, Swindon. (Page 42.)
4. ,, ,, Small specimen, Kimmeridge Clay, Wotton Basset, Wilts.
5. ,, ,, Portland Oolite, Isle of Portland.
6. ,, ,, Portland Oolite, Crookwood, near Devizes, Wilts.

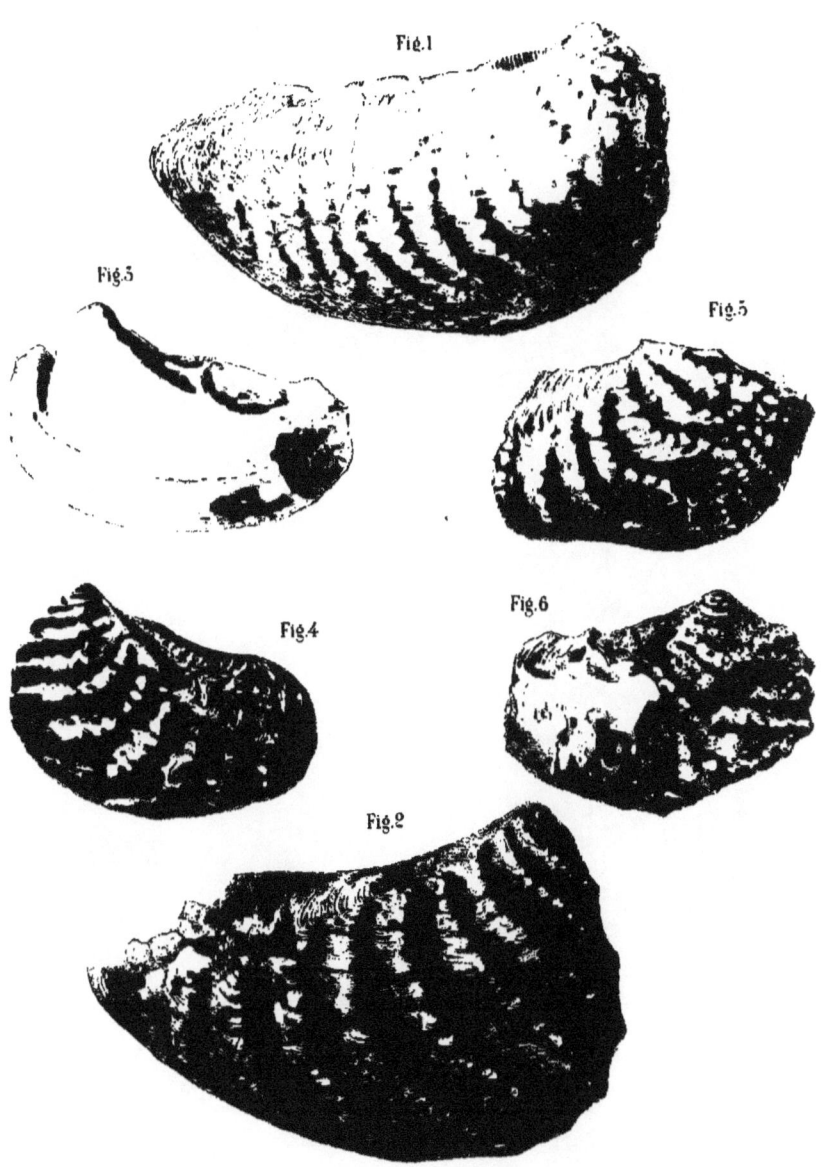

PLATE X.

Fig.
1, 2. *Trigonia Voltzii*, Ag. (*T. Thurmanni*, Cont.). Kimmeridge Clay, near Weymouth. (Page 20.)

3, 3, a. ,, *detrita*, Terq. and Jour. Forest Marble, near Cirencester, Gloucestershire. (Page 75.)

4. ,, ,, Cornbrash, Hilperton. (Page 75.)

5, 7, 8. ,, *conjungens*, Phil. Inferior Oolite (Millepore bed), Cloughton, near Scarborough. (Page 62.)

6. ,, *geographica*, Ag. Coral Rag, Pickering. (Page 69.)

Pl. 10.

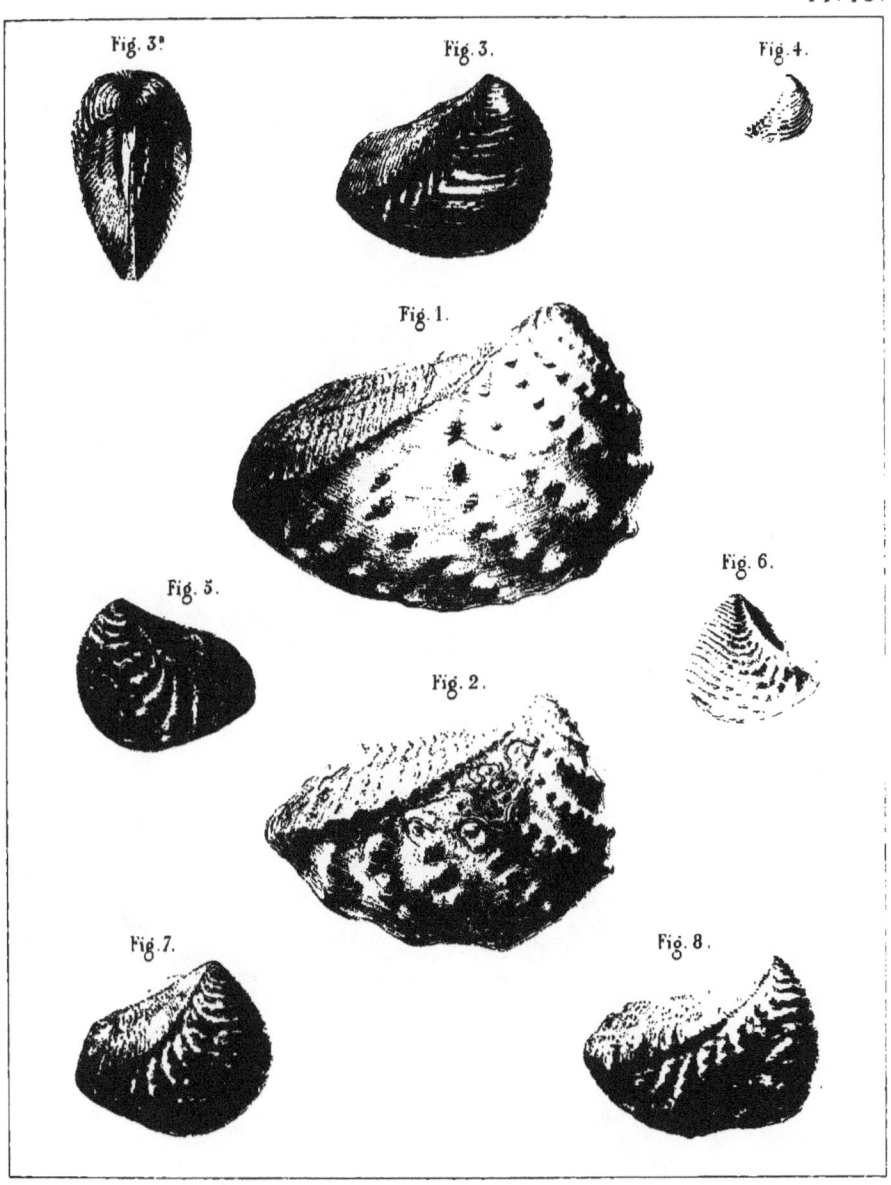

P. Lackerbauer ad nat.in lap. del. Imp. Becquet à Paris

PLATE XIII.

FIG.
1, 2. *Trigonia producta*, Lyc.   Inferior Oolite, Rodborough Hill, near Stroud. (Page 60.)

3, 4.   ,,   ,,   Valves of the same species, exhibiting the hinge. Inferior Oolite, Oxfordshire.   Coll. Royal School of Mines.

5.   ,,   *v.-costata*, Lyc.   Inferior Oolite, Cold Comfort, near Cheltenham; specimen with the costæ deformed.   (Page 66.)

6.   ,,   *conjungens*, Phil.   Inferior Oolite, Millepore bed, near Scarborough. Also Plate X, figs. 5, 7, 8.   (Page 62.)

Pl. 13.

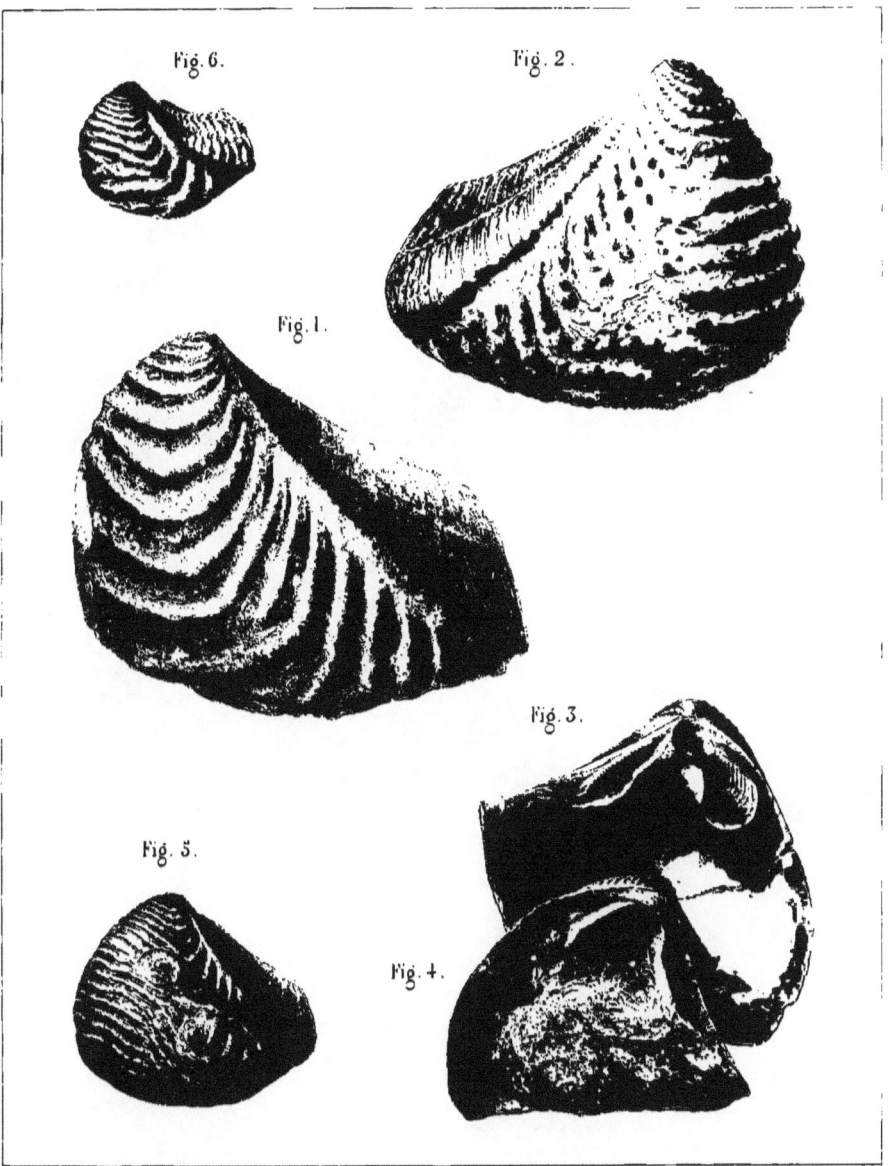

P. Lackerbauer ad nat.in lap del
Imp. Becquet à Paris.

PLATE XIV.

FIG.
1, 1, a, 2, 3, 4. *Trigonia literata*, Young and Bird. Upper Lias, Robin Hood's Bay, Yorkshire. (Page 64.)

5, 6. ,, *angulata*, Sow. Inferior Oolite, Rodborough Hill, Stroud. (Page 54.)

7, 8, 9, 10. ,, *flecta*, Lyc. Cornbrash or Forest Marble, Thornholme, Appleby, Lincolnshire. (Page 55.)

Pl. 14.

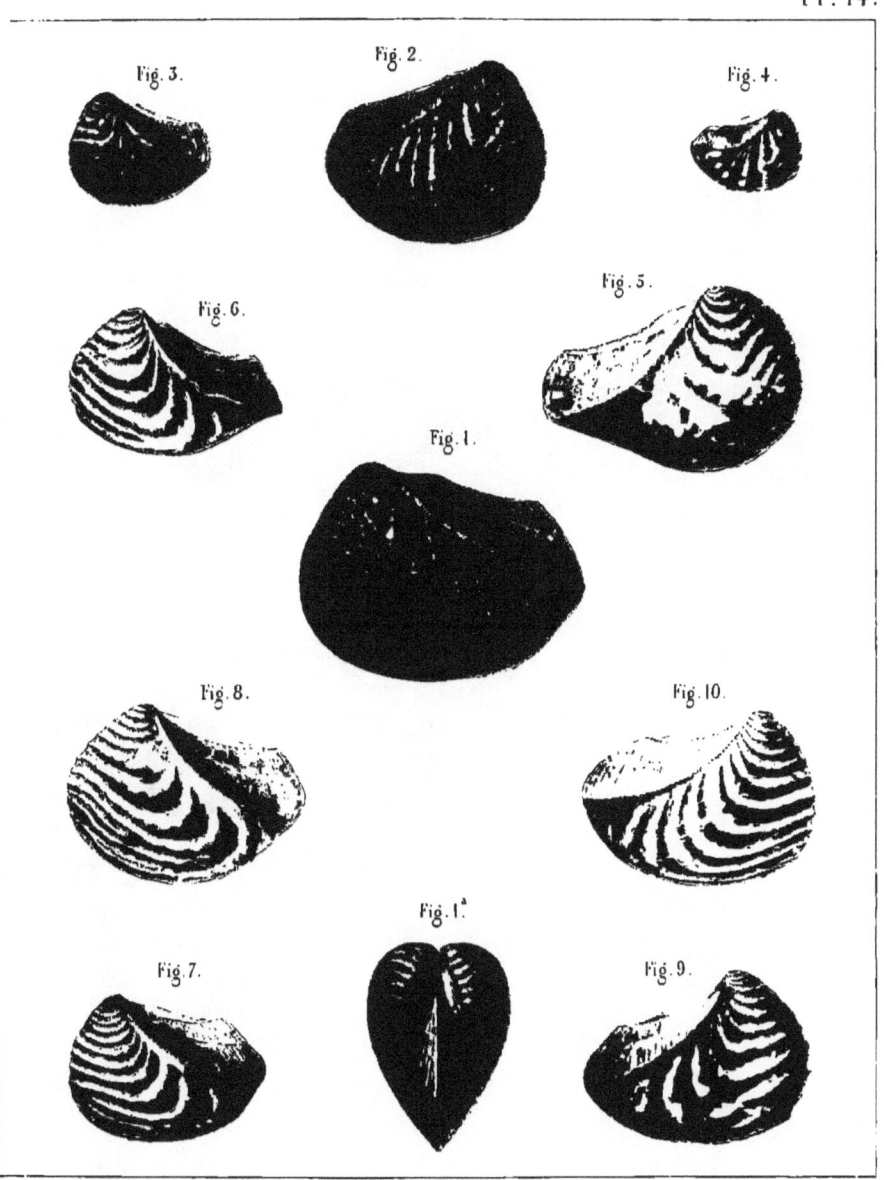

P. Lackerbauer ad nat. in lap. del. Imp. Becquet à Paris.

PLATE XV.

Fig.
1. *Trigonia v.-costata*, Lyc. Inferior Oolite, Rodborough Hill, Stroud. (Page 66.) Coll. Royal School of Mines.

2, 3, 4. ,, ,, Smaller examples, Inferior Oolite, Dogger, Blue Wyke, Yorkshire.

5, 6, 7. ,, *compta*, Lyc. Inferior Oolite, Collyweston slate. (Page 70.)

8, 10. ,, *costatula*, Lyc. Inferior Oolite, near Stroud. (Page 81.) Coll. Royal School of Mines.

9. ,, ,, A smaller specimen, from the same formation and locality.

11 ,, *Sharpiana*, Lyc. Inferior Oolite (ferruginous), near Northampton. (Page 79.)

Pl. 15.

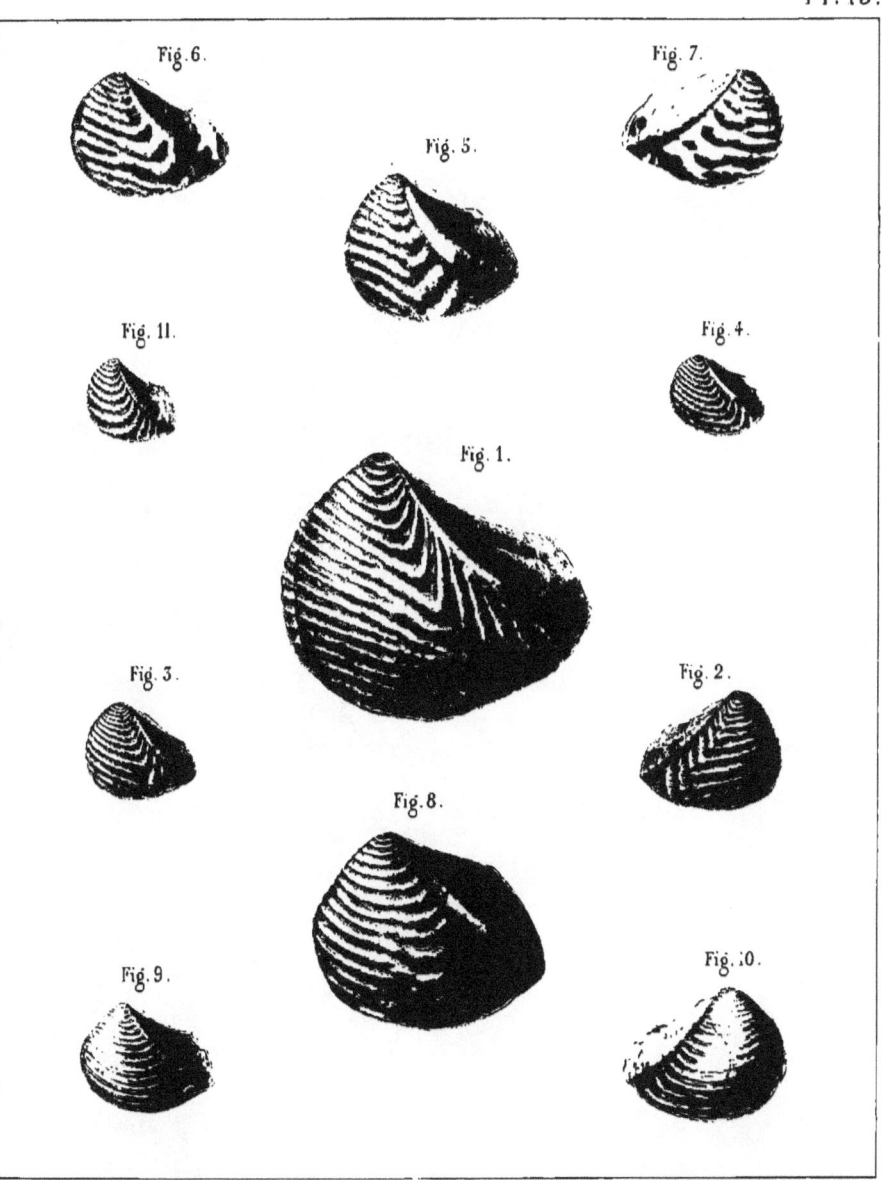

P. Lackerbauer ad nat. in lap. del. Imp. Becquet à Paris.

PLATE XVI.

FIG.
1. *Trigonia Leckenbyi*, Lyc. Supra-Liassic Sandstone, Robin Hood's Bay, Yorkshire. (Page 71.) Coll. Woodwardian Museum, Cambridge.

2. ,, ,, A smaller specimen, from the same formation and locality.

3. ,, *Sharpiana*. Dogger, Blue Wyke. (Page 79.)

4, 5, 6. ,, ,, Same species (ferruginous). Inferior Oolite, Northamptonshire.

7. ,, *paucicosta*, Lyc. Kelloway Rock, Cayton Bay, near Scarborough. (Page 57.)

8. ,, *Williamsoni*, Lyc. Kelloway Rock, Cayton Bay, near Scarborough. (Page 53.)

9, 10, 11. *undulata*, From. Great Oolite, near Bourn, Lincolnshire. (Page 77.)

Pl. 16.

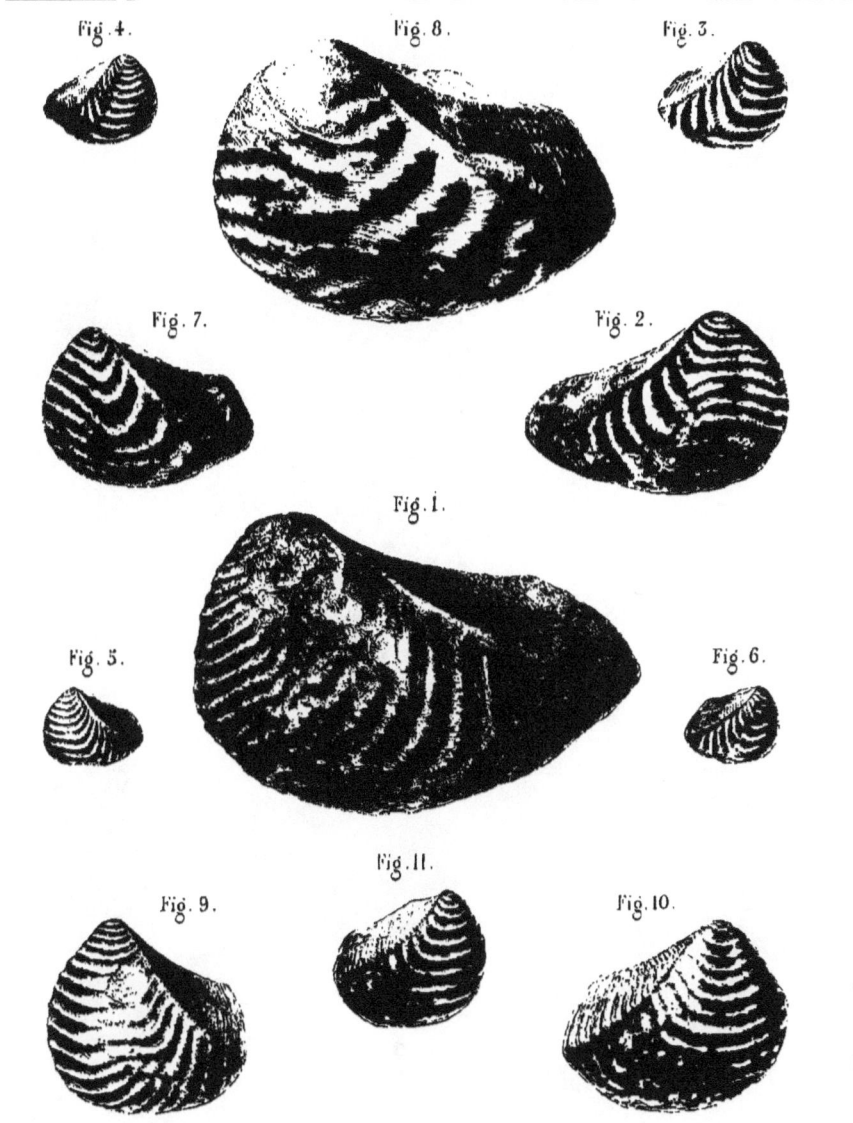

PLATE XVII.

Fig.
1. *Trigonia Woodwardi*, Lyc. Kimmeridge Clay, Dorsetshire. (Page 40.)
2, 3, 4. ,, *Beesleyana*, Lyc. Inferior Oolite, Combe Hill, Oxfordshire. (Page 90.)
5, 6. ,, *undulata*, From. Great Oolite, South Lincolnshire. (Page 77.)
7. ,, *clytia*, D'Orb. Great Oolite, Box, near Bath. (Page 76.)

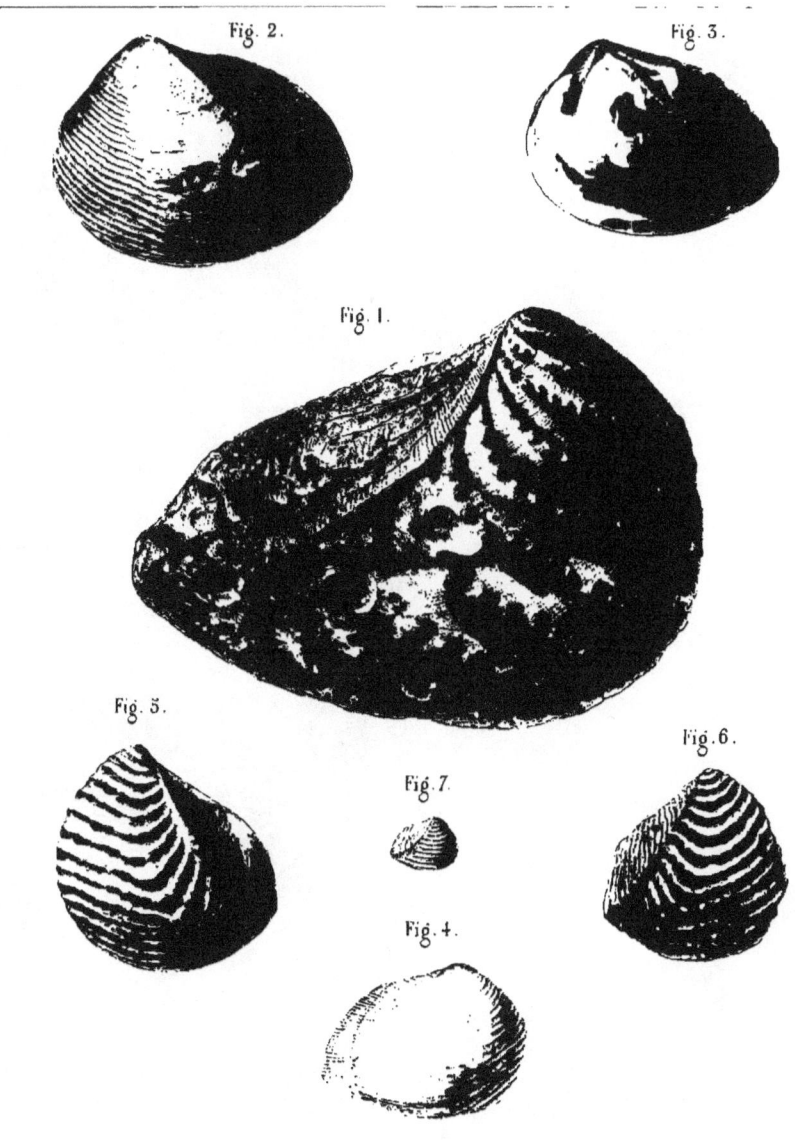

PLATE XVIII.

FIG.
1, 2, 2, a.   *Trigonia gibbosa*, Sow. Variety with prominent nodose costæ and wide ante-carinal space. Portland Oolite, Tisbury. Also Plate XXI, fig. 1. (Page 84.)

3.   „   *Damoniana*, De Lor. Variety with few large nodes upon the costæ; see also Plate XXI, figs. 2, 3, 4, 5. Portland Limestone, Portland. (Page 88.)

4.   „   *gibbosa*. Variety with the costæ small, partially plain or subtuberculated; see also Plate XIX, fig. 2. Portland Oolite, Tisbury. (Page 84.)

5, 6.   „   *gibbosa*. The typical form, longitudinal sulcations strongly defined. Portland Oolite, Tisbury. (Page 84.)

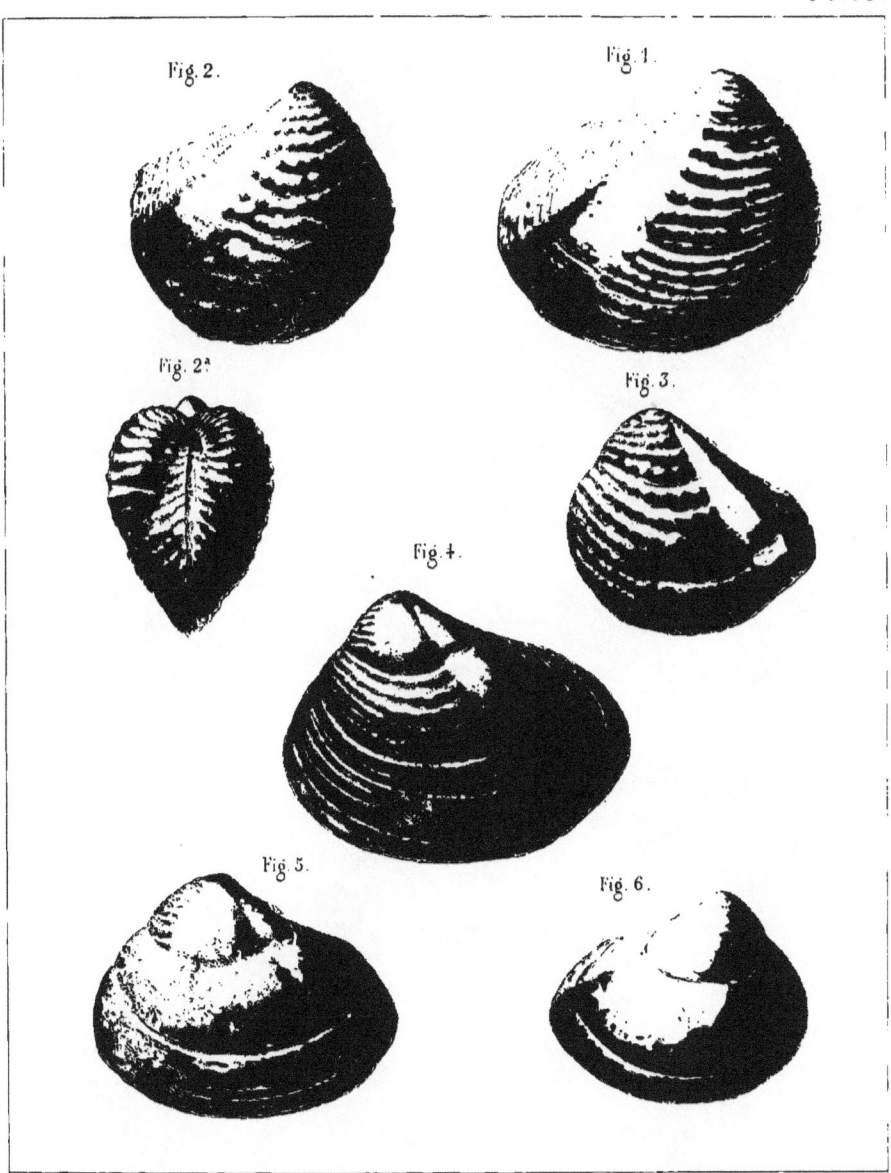

PLATE XIX.

Fig.
1, 1, a, b. *Trigonia Damoniana*, De Lor. Internal moulds. Portland Oolite, Swindon, Wilts. (Page 88.)

2. ,, *gibbosa*, Sow. Variety; see also fig. 4, Plate XVIII. Portland Oolite, Tisbury. (Page 84.)

3, 4, 4, a, b. ,, *Manseli*, Lyc. Portland Oolite, Isle of Portland. (Page 86.)

Pl. 19.

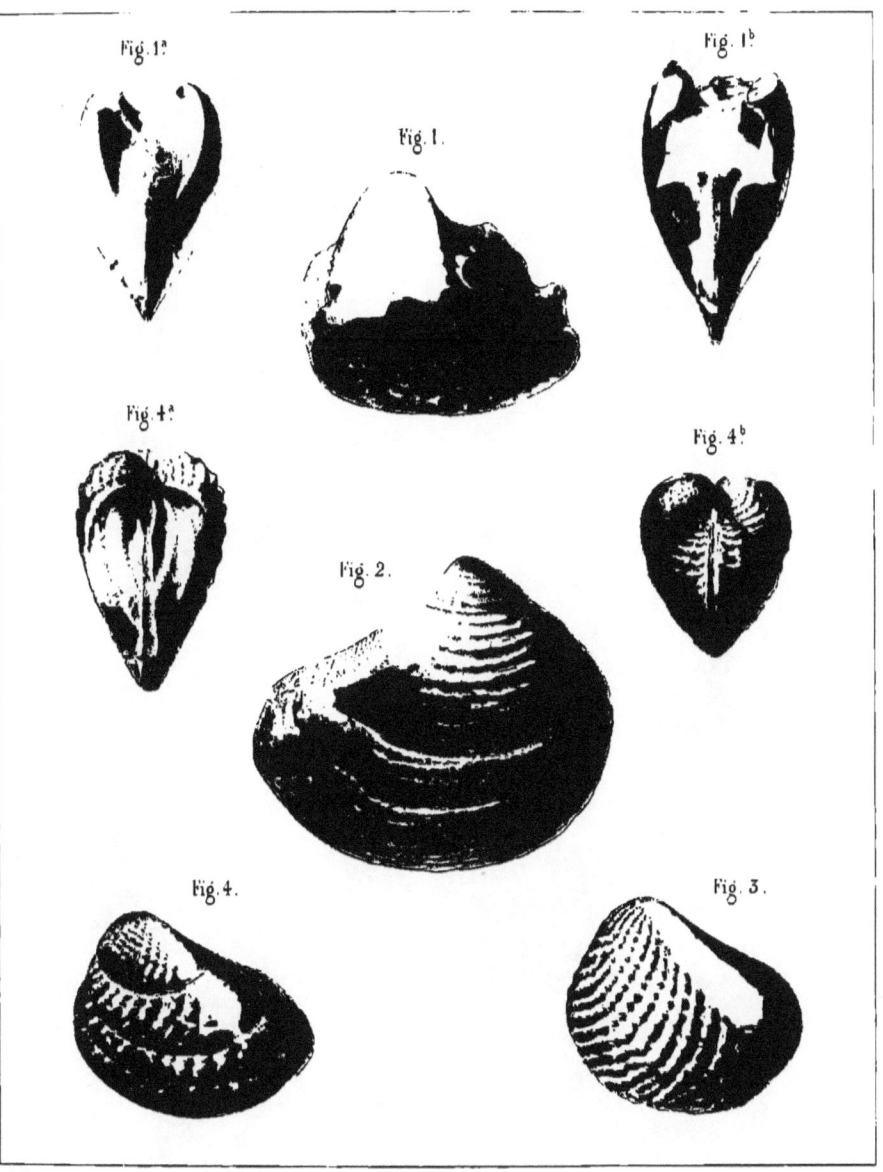

P. Lackerbauer ad nat. in lap. del.   Imp. Becquet à Paris.

PLATE XX.

Fig.
1. *Trigonia tenuitexta*, Lyc.   Portland Limestone, Isle of Portland.   (Page 90.)   My cabinet.

1 a.   „   „   A very young example enlarged one diameter.   Portland Limestone, Devizes.   Coll. Cunnington.

2.   „   *Joassi*, Lyc.   Mould with surface ornaments.   Lower Calcareous Grit, Brora, N.B.   (Page 82.)   My cabinet.

3.   „   „   Specimen in two portions from the same formation and locality.   Coll. British Museum.

4.   „   „   An entire specimen from the same formation and locality.   Coll. Rev. J. M. Joass.

5.   „   *excentrica*, Park.   Young example.   Blackdown.   (Page 94.)   Coll. British Museum.

6.   „   „   A very young example.   Greensand, near Collumpton.   Coll. Vicary.

7.   „   *Michelotti*, De Lor.   (Variety.)   Mould of external cast.   Portland Oolite, Devizes.   (Page 92.)   Coll. Cunnington.

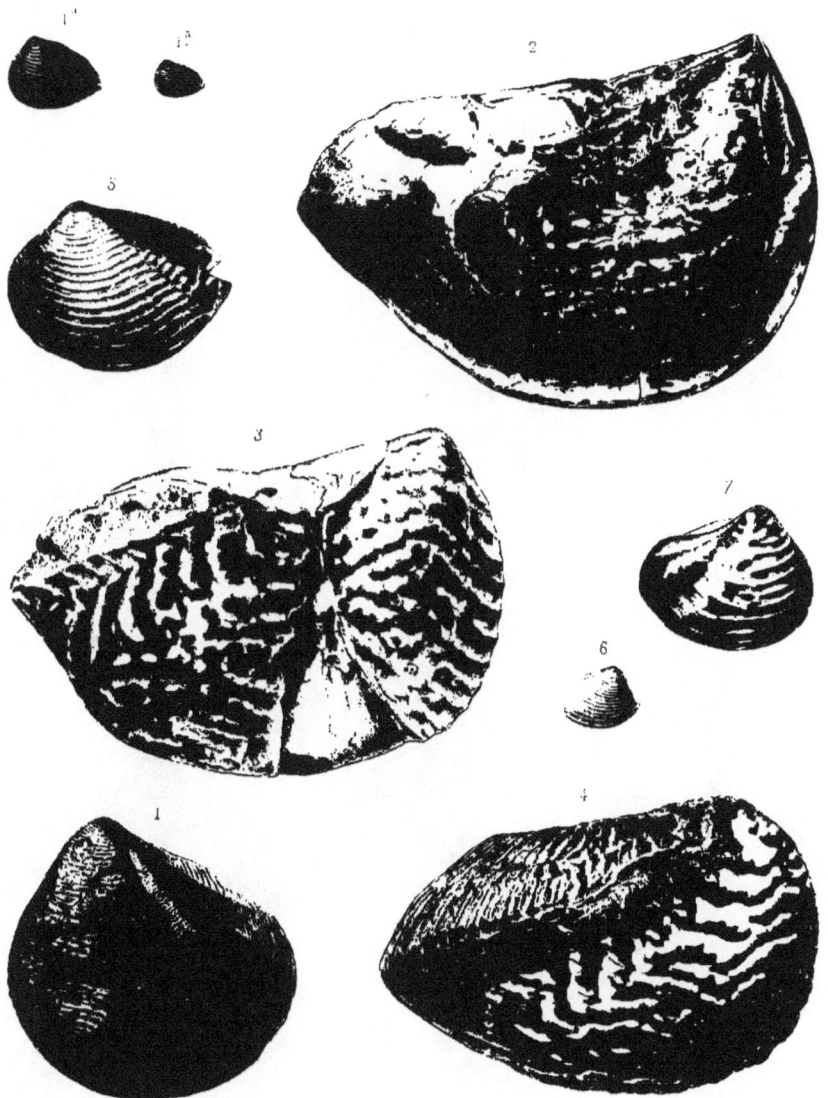

PLATE XXI.

Fig.
1. *Trigonia gibbosa*, Sow. (Variety.) Portland Oolite, Tisbury. (Page 84.) My cabinet.

2. ,, *Damoniana*, De Lor. Portland Oolite, Swindon. (Page 88.) My cabinet.

3. ,, ,, Another example. Portland Limestone, Isle of Portland. My cabinet.

4. ,, ,, Another specimen from the same formation and locality; 2 a, 2 b, upper and anteal surfaces of the same specimen. My cabinet.

5. ,, ,, Variety with small costæ and wide ante-carinal space. Portland Limestone, Isle of Portland. My cabinet.

6. ,, *excentrica*, Sow. Young example. Blackdown Greensand. (Page 94.) Coll. Royal School of Mines.

7. ,, ,, An example of fully developed growth. Blackdown Greensand. Coll. British Museum.

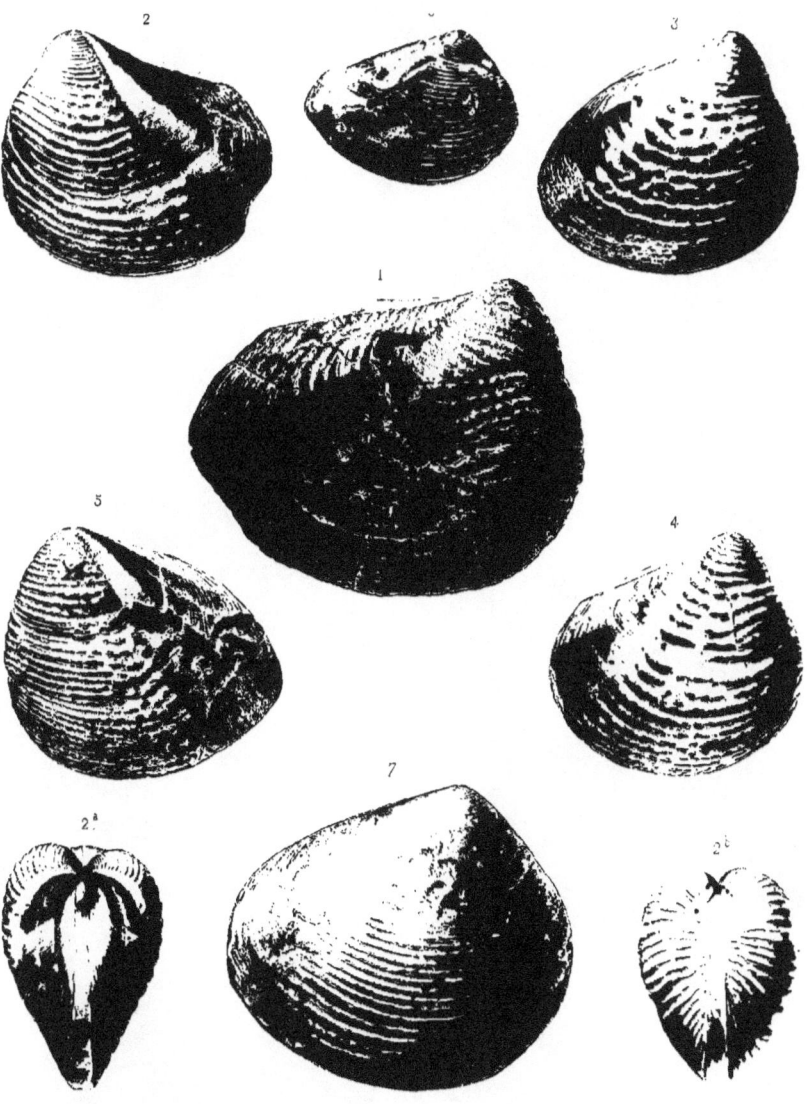

PLATE XXII.

Fig.
1, 2. *Trigonia Lingonensis,* Dum. Middle Lias, Eston Nab. (Page 98.) My cabinet.

1 a.   ,,   ,,   Area and escutcheon of the specimen fig. 1.

3.   ,,   ,,   Small specimen. Hob Hill Mine, near Saltburn, Middle Lias. My cabinet.

4.   ,,   ,,   Internal Mould, Middle Lias; Challoner Mine, near Guisborough. My cabinet.

5, 5 a.   ,,   *excentrica,* Sow. Greensand, Blackdown. (Page 94.) Coll. Royal School of Mines.

6.   ,,   *læviuscula,* Lyc. Greensand, Collumpton. (Page 96.) Coll. Vicary.

7.   ,,   *dædalea,* Park. Young example. Greensand, Blackdown. (Page 100.) My cabinet.

8.   ,,   ,,   A very young example, slightly magnified, from the same formation and locality. My cabinet. (*T. quadrata,* Sow.)

PLATE XXIII.

Fig.
1. *Trigonia dædalea*, Park., var. *confusa*. Greensand, Little Haldon. (Page 102.) Coll. Vicary.

2. ,, ,, Large example of the typical form. Greensand, Blackdown, near Collumpton. (Page 100.) My cabinet.

3. ,, ,, Another example from the same position and locality. My cabinet.

4, 4 a, 4 b. ,, *Fittoni*, Desh. Gault, Folkestone. (Page 132.) My cabinet.)

5. ,, ,, Internal mould from the same formation and locality. Coll. Rev. T. Wiltshire.

6. ,, *Meyeri*, Lyc. Chloritic Marl, near Sidmouth. (Page 125.) My cabinet.

7. ,, *Archiaciana*, D'Orb. Greensand, Blackdown. (Page 140.) Coll. Vicary.

8, 9. ,, *Upwarensis*, Lyc. Neocomian, Upware. (Page 143.) My cabinet.

10. ,, *Spinosa*, Park., var. Upper Greensand, Isle of Wight. (Page 136.) My cabinet.

11. ,, *Cunningtoni*, Lyc. Upper Greensand, Devizes. (Page 146.) Coll. Cunnington.

Lackerbauer (Karmansky) ad nap del.

PLATE XXIV.

Fig.
1, 1 a, 2.   *Trigonia nodosa*, Sow., var. *Orbignyana*. Neocomian, Isle of Wight. (Page 106.) My cabinet.

3.   ,,   ,,   Young specimen from the same formation and locality. My cabinet.

4.   ,,   *pennata*, Sow. Pebble bed, Upper Greensand, Haldon. (Page 133.) Coll. Vicary.

5.   ,,   ,,   Another specimen from the same formation. Coll. Cunnington.

6, 7.   ,,   *ornata*, D'Orb. Neocomian, Isle of Wight. (Page 139.) My cabinet.

8.   ,,   *spinosa*, Park. Greensand, Blackdown. (Page 136.) Coll. Rev. T. Wiltshire.

9.   ,,   ,,   Another example from the same formation and locality. Coll. Prof. J. C. Williamson.

10, 10 a, 10 b.   ,,   *Vectiana*, Lyc. Neocomian, Isle of Wight. (Page 123.) My cabinet.

11.   ,,   ,,   A large specimen from the same formation and locality. My cabinet.

PLATE XXV.

Fig.
1. *Trigonia nodosa*, Sow. Neocomian, Sandown. (Page 106.) My cabinet.
2. ,, ,, Specimen from ferruginous pisolite, Tealby. My cabinet.
3, 3 a. ,, *aliformis*, Park. Greensand, Blackdown. (Page 116.) Coll. Royal School of Mines.
4, 4 a. ,, ,, Greensand, Blackdown. My cabinet.
5, 6. ,, ,, variety, *attenuata*. Upper Greensand, Isle of Wight. (Page 118.) My cabinet.
7. ,, *Vectiana*. Internal mould, Neocomian, Isle of Wight. (Page 123.) My cabinet.
8. ,, *Vicaryana*, Lyc. Upper Greensand, near Sidmouth. (Page 141.) My cabinet.
9. ,, ,, Another specimen, having the costæ more widely separated. Pebble bed above the Greensand, Haldon. Coll. Vicary.
10. ,, *Archiaciana*, D'Orb. Specimen deprived of the test. Upper Greensand, Isle of Wight. (Page 140.) My cabinet.

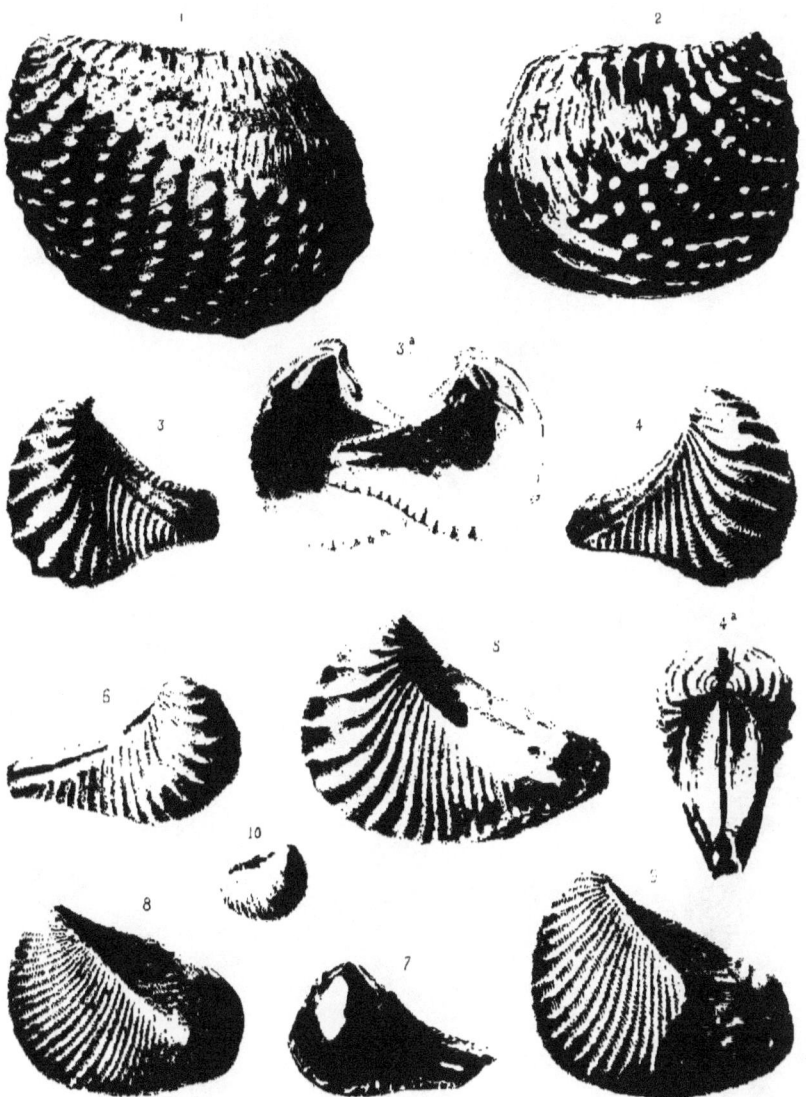

Lackerbauer (Karmanski) ad lap del. Imp Becquet à Paris

PLATE XXVI.

Fig.
1, 2, 3. *Trigonia spectabilis*, Sow. Greensand, Blackdown. (Page 112.) My cabinet.

4. ,, ,, Young example from the same formation and locality. Coll. Vicary.

5, 6, 6 *a*, 6 *b*. *caudata*, Ag. Neocomian, Isle of Wight. (Page 129.) My cabinet.

7. ,, ,, Young specimen from the same formation and locality. My cabinet.

8. ,, *sulcataria*, Lam. Pebble bed overlying Greensand, Great Haldon. (Page 135.) Coll. Vicary.

PLATE XXVII.

Fig.
1, 1 a, 1 b, 2. *Trigonia Etheridgei*, Lyc. Neocomian, Perna bed, Isle of Wight. (Page 127.) My cabinet.

3, 3 a. ,, ,, Young example from the same formation and locality. My cabinet.

4, 5, 5 a, 5 b. ,, *scabricola*, Lyc. Greensand, Blackdown. (Page 130.) My cabinet.

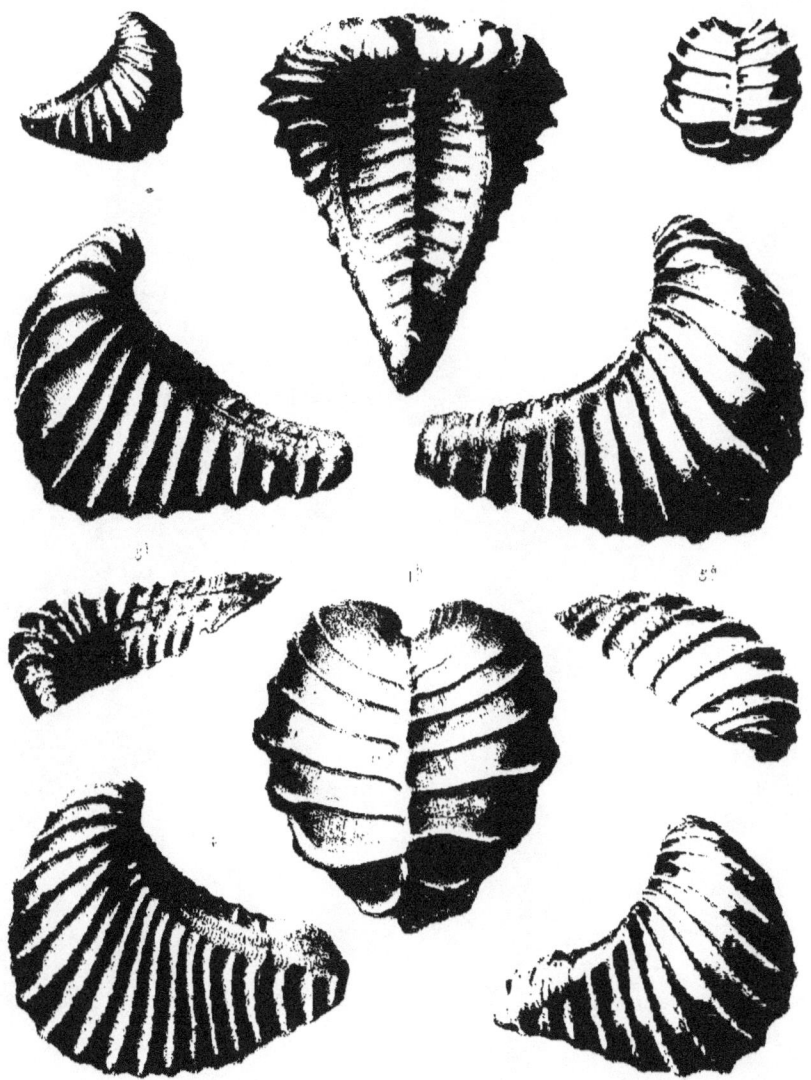

PLATE XXVIII.

Fig.
1, 2.   *Trigonia spinosa* var. *subovata*, Lyc. Chloritic Sands, Warminster. See also Pl. XXIII, fig. 10. (Page 136.) My collection.
3.   ,,   *sulcataria*, Lam. Pebble bed, Haldon. See also Pl. XXVI, fig. 8. (Page 135.) Coll. W. Vicary, Esq.
4, 4 a.   *Vicaryana*, Lyc. Portion of the surface magnified. See also Pl. XXV, figs. 8, 9. Pebble bed, Haldon. (Page 141.) Coll. W. Vicary, Esq.
5, 5 a.   ,,   *aliformis*, Park. Also Pl. XXV, figs. 3, 3 a, 4, 4 a. Greensand, Blackdown. (Page 116.) Coll. W. Vicary, Esq.
6, 6 a, 9, 10.   ,,   *excentrica*, Park. Specimens exhibiting the change from *T. sinuata* to *T. excentrica*. Greensand, Blackdown. (Page 94.) Coll. W. Vicary, Esq.
7.   ,,   *Tealbyensis*, Lyc. Middle Neocomian Formation, Tealby. (Page 114.) Woodwardian Museum, Cambridge.
8.   ,,   *dædalæa*, Park. Inner surface. See also Pl. XXII, figs. 7, 8, and Pl. XXIII, figs. 2, 3. (Page 100.) Greensand, Blackdown. My cabinet.

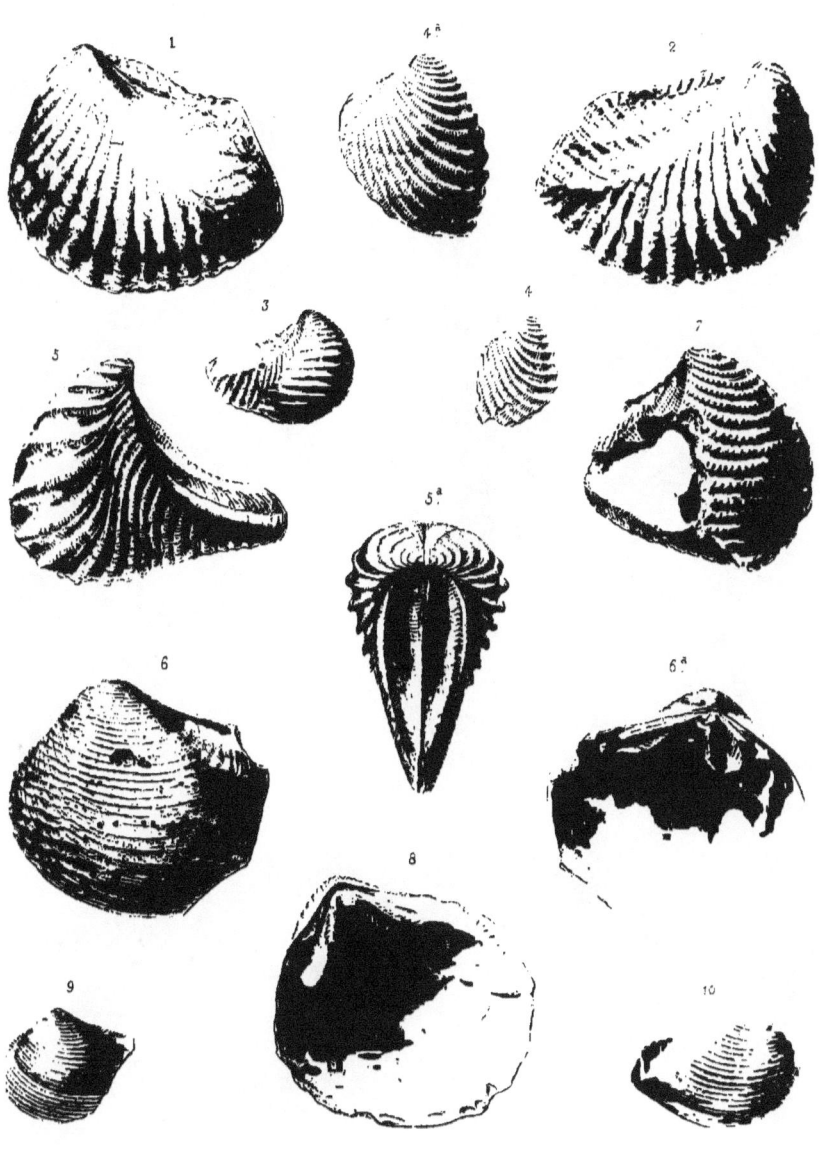

PLATE XXIX.

Fig.
1, 2, 3. *Trigonia denticulata*, Ag. Inf. Oolite, grey limestone. Cloughton, near Scarborough. (Page 152.) My collection.

4. ,, ,, Id. Small variety. Great Oolite, South Lincolnshire. My collection.

5, 6, 7. 8. ,, *costata*, Sow. The typical form. Inf. Oolite, Bradford Abbas. (Page 147.) My collection.

9, 10. ,, ,, var. *lata*, Lyc. Inf. Oolite. (Page 149.) Cotteswold Hills. My collection.

11, 12. ,, *formosa* var. *lata*, Lyc. Inf. Oolite, Bradford Abbas. See also Pl. XXXV, fig. 7. (Pages 35, 202.) My collection.

PLATE XXX.

Fig.
1, 1 a, 2. *Trigonia elongata* var. *angustata*, Lyc. Cornbrash, Scarborough. (Page 154.) My collection.
3, 3 a, 3 b, 6. ,, ,, The typical form. Oxford Clay, Weymouth. (Page 154.) My collection.
4, 5. ,, ,, var. *lata*, Lyc. Cornbrash, Scarborough. Also South Lincolnshire. (Page 154.) My collection.

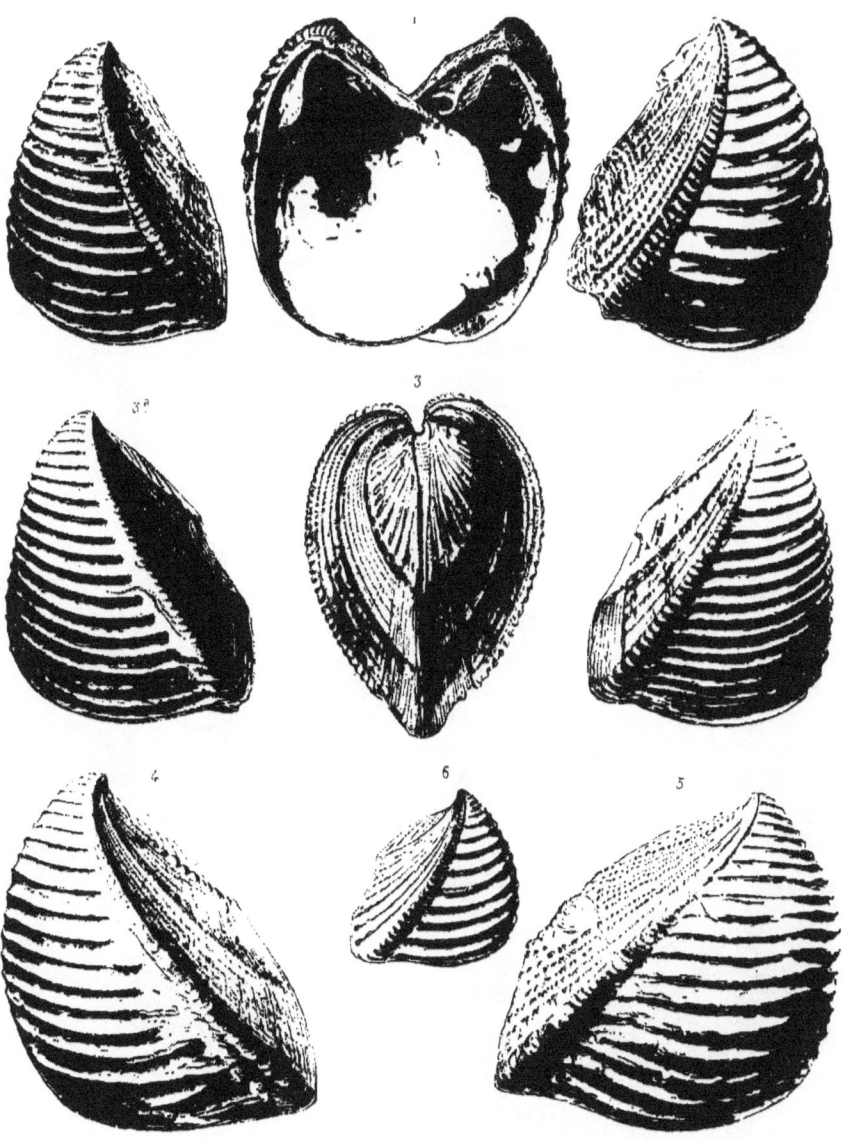

PLATE XXXI.

Fig.
1, 1 *a*, 2, 3, 10. *Trigonia monilifera*, Ag. Kimmeridge Clay, Weymouth. (Page 165.) My collection.
2 *a*. ,, ,, Portion of the surface magnified.
4, 5, 7, 8. ,, *hemisphærica*, Lyc. Inf. Oolite, Santon Bridge, near Appleby. (Page 174.) My collection.
6. ,, ,, Specimen with the costæ fewer and larger. Inf. Oolite, Santon Bridge. My collection.
9, 9 *a*. ,, *Cülleni*, Lyc. Inf. Oolite, Millepore bed, Cloughton. (Page 173.) My collection.

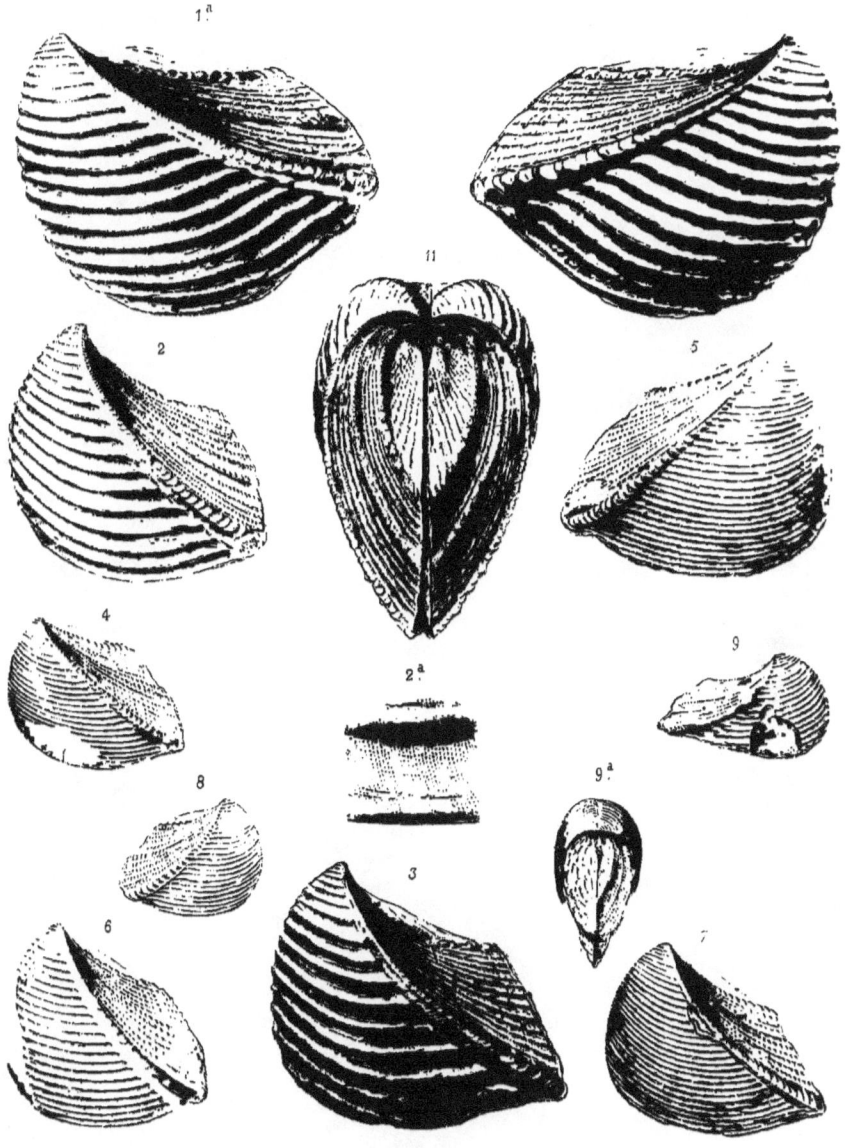

PLATE XXXII.

FIG.
1, 2, 3, 4, 5. *Trigonia Cassiope*, D'Orb. Cornbrash, Scarborough. (Page 170.) collection.
6, 7. ,, *bella*, Lyc. Bradford Abbas. Inf. Oolite. (Page 162.) My collection.
8, 8 *a*. ,, ,, Small specimen. The same locality. Coll. Col. Mansel-Pleydell.
9. ,, *geographica*, Ag. Also Pl. X, fig. 6. Coral Rag, Pickering. (Page 69.) My collection.

PLATE XXXIII.

Fig.
1, 2. *Trigonia Meriani*, Ag.   Coral Rag, Pickering. (Page 167.)   My collection.
3.   ,,   ,,   Calcareous Grit, Weymouth.   My collection.
8.   ,,   *tenuicosta*, Lyc.   Inf. Oolite, Walditch. (Page 160.)   Coll. Col. Mansel-Pleydell.
7, 9, 9 a. ,,   ,,   Inf. Oolite, Bradford Abbas.   My collection.
4, 5, 6.   ,,   *hemisphærica* var. *gregaria*, Lyc.   Appleby, N. W. Lincolnshire. (Page 174.)   Coll. Rev. J. E. Cross.   Also my collection.

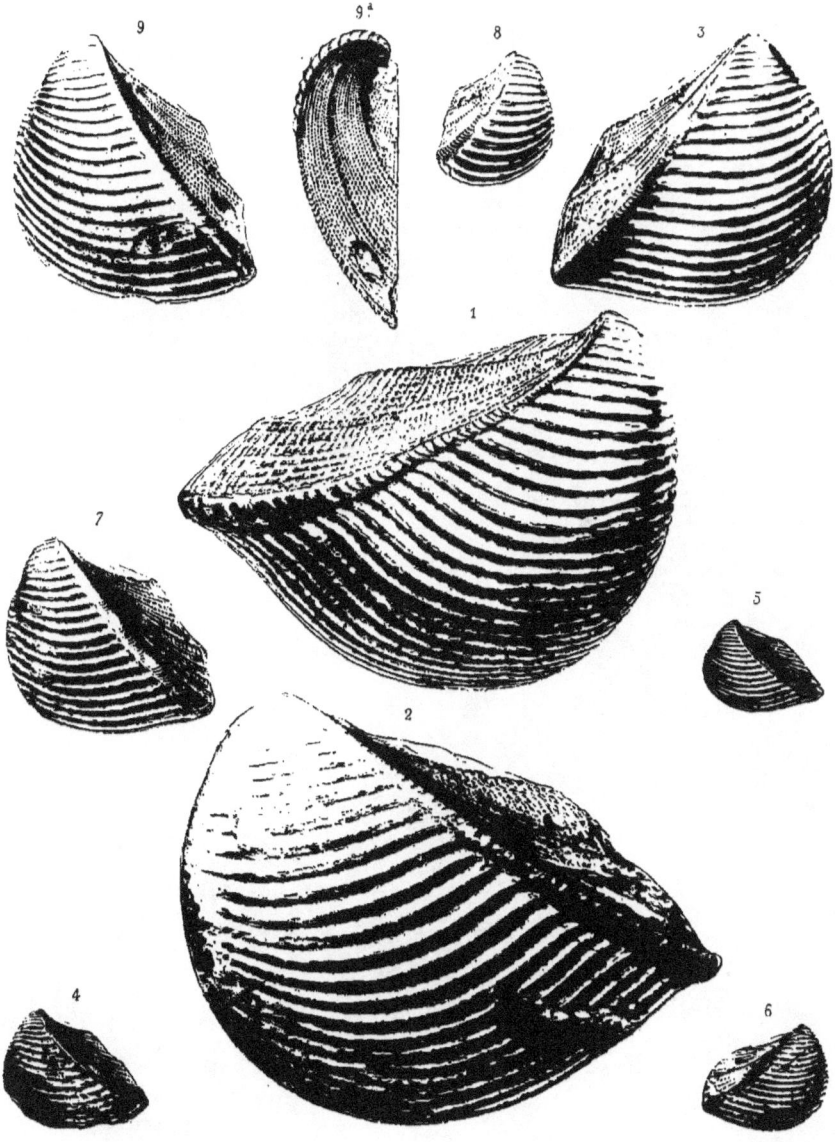

Lackerbauer (Karmanski) ad.lap del. Imp. Becquet à Paris

PLATE XXXIV.

Fig.
1, 2, 2 a. *Trigonia sculpta*, Lyc.   Inf. Ool., near Stroud.   (Page 157.)   My collection.
3.   ,,   variety *Cheltensis*, Inf. Ool., Cotteswold Hills.   (Page 159.) My collection.
4.   ,,   ,,   variety *Rolandi*, Cross.   Cornbrash, North Lincolnshire; also Cornbrash, Hilperton, Wilts.   (Page 159.)   My collection.
5.   ,,   *Hudlestoni*, Lyc.   Coral Rag, Headington, near Oxford.   See also Pl. XXXIX, figs. 1 a, 2.   (Page 194)   University Museum, Oxford.
6.   ,,   ,,   Elsworth Rock, Cambridge.   Coll. J. F. Walker, York.
7, 7 a.   ,,   *pullus*, Sow.   Cornbrash, Hilperton, Wilts.   (Page 164.)   My collection.
8, 9.   ,,   ,,   Great Oolite, Minchinhampton.   My collection.

PLATE XXXV.

Fig.
1, 2. *Trigonia Keepingi*, Lyc. Middle Neocomian Formation, Tealby, Lincolnshire. (Page 196.) Coll. Woodwardian Museum, Cambridge.
3. ,, *carinata*, Ag. Young specimen. Upper Greensand, Isle of Wight, Ventnor. (Page 179.) My collection.
4, 4 a. ,, ,, Young specimen. Neocomian Formation, Atherfield, Isle of Wight. My collection.
5. ,, ,, Specimen of more fully developed growth. Atherfield.
5 a. ,, ,, Anterior side, exhibiting the byssal aperture. Atherfield, Isle of Wight. My collection.
6, 6 a. ,, ,, Atherfield, Isle of Wight. My collection.
7. ,, *formosa* var. *lata*, Lyc. Inf. Ool., Bradford Abbas. See also Pl. XXIX, figs. 11, 12. (Pages 35, 202). My collection.
8. ,, *Brodiei*, Lyc. Inf. Ool., Northampton Sands, Milcombe Hill, Oxon. (Page 195.) Coll. Rev. P. B. Brodie.
9. ,, ,, My collection.

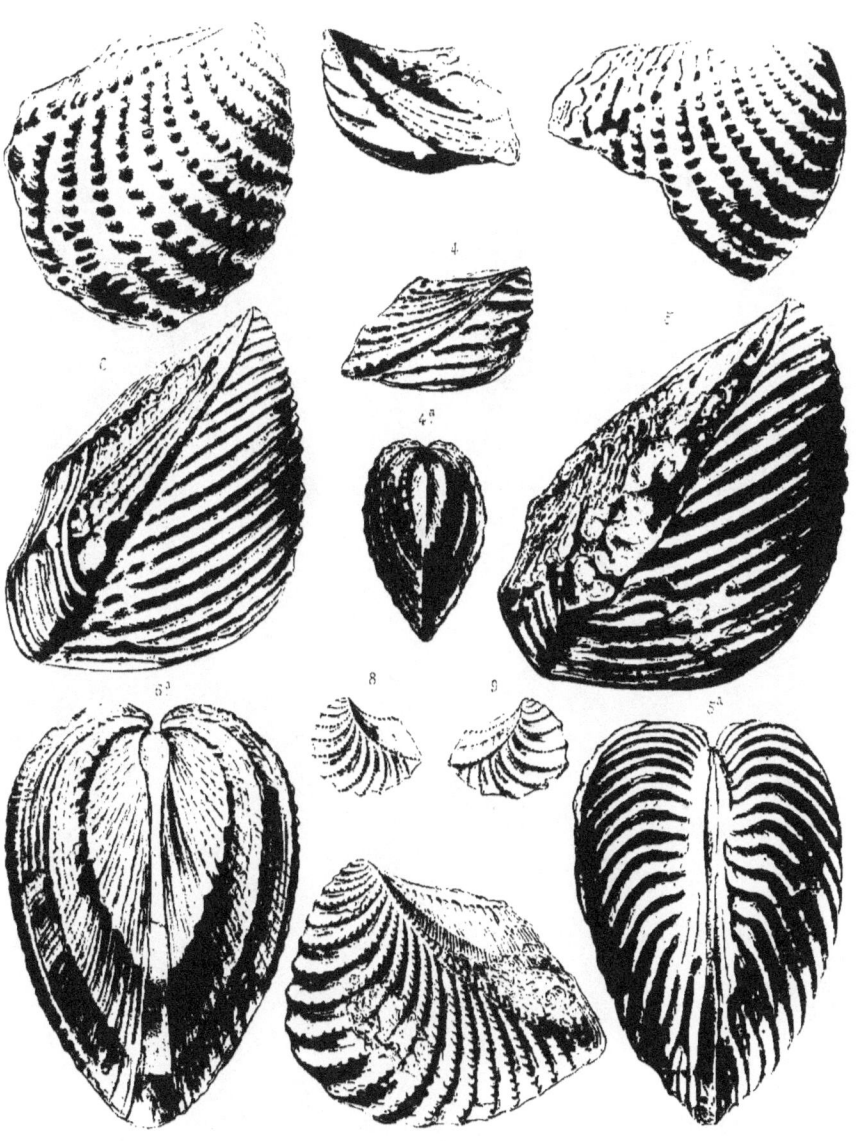

PLATE XXXVI.

FIG.
1, 2, 3, 4. *Trigonia Rupellensis*, D'Orb. Kelloway Rock. Also Pl. VIII, fig. 4. Cayton Bay, Scarborough. (Page 28.) My collection.
7.   ,,   *triquetra*, Sceb. Also Pl. VI, figs. 1, 1 *a*, 2. Coral Rag, Filey Point. (Page 26.) My collection.
5, 6.   ,,   *ingens*, Lyc. Young specimens. Also Pl. VIII, figs. 1, 3. Middle Neocomian formation, Tealby, Lincolnshire. (Page 24.) My collection.
9, 10.   ,,   *imbricata*, Sow. Also Pl. VI, fig. 5 *a*, *b*. Great Oolite. (Page 33.) British Museum.
8.   ,,   *parcinoda*, Lyc. Inf. Oolite. (Page 46.) British Museum.
11, 11 *a*. ,,   *Griesbachi*. Also Pl. III, figs. 10 *a*, *b*. (Page 34.) Coll. Rev. A. W. Griesbach.

PLATE XXXVII.

FIG.
1, 2. *Trigonia producta*, Lyc. Also Pl. XIII, figs. 1, 2, 3, 4. Inferior Oolite, Hook Norton, Oxon. (Page 60.) Coll., Royal School of Mines.
3. ,, *paucicosta*, Lyc. Kelloway Rock, Scarborough. See also Pl. XI, figs. 8, 9, and Pl. XVI, fig. 7. (Page 57.) My collection.
4, 4 a. ,, *pennata*, Sow. 4 a, magnified. Also Pl. XXIV, figs. 4, 5. Chloritic Marl, South Devon. (Page 133.) Coll. W. Vicary, Esq.
5, 5 a. ,, *nodosa*, Sow. Specimen less developed than the figures upon Pls. XXIV and XXV. (Page 106.) Neocomian formation, Atherfield. My collection.
6. ,, ,, Young specimen; the costæ had not acquired nodes. Neocomian formation, Atherfield. My collection.
7, 8. ,, *angulata*, Sow. Specimens of advanced stages of growth. See also Pl. XIV, figs. 5, 6. Inferior Oolite, near Stroud. (Page 54.) Coll. E. Witchell, Esq.
9. ,, ,, Young specimen having costæ without angularity. Inf. Oolite, Stroud. Coll. E. Witchell, Esq.
10. ,, *formosa*, var. Inferior Oolite, Bradford Abbas. See also Pl. V, figs. 4, 5, 6. (Page 35.) My collection.

PLATE XXXVIII.

Fig.
1. *Trigonia Cymba*, Cont. (Mould). Portland Sand, Dorset. (Page 192.) Coll. Col. J. C. Mansel-Pleydell.
2. ,, *exaltata*, Lyc. Middle Neocomian formation, Norfolk. (Page 184.) Coll. British Museum.
3. ,, *Alina*, Cont. Portland Limestone, Shotover Hill. (Page 193.) See also Pl. IX, fig. 2; there mentioned as a variety of *T. incurva*. Coll. Museum, Oxon.
4. ,, *compta*, Lyc. Slate of Collyweston. Specimen unusually large, having the postcal terminal tubercle of each costa much developed. See also Pl. XV, figs. 5, 6, 7. (Page 70.) Coll. S. Sharp, Esq.
6. ,, *scapha*, Ag. Middle Neocomian formation, Norfolk. (Page 183.) Woodwardian Museum, Cambridge.
7. ,, *Williamsoni*, Lyc. Kelloway Rock, Scarborough. See also Pl. XVI, fig. 8. (Page 53.) My collection.
8, 9. ,, *Witchelli*, Lyc. Fullers' Earth, Stroud. (Page 197.) Coll. E. Witchell, Esq.
10, 11, 12, 12 a. ,, *pulchella*, Ag. Upper Lias, Lincoln. (Page 185.) My collection.

PLATE XXXIX.

FIG.
1. *Trigonia* **Pellati**. Mun. Chal. . Oxford Clay, St. Ives. See also Pl. VII, figs. 1, 2 *a*, *b*, and Pl. XI, fig. 1. (Page 41.) Coll. J. F. Walker, Esq.

1 *a*, 2. ,, *Hudlestoni*, Lyc. Coral Rag, Cawkley. Also Pl. XXXIV, figs. 5, 6. (Page 194.) Coll. W. H. Hudleston, Esq.

3. ,, *irregularis*, Seeb. Var. Kimmeridge Clay, Wotton Basset. The costæ are nearly without the irregularity of the figures upon Pl. V. (Page 39.) My collection.

4. ,, *Upwarensis*, Lyc. Neocomian formation, Upware, Cambridgeshire. See also Pl. XXIII, figs. 8, 9. (Page 143.) Coll. J. F. Walker, Esq.

Lackerbauer (Karmanski:) ad lap del.

PLATE XL.

Fig.
1, 1 a, 1 b, 7, 9, 9 a. *Trigonia crenulifera*, Lyc. Chloritic Marls, Dunscomb Cliffs. (Page 189.) Coll. Meÿer.
2. „ *affinis*, Sow. Pebble bed, Great Haldon. (Page 187.) Coll. Vicary.
3, 4. „ *Vicaryana*, Lyc. Chloritic Marls, Dunscomb Cliffs, near Sidmouth. (Page 203.) Coll. Meÿer.
5, 6. „ *Dunscombensis*, Lyc. Chloritic Marls, Sidmouth. (Page 188.) Coll. Meÿer.
8, 8 a. „ *debilis*, Lyc. Chloritic Marls, Dunscomb Cliffs. (Page 189.) Coll. Meÿer.

PLATE XLI.

Fig.
1, 1 a, 2, 3. *Trigonia Snaintonensis*, Lyc. Passage Beds of the Lower Calcareous Grit, Snainton, near Scarborough. (Page 198.) Coll. Hudleston.
4.   "   *Blakei*, Lyc. Passage Beds of the Lower Calcareous Grit, Snainton, near Scarborough. (Page 205.) Coll. Hudleston.
5.   "   *debilis*, Lyc. Chloritic Marls, Dunscombe Cliffs. See also Plate XL, fig. 8. (Page 189.) My collection.
6, 8.   "   *Myophoria postera*, Quenst. The left valve, natural size. Rhœtic Limestone, near Ilminster. (Page 215.) Coll. Moore.
6 a, 8 a.   "   "   "   The same valve, enlarged.
7, 9,   "   "   "   The right valve, natural size. Coll. Moore.
7 a, 9 a.   "   "   "   The right valve, enlarged.
10, 11, 12.   "   *imbricata*, Sow. Fullers Earth, Stroud. See also Plate VI, figs. 5 a, b; and Plate XXXVI, figs. 9, 10. (Page 33.) Coll. Witchell.
13, 13 a.   "   ? *modesta*, Tate. Natural size and enlarged. Lower Lias, Warter's Bay, near Whitby. (Page 212.) Coll. Rev. J. F. Blake.
14.   "   *Dunscombensis*, Lyc. Chloritic Marls, Dunscombe Cliffs, South Devon. See also Plate XL, fig. 5. (Page 188.) My collection.
15, 16.   "   *Meyeri*, Lyc. Chloritic Marls, Dunscombe Cliffs. See also Plate XXIII, fig. 6. (Page 125.) My collection.
17.   "   *costigera*, Lyc. Chloritic Marls, Dunscombe Cliffs. (Page 205.) My collection.

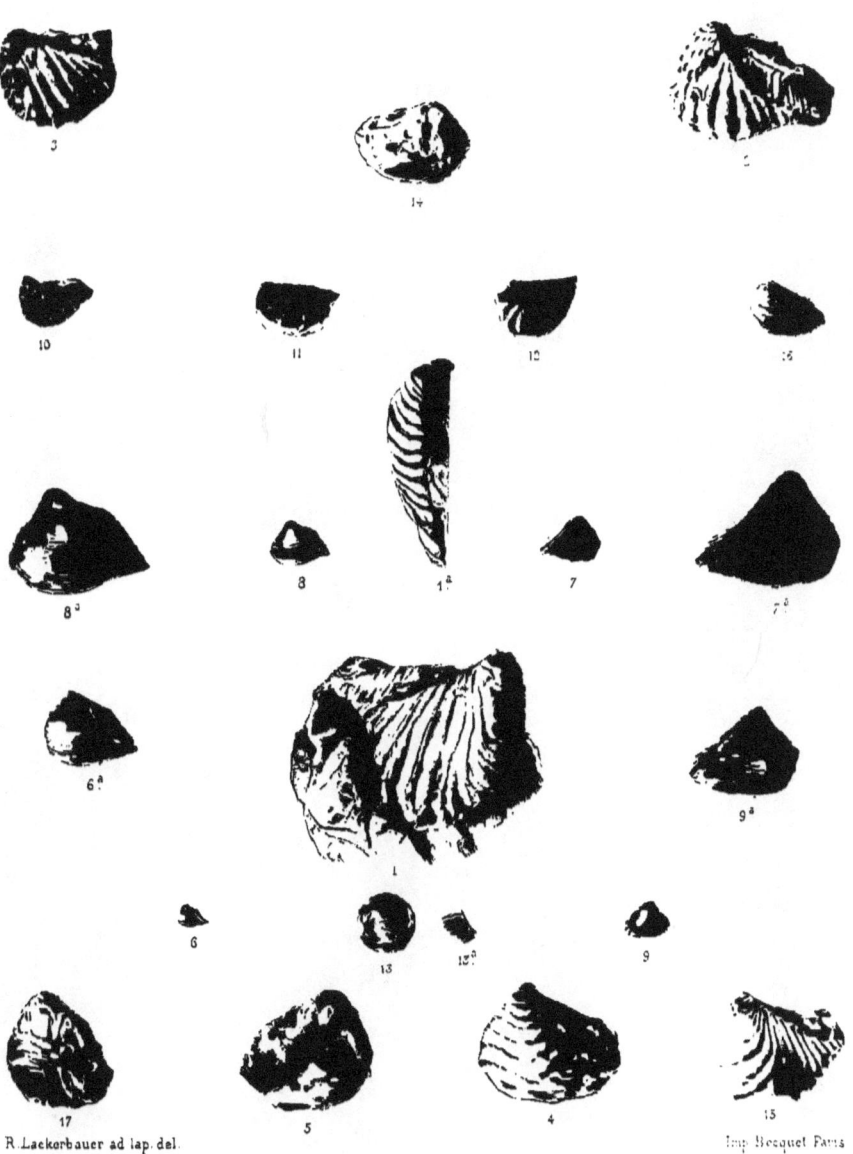

www.ingramcontent.com/pod-product-compliance
Lightning Source LLC
Chambersburg PA
CBHW022112290426
44112CB00008B/650